ENCYCLOPEDIA OF MATHEMATICS AND ITS APPLICATIONS

EDITED BY G.-C. ROTA

Editorial Board
R. S. Doran, M. Ismail, T.-Y. Lam, E. Lutwak, R. Spigler

Volume 82

The Foundations of Mathematics in The Theory of Sets

ENCYCLOPEDIA OF MATHEMATICS AND ITS APPLICATIONS

4   W. Miller, Jr. *Symmetry and separation of variables*
6   H. Minc *Permanents*
11 W. B. Jones and W. J. Thron *Continued fractions*
12 N. F. G. Martin and J. W. England *Mathematical theory of entropy*
18 H. O. Fattorini *The Cauchy problem*
19 G. G. Lorentz, K. Jetter and S. D. Riemenschneider *Birkhoff interpolation*
21 W. T. Tutte *Graph theory*
22 J. R. Bastida *Field extensions and Galois theory*
23 J. R. Cannon *The one-dimensional heat equation*
25 A. Salomaa *Computation and automata*
26 N. White (ed.) *Theory of matroids*
27 N. H. Bingham, C. M. Goldie and J. L. Teugels *Regular variation*
28 P. P. Petrushev and V. A. Popov *Rational approximation of real functions*
29 N. White (ed.) *Combinatorial geometrics*
30 M. Pohst and H. Zassenhaus *Algorithmic algebraic number theory*
31 J. Aczel and J. Dhombres *Functional equations containing several variables*
32 M. Kuczma, B. Chozewski and R. Ger *Iterative functional equations*
33 R. V. Ambartzumian *Factorization calculus and geometric probability*
34 G. Gripenberg, S.-O. Londen and O. Staffans *Volterra integral and functional equations*
35 G. Gasper and M. Rahman *Basic hypergeometric series*
36 E. Torgersen *Comparison of statistical experiments*
37 A. Neumaier *Intervals methods for systems of equations*
38 N. Korneichuk *Exact constants in approximation theory*
39 R. A. Brualdi and H. J. Ryser *Combinatorial matrix theory*
40 N. White (ed.) *Matroid applications*
41 S. Sakai *Operator algebras in dynamical systems*
42 W. Hodges *Model theory*
43 H. Stahl and V. Totik *General orthogonal polynomials*
44 R. Schneider *Convex bodies*
45 G. Da Prato and J. Zabczyk *Stochastic equations in infinite dimensions*
46 A. Bjorner, M. Las Vergnas, B. Sturmfels, N. White and G. Ziegler *Oriented matroids*
47 E. A. Edgar and L. Sucheston *Stopping times and directed processes*
48 C. Sims *Computation with finitely presented groups*
49 T. Palmer *Banach algebras and the general theory of *-algebras*
50 F. Borceux *Handbook of categorical algebra I*
51 F. Borceux *Handbook of categorical algebra II*
52 F. Borceux *Handbook of categorical algebra III*
54 A. Katok and B. Hassleblatt *Introduction to the modern theory of dynamical systems*
55 V. N. Sachkov *Combinatorial methods in discrete mathematics*
56 V. N. Sachkov *Probabilistic methods in discrete mathematics*
57 P. M. Cohn *Skew Fields*
58 Richard J. Gardner *Geometric tomography*
59 George A. Baker, Jr. and Peter Graves-Morris *Padé approximants*
60 Jan Krajicek *Bounded arithmetic, propositional logic, and complex theory*
61 H. Gromer *Geometric applications of Fourier series and spherical harmonics*
62 H. O. Fattorini *Infinite dimensional optimization and control theory*
63 A. C. Thompson *Minkowski geometry*
64 R. B. Bapat and T. E. S. Raghavan *Nonnegative matrices and applications*
65 K. Engel *Sperner theory*
66 D. Cvetkovic, P. Rowlinson and S. Simic *Eigenspaces of graphs*
67 F. Bergeron, G. Labelle and P. Leroux *Combinatorial species and tree-like structures*
68 R. Goodman and N. Wallach *Representations of the classical groups*
69 T. Beth, D. Jungnickel and H. Lenz *Design Theory volume I 2 ed.*
70 A. Pietsch and J. Wenzel *Orthonormal systems and Banach space geometry*
71 George E. Andrews, Richard Askey and Ranjan Roy *Special Functions*
72 R. Ticciati *Quantum field theory for mathematicians*
76 A. A. Ivanov *Geometry of sporadic groups I*
78 T. Beth, D. Jungnickel and H. Lenz *Design Theory volume II 2 ed.*
80 O. Stormark *Lie's Structural Approach to PDE Systems*

ENCYCLOPEDIA OF MATHEMATICS AND ITS APPLICATIONS

The Foundations of Mathematics in the Theory of Sets

J. P. MAYBERRY

University of Bristol

CAMBRIDGE
UNIVERSITY PRESS

CAMBRIDGE UNIVERSITY PRESS
Cambridge, New York, Melbourne, Madrid, Cape Town, Singapore,
São Paulo, Delhi, Dubai, Tokyo, Mexico City

Cambridge University Press
The Edinburgh Building, Cambridge CB2 8RU, UK

Published in the United States of America by Cambridge University Press, New York

www.cambridge.org
Information on this title: www.cambridge.org/9780521172714

© Cambridge University Press 2000

This publication is in copyright. Subject to statutory exception
and to the provisions of relevant collective licensing agreements,
no reproduction of any part may take place without the written
permission of Cambridge University Press.

First published 2000
First paperback edition 2010

A catalogue record for this publication is available from the British Library

Library of Congress Cataloguing in Publication data

Mayberry, John P.
 The foundations of mathematics in the theory of sets / J.P. Mayberry.
  p. cm (Encyclopedia of mathematics and its applications; v. 82)
 Includes bibliographical references and index.
 ISBN 0 521 77034 3
 1. Set theory. I. Title. II. Series.
 QA248.M375 2001
 511.3'22-dc21   00-45435 CIP

ISBN 978-0-521-77034-7 Hardback
ISBN 978-0-521-17271-4 Paperback

Cambridge University Press has no responsibility for the persistence or
accuracy of URLs for external or third-party internet websites referred to in
this publication, and does not guarantee that any content on such websites is,
or will remain, accurate or appropriate.

To Mary Penn Mayberry and Anita Kay Bartlett Mayberry

Contents

Preface		*page* x
Part One: Preliminaries		1
1	**The Idea of Foundations for Mathematics**	3
	1.1 Why mathematics needs foundations	3
	1.2 What the foundations of mathematics consist in	8
	1.3 What the foundations of mathematics need not include	10
	1.4 Platonism	14
2	**Simple Arithmetic**	17
	2.1 The origin of the natural numbers	17
	2.2 The abstractness of the natural numbers	19
	2.3 The original conception of number	21
	2.4 Number words and ascriptions of number	24
	2.5 The existence of numbers	29
	2.6 Mathematical numbers and pure units	39
	2.7 Ascriptions of number: Frege or Aristotle?	45
	2.8 Simple numerical equations	52
	2.9 *Arithmetica universalis*	59
Part Two: Basic Set Theory		65
3	**Semantics, Ontology, and Logic**	67
	3.1 Objects and identity	67
	3.2 *Arithmoi* and their units	70
	3.3 Sets	74
	3.4 Global functions	78
	3.5 Species	85

	3.6	Formalisation	94
	3.7	Truth and proof in mathematics	99
4	**The Principal Axioms and Definitions of Set Theory**		**111**
	4.1	The Axiom of Comprehension and Russell's Theorem	111
	4.2	Singleton selection and description	114
	4.3	Pair Set, Replacement, Union, and Power Set	115
	4.4	The status of the principal axioms of set theory	118
	4.5	Ordered pairs and Cartesian products	124
	4.6	Local functions and relations	130
	4.7	Cardinality	134
	4.8	Partial orderings and equivalence relations	137
	4.9	Well-orderings and local recursion	139
	4.10	Von Neumann well-orderings and ordinals	144
	4.11	The Principle of Regularity	146

Part Three: Cantorian Set Theory 151

5	**Cantorian Finitism**		**153**
	5.1	Dedekind's axiomatic definition of the natural numbers	153
	5.2	Cantor's Axiom	161
	5.3	The Axiom of Choice	162
	5.4	The extensional analysis of sets	164
	5.5	The cumulative hierarchy of sets	170
	5.6	Cantor's Absolute	176
	5.7	Axioms of strong infinity	185
6	**The Axiomatic Method**		**191**
	6.1	Mathematics before the advent of the axiomatic method	191
	6.2	Axiomatic definition	195
	6.3	Mathematical logic: formal syntax	207
	6.4	Global semantics and localisation	213
	6.5	Categoricity and the completeness of theories	221
	6.6	Mathematical objects	225
7	**Axiomatic Set Theory**		**237**
	7.1	The Zermelo–Fraenkel axioms	237
	7.2	Axiomatic set theory and Brouwer's Principle	242
	7.3	The localisation problem for second order logic	251

Contents

Part Four: Euclidean Set Theory — 259

8 Euclidean Finitism — 261
    8.1   The serpent in Cantor's paradise — 261
    8.2   The problem of non-Cantorian foundations — 270
    8.3   The Axiom of Euclidean Finiteness — 276
    8.4   Linear orderings and simple recursion — 282
    8.5   Local cardinals and ordinals — 291
    8.6   Epsilon chains and the Euclidean Axiom of Foundation — 294

9 The Euclidean Theory of Cardinality — 300
    9.1   Arithmetical functions and relations — 300
    9.2   Limited recursion — 310
    9.3 S-ary decompositions and numerals — 317

10 The Euclidean Theory of Simply Infinite Systems — 325
    10.1   Simply infinite systems — 325
    10.2   Measures, scales, and elementary arithmetical operations — 330
    10.3   Limited recursion — 333
    10.4   Extending simply infinite systems — 336
    10.5   The hierarchy of S-ary extensions — 350
    10.6   Simply infinite systems that grow slowly in rank — 352
    10.7   Further axioms — 362

11 Euclidean Set Theory from the Cantorian Standpoint — 369
    11.1   Methodology — 369
    11.2   Cumulation models — 372

12 Envoi — 381
    12.1   Euclid or Cantor? — 381
    12.2   Euclidean simply infinite systems — 382
    12.3   Speculations and unresolved problems — 387

Appendix 1   **Conceptual Notation** — 396
    A1.1   Setting up a conceptual notation — 396
    A1.2   Axioms, definitions, and rules of inference — 398
    A1.3   Global propositional connectives — 406

Appendix 2   **The Rank of a Set** — 411

Bibliography — 415

Index — 421

Preface

Dancing Master: *All the troubles of mankind, all the miseries which make up history, all the blunders of statesmen, all the failures of great captains – all these come from not knowing how to dance.*
Le Bourgeois Gentilhomme, Act 1, Scene 2

The importance of set-theoretical foundations

The discovery of the so-called "paradoxes" of set theory at the beginning of the twentieth century precipitated a profound crisis in the foundations of mathematics. This crisis was the more serious in that the then new developments in the theory of sets had allowed mathematicians to solve earlier difficulties that had arisen in the logical foundations of geometry and analysis. More than that, the new, set-theoretical approach to analysis had completely transformed that subject, allowing mathematicians to make rapid progress in areas previously inaccessible (in the theory of measure and integration, for example).

All of these advances seemed to be placed in jeopardy by the discovery of the paradoxes. Indeed, it seemed that mathematics itself was under threat. Clearly a retreat to the *status quo ante* was not an option, for serious difficulties once seen cannot just be ignored. But without secure foundations – clear concepts that can be employed without prior definition and true principles that can be asserted without prior justification – the very notion of *proof* is undermined. And, of course, it is the demand for rigorous proof that, since the time of the Greeks, has distinguished mathematics from all of the other sciences.

This crisis profoundly affected some mathematicians' attitudes to their subject. Von Neumann, for example, confessed in a brief autobiographical essay that the existence of the paradoxes of set theory cast a blight on his entire career, and that whenever he encountered technical difficulties in his research he could not suppress the discouraging thought that the problems in the foundations of mathematics doomed the whole mathematical enterprise to failure, in any case.

Mathematics, however, has passed through this crisis, and it is unlikely that a contemporary mathematician would suffer the doubts that von Neumann suffered. Indeed, mathematicians, in general, do not *worry* about foundational questions now, and many, perhaps most, of them are not even interested in such matters. It is surely natural to ask what is the cause of this complacency and whether it is justified.

Of course, every mathematician must master some of the facts about the foundations of his subject, if only to acquire the basic tools and techniques of his trade. But these facts, which are, essentially, just the elements of set theory, can be, and usually are, presented in a form which leaves the impression that they are just definitions or even mere notational conventions, so that their existential content is overlooked. What is more, the exposition of such foundational matters typically begins *in medias res*, so to speak, with the natural numbers and real numbers simply regarded as *given*, so that the beginner is not even aware that these things require proper mathematical definitions, and that those definitions must be shown to be both logically consistent and adequate to characterise the concepts being defined.

These fundamental number systems are nowadays defined using the axiomatic method. But there is a surprisingly widespread misunderstanding among mathematicians concerning the underlying logic of the axiomatic method. The result is that many of them regard the foundations of mathematics as just a branch of mathematical logic, and this encourages them to believe that the foundations of their subject can be safely left in the hands of expert colleagues. But formal mathematical logic itself rests on the same assumptions as do the other branches of mathematics: it, too, stands in need of foundations. Indeed, mathematical logicians are as prone to confusion over the foundations of the axiomatic method as their colleagues.

But this complacency about foundations does have a certain practical justification: modern mathematics does, indeed, rest on a solid and safe foundation, more solid and more safe than most mathematicians realise. Moreover, since mathematics is largely a technical, as opposed to philosophical, discipline, it is not unreasonable that mathematicians should, in the main, get on with the business of pursuing their technical specialities without worrying unduly about foundational questions. But that does not give them licence to pronounce upon matters on which they have not seriously reflected and are ignorant, or to assume that expertise in some special branch of their subject gives them special insight into its foundations.

However, even though it is not, strictly speaking, always necessary for mathematicians to acquire more than a basic knowledge of the foundations of their subject, surely it is desirable that they should do so. Surely the practitioners of a subject the very essence of which is proof and definition ought to be curious about the concepts and principles on which those activities rest.

Philosophers too have an important stake in these questions. Indeed, it is the fundamental role accorded to questions in the philosophy of mathematics that is the characterising feature of western philosophy, the feature that sharply distinguishes it from the other great philosophical traditions.

Problems relating to mathematics and its foundations are to be found everywhere in the writings of Plato and Aristotle, and every major modern philosopher has felt compelled to address them[1]. The subjects that traditionally constitute the central technical disciplines of philosophy – logic, epistemology, and metaphysics – cannot be studied in any depth without encountering problems in the foundations of mathematics. Indeed, the deepest and most difficult problems in those subjects often find their most perspicuous formulations when they are specialised to mathematics and its foundations. Even theology must look to the foundations of mathematics for the clearest and most profound study yet made of the nature of the infinite.

Unfortunately, the complacency, already alluded to, among mathematicians concerning the foundations of their subject has had a deleterious effect on philosophy. Deferring to their mathematical colleagues' technical competence, philosophers are sometimes not sufficiently critical of received opinions even when those opinions are patently absurd.

The mathematician who holds foolish philosophical opinions – about the nature of truth or of proof, for example – is protected from the consequences of his folly if he is prepared to conform to the customs and

---

[1] This is notoriously the case with Descartes, Leibniz, Kant, and, of course, Frege, who is the founder of the modern analytic school of philosophy; but it is no less true of Berkeley, Hume, and Schopenhauer. Among twentieth century philosophers, Husserl, Russell, and Wittgenstein come to mind.

mores of his professional tribe. But the philosopher who follows him in adopting those opinions does not have that advantage.

In any case, it is one thing to flirt with anarchist views if one lives in a settled, just, and well-policed society, but quite another if one is living in a society in which the institutions of law and justice threaten to collapse. Twice in the last two hundred years mathematicians have been threatened with anarchy – during the early nineteenth century crisis in the foundations of analysis and the early twentieth century crisis in the foundations of set theory – and in both of these crises some of the best mathematicians of the day turned their attention to re-establishing order.

The essential elements of the set-theoretical approach to mathematics were already in place by the early 1920s, and by the middle of the century the central branches of the subject – arithmetic, algebra, geometry, analysis, and logic – had all been recast in the new set-theoretical style. The result is that set theory and its methods now permeate the whole of mathematics, and the idea that the foundations of all of mathematics, including mathematical logic and the axiomatic method, now lie in the theory of sets is not so much a theory as it is a straightforward observation.

Of course that, on its own, doesn't mean that set theory is a *suitable* foundation, or that it doesn't require justification. But it does mean that any would-be reformer had better have something more substantial than a handful of new formalised axioms emblazoned on his banner. And he had better take it into account that even mathematical logic rests on set-theoretical foundations, and so is not available to him unless he is prepared to reform *its* foundations.

The point of view embodied in this book

My approach to set theory rests on one central idea, namely, that the modern notion of *set* is a refined and generalised version of the classical Greek notion of *number* (*arithmos*), the notion of number found in Aristotle and expounded in Book VII of Euclid's *Elements*. I arrived at this view of set theory more than twenty years ago when I first read *Greek Mathematical Thought and the Origin of Algebra* by the distinguished philosopher and scholar Jacob Klein.

Klein's aim was to explain the rise of modern algebra in the sixteenth and seventeenth centuries, and the profound change in the traditional concept of number that accompanied it. But it struck me then with the force of revelation that the later, nineteenth century revolution in the foundations of mathematics, rooted, as it was, in Cantor's new theory of transfinite numbers, was essentially a return to Greek arithmetic as Klein had described it, but in a new, non-Euclidean form.

As Klein points out, in Greek mathematics a number was defined to be a finite plurality composed of units, so what the Greeks called a number (*arithmos*) is not at all like what *we* call a number but more like what we call a set. It is having a finite size (cardinality) which makes a plurality a "number" in this ancient sense. But what is it for a plurality to have a finite size? *That* is the crucial question.

The Greeks had a clear answer: for them a definite quantity, whether continuous like a line segment, or discrete – a "number" in their sense – must satisfy the axiom that *the whole is greater than the part*[2]. We obtain the modern, Cantorian notion of set from the ancient notion of number by abandoning this axiom and acknowledging as finite, in the root and original sense of "finite"– "limited", "bounded", "determinate", "definite" – certain pluralities (most notably, the plurality composed of all natural numbers, suitably defined) which on the traditional view would have been deemed infinite.

By abandoning the Euclidean axiom that the whole is greater than the part, Cantor arrived at a new, *non-Euclidean arithmetic*, just as Gauss, Lobachevski, and Bolyai arrived at *non-Euclidean geometry* by abandoning Euclid's Axiom of Parallels. Cantor's innovation can thus be seen as part of a wider nineteenth century program of correcting and generalising Euclid.

Cantor's non-Euclideanism is much more important even than that of the geometers, for his new version of classical arithmetic that we call *set theory* serves as the foundation for the whole of modern mathematics, including geometry itself. The set-theoretical approach to mathematics is now taken by the overwhelming majority of mathematicians: it is embodied in the mathematical curricula of all the major universities and is reflected in the standards of exposition demanded by all the major professional journals.

Since the whole of mathematics rests upon the notion of set, this view of set theory entails that the whole of mathematics is contained in *arithmetic*, provided that we understand "arithmetic" in its original and historic sense, and adopt the Cantorian version of finiteness. In set theory, and the mathematics which it supports and sustains, we have

---

[2] This is Common Notion 5 in Book I of the *Elements*.

made real the seventeenth century dream of a *mathesis universalis*, in which it is possible to express the exact part of our thought[3].

But what are the practical consequences of this way of looking at set theory for mathematics and its foundations? They are, I am convinced, profound and far-reaching, both for orthodox set-theoretical foundations, and for the several dissenting and heterodox schools that go under various names – "constructivism", "intuitionism", "finitism", "ultra-intuitionism", etc. – but whose common theme is the rejection of the great revolution in mathematical practice that was effected by Cantor and his followers.

For orthodox foundations the principal benefit of looking at things in this way is that it enables us to see that the central principles – axioms – of set theory are really *finiteness principles* which, in effect, assert that certain multitudes (pluralities, classes, species) are finite in extent and *for that reason* form sets.

Taking finitude (in Cantor's new sense) to be the defining characteristic of sets, as the Greeks took it (in their sense) to be the defining characteristic of numbers (*arithmoi*), allows us to see why the conventionally accepted axioms for set theory – the Zermelo–Fraenkel axioms – are both natural and obvious, and why the unrestricted comprehension principle, which is often *claimed* as natural and obvious (though, unfortunately, self-contradictory), is neither.

This is a matter of considerable significance, for there is a widespread view that all existing axiomatisations of set theory are more or less *ad hoc* attempts to salvage as much of the "natural" unrestricted comprehension principle – the principle that the extension of any well-defined property is a set – as is consistent with avoiding outright self-contradiction[4]. On this view set theory is an unhappy compromise, a botched job at best.

Hence the widespread idea that set theory must be presented as an axiomatic theory, indeed, as an axiomatic theory *formalised* in first order mathematical logic. It is felt that the very formalisation itself somehow confers mathematical respectability on the theory formalised. But this is a serious confusion, based on a profound misunderstanding of the logical and, indeed, *ontological* presuppositions that underlie the axiomatic method, formal or informal.

The mathematician's "set" is the mathematical logician's "domain of discourse", so conventional ("classical") mathematical logic is, like every

---

[3] Perhaps we might more appropriately describe the theory as an *arithmetica universalis*, a *universal arithmetic* which encompasses the whole of mathematics.

[4] See Quine's *Set Theory and its Logic*, for example.

other branch of mathematics, based on set theory[5]. This means, among other things, that we cannot use the standard axiomatic method to establish the theory of sets, on pain of a circularity in our reasoning.

Moreover, on the arithmetical conception of set the totality of all sets, since it is easily seen not to be a set, is not a conventional domain of discourse either. Hence quantification over that non-conventional domain (which is *absolutely infinite* in Cantor's terminology) cannot simply be *assumed* to conform to the conventional, "classical" laws.

As Brouwer repeatedly emphasised, since classical logic is the logic of the finite, the logic of infinite domains must employ different laws. And, of course, in the present context "finite domain" simply means "set". The consequences of this view for the *global* logic of set theory are discussed at length in Section 3.5 and Section 7.2.

But what are the consequences of this *arithmetical* conception of set for those who reject Cantor's innovations – the intuitionists, finitists, constructivists, etc., of the various schools?

Klein's profound scholarship is very much to the point here. For the one thing on which all these schools agree is the central importance of the system of natural numbers as the basic *datum* of mathematics. But Klein shows us that, on the contrary, the natural numbers are a recent invention: the oldest mathematical concept we have is that of *finite plurality* – the Greek notion of *arithmos*. This is so important a matter that I have devoted an entire chapter (Chapter 2) to its dicussion.

When the natural number system is taken as a primary *datum*, something simply "given", it is natural to see the principles of proof by mathematical induction and definition by recursion along that system as "given" as well. We gain our knowledge of these numbers when we learn to count them out and to calculate with them, so we are led to see these *processes* of counting out and calculating as *constitutive* of the very notion of natural number. The natural numbers are thus seen as what we arrive at in the *process* of counting out: 0, 1, 2, ..., where the dots of ellipsis, "...", are seen as somehow self-explanatory – after all, we all know how to continue the count no matter how we have taken it. But those dots of ellipsis contain the whole mystery of the notion of natural number!

If, however, we see the notion of natural number as a secondary

---

[5] Thus set theory stands the "logicist" view of Frege and Russell on its head: arithmetic isn't a branch of logic, logic is a branch of arithmetic, the non-Euclidean arithmetic of Cantor that we call set theory.

growth on the more fundamental notion of *arithmos* – finite plurality, in the original Greek sense of "finite" – then the principles of proof by induction and definition by recursion are no longer just "given" as part of the raw data, so to speak, but must be established from more fundamental, set-theoretical principles.

Nor are the *operations* of counting out or calculating to be taken as primary data: they too must be analysed in terms of more fundamental notions. We are thus led to reject the *operationalism* that all the anti-Cantorian schools share.

For us moderns numbers take their being from what we can *do* with them, namely count and calculate; but Greek "numbers" (*arithmoi*) were objects in their own right with simple, intelligible natures. Our natural numbers are things that we can (in principle) *construct* (by counting out to them); Greek numbers were simply "there", so to speak, and it would not have occurred to them that their numbers had to be "constructed" one unit at a time[6].

I am convinced that this operationalist conception of natural number is the central fallacy that underlies *all* our thinking about the foundations of mathematics. It is not confined to heretics, but is shared by the orthodox Cantorian majority. This *operationalist fallacy* consists in the assumption that the mere *description* of the natural number system as "what we obtain from zero by successive additions of one" suffices *on its own* to define the natural number system as a unique mathematical structure – the assumption that the operationalist description of the natural numbers is itself what provides us with a *guarantee* that the system of natural numbers has a unique, fixed structure.

Let me not be mistaken here: the existence of a unique (up to isomorphism) natural number system is a *theorem* of orthodox, Cantorian mathematics. The fallacy referred to thus does *not* consist in supposing that *there is* a unique system of natural numbers, but rather in supposing that the existence of this system, and its uniqueness, are immediately given and do not need to be *proved*. And if we abandon Cantorian orthodoxy we thereby abandon the means with which to prove these things.

---

[6] Oswald Spengler, who thought that the mathematics of a civilization held a clue to its innermost nature, contrasted the *Apollonian* culture of classical Greece, which was static and contemplative, with the *Faustian* culture of modern Europe, which is dynamic and active. Whatever the virtues of his general thesis, he seems to have got it right about the mathematics. The "operationalism" to which I refer here seems to be quintessentially Faustian in his sense, which perhaps explains its grip on our imaginations.

But if we acknowledge that the natural numbers are not *given* to us, the alternative, if we decide to reject Cantor's radical new version of finitude, is to return to arithmetic as practiced by the mathematicians of classical Greece, but equipped now with the more powerful and more subtle techniques of modern set theory. If we should decide to do this we should be going back to the very roots of our mathematical culture, back before Euclid and Eudoxus to its earliest Pythagorean origins. We should have to rethink our approach to geometry and the Calculus. It is a daunting prospect, though an exciting one.

The resulting theory, which I call *Euclidean set theory* by way of contrast with *Cantorian set theory*, the modern orthodoxy, is very like its Cantorian counterpart, except that Cantor's assumption that the species of natural numbers forms a set is replaced by the traditional Euclidean assumption that every set is strictly larger than any of its proper subsets.

This theory, not surprisingly, constitutes a radical departure from Cantorian orthodoxy. But it stands in even sharper contrast to the various operationalist theories which have been put forward as alternatives to that orthodoxy. So far from taking the natural numbers as given, Euclidean set theory forces us to take seriously the possibility that there is no unique natural number system, and that the various ways of attempting to form such a system lead to "natural number systems" of differing lengths.

But *should* we abandon Cantorian orthodoxy? There is obviously a *prima facie* case against the Cantorian account of finiteness, and, indeed, that case was made by some of his contemporaries. But against that there is the experience of more than one hundred years during which Cantor's ideas have been the engine driving a quite astonishing increase in the subtlety, power, and scope of mathematics.

Perhaps I should come clean with the reader and admit that I am attracted to the anti-Cantorian position. I put it no stronger than that because the issue is by no means clear-cut, and we do not yet know enough to be sure that the Cantorian conception of finiteness should be rejected.

Indeed, it seems to me that the common failing of all the advocates of the various alternatives to Cantorian orthodoxy is that they fail to appreciate how simple, coherent, and plausible are the foundational ideas that underlie it. These enthusiasts rush forward with their proposed cures without having first carried out a proper diagnosis to determine the nature of the disease, or even whether there *is* a disease that requires their ministrations.

Accordingly, I shall devote much of my attention to a careful, sympathetic, and detailed treatment of the Cantorian version of the theory. This is of interest in its own right, for this is the theory on which all of current mathematics rests. But it is also essential for those who are dissatisfied (or who fancy themselves dissatisfied) with the current orthodoxy, to discover what principles that orthodoxy really rests on, and to determine exactly where its strengths and weaknesses lie.

I have divided my exposition into four parts. *Part One* deals with the criteria which any attempt to provide foundations for mathematics must meet, and with the significance of the Greek approach to arithmetic for modern foundations.

Part Two is an exposition of the elements of set theory: the basic concepts of set theory, which neither require, nor admit of, definition, but in terms of which all other mathematical concepts are defined; and the basic truths of set theory, which neither require, nor admit of, proof, but which serve as the ultimate assumptions on which all mathematical proofs ultimately rest. The theory presented in *Part Two* is common to both the Cantorian and the Euclidean versions of set theory.

Part Three is an exposition of the Cantorian version of the theory and *Part Four* of the Euclidean. I have also included an appendix which deals with logical technicalities.

This, then, is the point of view embodied in this book: all of mathematics is rooted in *arithmetic*, for the central concept in mathematics is the concept of a plurality limited, or bounded, or determinate, or definite – in short, *finite* – in size, the ancient concept of *number* (*arithmos*).

From this it follows that there are really only two central tasks for the foundations of mathematics:

1. To determine what it is to be *finite*, that is to say, to discover what basic principles apply to finite pluralities *by virtue of their being finite*.
2. To determine what logical principles should govern our reasoning about *infinite* and *indefinite* pluralities, pluralities that are *not* finite in size.

On this analysis, all disputes about the proper foundations for mathematics arise out of differing solutions to these two central problems.

Such a way of looking at things is not easily to assimilate to any of the well-known "isms" that have served to describe the various approaches to the study of mathematical foundations in the twentieth century. But to my mind it has a certain attractive simplicity. Moreover, it is rooted in the history of mathematics and, indeed, takes as its starting point the oldest mathematical concept that we possess.

Acknowledgements

I am especially grateful to Moshé Machover, T.E. Conlon, and Dickon Lush, all of whom read drafts of my manuscript and made many useful suggestions on matters of both style and substance. H.E. Rose and J.C. Shepherdson gave me helpful advice on the material in Chapters 8 and 9.

Alberto Peruzzi invited me to spend a study leave at the University of Florence, during which I started the book, and my many stimulating conversations with him, then and since, have proved invaluable. I shall also always be grateful to him and his family for their wonderful hospitality which made my stay there so pleasant and productive.

My ideas on the concept of *arithmos* were developed in discussions over a number of years with C.J. Rowe and Paul Pritchard, both whom gave me valuable advice in drafting Chapter 2.

I must also acknowledge my debt to my former students Maria del Camino Cañon Loyes, Vincent Homolka, Kevin Lano, Peter Fletcher, Martyn Prigmore, and Nicholas Bamber for many stimulating and fruitful conversations on both technical and philosophical matters.

I am especially indebted to my former student S. J. Popham. The key results in Section 11.2 are due to him, as is the concept of a cumulation model (Definition 11.2.2) upon which they depend. These results are of fundamental importance for the whole approach to foundations taken in Part Four, for they provide crucial information on what *cannot* be proved from the most obvious and natural axioms.

Moreover, it was he who called my attention to the importance of the Ackermann simply infinite system treated in Section 10.6, and who, with the help of a suggestion by Robin Gandy, laid down a definition of that system in the special case of pure sets. Without his work this would have been a very different book.

Finally, I want to thank my wife, Anita, for her help and encouragement. She has given me invaluable advice on the prose composition, and her gentle but persistent criticism has greatly improved the exposition.

Part One
Preliminaries

It is the mark of an educated man to look for precision in each kind of enquiry just to the extent that the nature of the subject allows.

Aristotle

1

The Idea of Foundations for Mathematics

1.1 Why mathematics needs foundations

Mathematics differs from all the other sciences in requiring that its propositions be proved. Certainly no one will deny that proof is the goal of mathematics, even though there may be disagreement over whether, or to what extent, that goal is achieved. But you cannot prove a proposition unless the concepts employed in formulating it are clear and unambiguous, and this means that the concepts used in a proof either must be basic concepts that can be grasped directly and can be seen immediately to be clear and unambiguous, or must be rigorously *defined* in terms of such basic concepts. Mathematics, therefore, since it is about proof is also about definition.

Now definition and proof are both species of the genus *explanation*: to define something is to explain *what* it is; to prove something is to explain *why* it is true. All scholars and scientists, of course, deal in explanation. But mathematicians are unique in that they intend their explanations to be complete and final: that must be their aim and ideal, even if they fail to realise it in full measure. From these simple observations many consequences flow.

Perhaps the most important of them concerns the mathematician's claims to truth. Because he deals in proof, those claims must be absolute and unqualified. Whether they are justified, either in general, or in particular cases, is, of course, quite another matter: but that they are, in fact, made cannot be denied without stripping the word "proof" of all meaning. To claim to have proved something is to claim, among other things, that it is true, that its truth is an objective fact, and that its being so is independent of all authority and of our wishes, customs, habits, and interests. Where there are no truth and falsehood, objectively determined, there can be no proof; and where there is no proof there can be no mathematics.

No doubt all of this is at odds with the *Zeitgeist*[1]: it would seem that we must come to terms with the fact that when there is disagreement about a genuine mathematical proposition, someone must be right and someone must be wrong. But the requirement that we must lay unqualified claims to truth in mathematics is quite compatible with our maintaining a prudent and healthy scepticism about such claims: what it rules out is dogmatic or theoretical scepticism.

You may, as a mathematician, reasonably doubt that such and such a theorem is true, or that such and such a proof is valid: indeed, there are many occasions on which it is your professional duty to do this, even to the point of struggling to maintain doubt that is crumbling under the pressure of argument: for it is precisely when you begin to settle into a conviction that you are most liable to be taken in by a specious but plausible line of reasoning. When your business is judging proofs you must become a kind of professional sceptic. But scepticism, properly understood, is an attitude of mind, not a theory, and you cannot systematically maintain that there is no such thing as a true proposition or a valid argument and remain a mathematician.

A proof, to be genuine, must still all reasonable doubts as to the truth of the proposition proved. But the doubts to be stilled are those that pertain to that proposition: a proof need not, indeed cannot, address general sceptical doubts. Anyone who proposes to pass judgement on the validity of an intended proof must address his attention to the propositions and inferences contained in the argument actually presented. It won't do to object to a particular argument on the ground that all argument is suspect. The fact, for example, that people often make mistakes in calculating sums does not provide grounds for concluding that any particular calculation is incorrect, or even uncertain: each must be judged separately, on its own merits.

In the final analysis, there are only two grounds upon which you may reasonably call the efficacy of a purported proof into question: you may dispute the presuppositions upon which the argument rests, or you may dispute the validity of one or more of the inferences by means of which the argument advances to its conclusion. If, after careful, and perhaps prolonged, reflection, you cannot raise an objection to an argument on

---

[1] Cantor complained of the "Pyrrhonic and Academic scepticism" that prevailed in his day. *Plus ça change* ...

1.1 Why mathematics needs foundations

either of these two grounds, then you should accept it as valid and its conclusion as true.

Here we must include among the presuppositions of a proof not only the truth of the propositions that are taken as unproved starting points of the argument, but also the clarity, unambiguity, and unequivocality of the concepts in which the propositions employed in the argument are couched.

Of course in practice, actual proofs start from previously established theorems and employ previously defined concepts. But if we persist in our analysis of a proof, always insisting that, where possible, assertions should be justified and concepts defined, we shall eventually reach the ultimate presuppositions of the proof: the propositions that must be accepted as true without further argument and the concepts that must be understood without further definition. Of course when I say that these things *must* be accepted without proof or understood without definition I mean that they must be so accepted and so understood if the given proof is to be judged valid and its conclusion true.

If we were to carry out such a complete analysis on all mathematical proofs, the totality of ultimate presuppositions we should then arrive at would obviously constitute the foundations upon which mathematics rests. Naturally, I'm not planning to embark on the enterprise of analysing actual proofs to discover those foundations. My point here is rather that solely in virtue of the fact that mathematics is about proof and definition it must of necessity *have* foundations, ultimate presuppositions – unproved assertions and undefined concepts – upon which its proofs and definitions rest.

Of course that observation is compatible with there being a motley of disparate principles and concepts underlying the various branches of the subject, with no overarching ideas that impose unity on the whole. The question thus arises whether it is possible to discover a small number of clear basic concepts and true first principles from which the whole of mathematics can be systematically developed: that is, I suspect, what most mathematicians have in mind when they speak of providing foundations for mathematics.

From the very beginnings of the subject, that is to say, from the time when proof became central in mathematics, mathematicians and philosophers have been aware of the need to provide for foundations in the ideal and general sense just described. But there are particular, and pressing, practical reasons why present day mathematics needs foundations in this sense. Mathematics today is, for mathematicians, radically different from what it was in the relatively recent past, say one hundred and fifty years ago, and, indeed, come to that, from what it is now for professional *users* of mathematics, such as physicists, engineers, and economists. The difference lies in the greatly enhanced role that definition now plays. Present day mathematics deals with rigorously defined mathematical structures: groups, rings, topological spaces, manifolds, categories, etc. Traditional mathematics, on the other hand, was based on geometrical and kinematical intuition. Its objects were idealised shapes and motions. They could be imagined – pictured in the mind's eye – but they could not be rigorously defined.

Now it is precisely in our possession of powerful and general methods of rigorous definition that we are unquestionably superior to our mathematical predecessors. However, this superiority does not consist primarily in our basic definitions being more certain or more secure – although, indeed, they are more certain and secure, as are the proofs that employ them – but rather in the fact that they can be generalised and modified to apply in circumstances widely remote from those in which they were originally conceived.

There is a certain irony here. For although the earliest pioneers of modern rigour – Weierstrass for example – set out in search of safer, more certain methods of definition and argument by cutting mathematics free of its former *logical* dependence on geometrical and kinematical intuitions, they have, paradoxically, enormously enlarged the domain in which those intuitions can be applied.

When we give a rigorous "analytic" (i.e. non-geometrical, non-kinematical) definition of "limit" or "derivative" we do, undoubtedly, attain a greater certainty in our proofs. But, what is just as important, we can *generalise* a rigorous, analytic definition, while a definition based on geometrical or kinematical intuition remains tied to what we can actually visualise. By purging our definitions of their *logical* dependence on geometrical and kinematical intuition, we clear the way for transferring our insights based on that intuition to "spaces", for example, infinite dimensional ones, in which intuition, in the Kantian sense of sensual intuition – images in the mind's eye – is impossible. The mathematicians of the nineteenth century noticed that by a novel use of definition they could convert problems in geometry into problems in algebra and set theory, which are more amenable to rigorous treatment[2]. What they didn't

---

[2] Descartes saw that problems in geometry could be converted into problems in algebra. But his algebra, the algebra of real numbers, rested logically on geometrical conceptions.

1.1 Why mathematics needs foundations

foresee – how could they have foreseen it? – was the enormous increase in the scope of mathematics that these new methods made possible. By banishing "intuitive" ("*anschaulich*") geometry from the logical foundations of mathematics, they inadvertently, and quite unintentionally, gave that geometry a new lease of life.

But it was the technique of axiomatic definition that made the transition from traditional to modern mathematics possible. Naively, an axiomatic definition defines a *kind* or *species* of mathematical structure (e.g. *groups, rings, topological spaces, categories,* etc.) by laying down conditions or *axioms* that a structure must satisfy in order to be of that kind. Axiomatic definition is the principal tool employed in purging the foundations of mathematics of all *logical* dependence on geometrical and kinematical intuition. It follows that if we wish to understand *how* geometry has disappeared from the logical foundations of mathematics, we must understand the logical underpinnings of axiomatic definition. To understand those underpinnings is to understand how set theory provides the foundations for all mathematics.

Here we come to the central reason why modern mathematics especially stands in need of a careful examination and exposition of its foundations. For there is widespread confusion concerning the very nature of the modern axiomatic method and, in particular, concerning the essential and ineliminable role set theory plays in that method[3]. I shall discuss this critical issue later in some detail[4]. But for now, suffice it to say that the *logical* dependence of axiomatics on the set-theoretical concept of mathematical structure requires that set theory already be in place before an account of the axiomatic method, understood in the modern sense of axiomatic *definition*, can be given. It follows necessarily, therefore, that *we cannot use the modern axiomatic method to establish the theory of sets*. We cannot, in particular, simply employ the machinery of modern logic, modern *mathematical* logic, in establishing the theory of sets.

There is, to be sure, such a thing as "axiomatic set theory"; but although this theory is of central importance for the study of the foundations of mathematics, *it is a matter of logic* that it cannot itself, as an axiomatic theory in the modern sense, serve as a foundation for mathematics. Set theory, as a foundational theory, is, indeed, an axiomatic

---

The novelty introduced by later mathematicians was to base the algebra of real numbers on set theory, using the technique of axiomatic definition.

[3] I have discussed this matter at some length in my article "What is required of a foundation for mathematics?" to which I refer the interested reader.

[4] Chapter 6, especially Sections 6.2, 6.3 and 6.4.

theory, but in the original sense of "axiomatic" that applies to traditional Euclidean geometry as traditionally understood. The axioms of set theory are not conditions that single out a class of interpretations, as are, for example, Hilbert's axioms for geometry. On the contrary, they are fundamental truths expressed in a language whose fundamental vocabulary must be understood *prior* to the laying down of the axioms. That, in any case, must be the view taken of those axioms by anyone who embarks on the enterprise of expounding the set-theoretical foundations of mathematics. Whether, or to what extent, any such enterprise is successful, whether, or to what extent, the axioms can legitimately be regarded in this manner, is, of course, a matter for judgement. But it will be a central part of my task to show that they can be so regarded.

1.2 What the foundations of mathematics consist in

As I have just explained, the foundations of mathematics comprise those ideas, principles, and techniques that make rigorous proof and rigorous definition possible. To expound those foundations systematically, one must provide three things: an account of the *elements* of mathematics, an account of its *principles*, and an account of its *methods*.

The *elements* of mathematics are its basic notions: the fundamental *concepts* of mathematics, the *objects* that fall under those concepts, and the fundamental *relations* and *operations* that apply to them. These basic notions are those that neither require, nor admit of, proper mathematical definition, but in terms of which all other mathematical notions are ultimately defined. Insofar as these basic notions of mathematics are clear and unambiguous, the basic propositions of mathematics, which employ them, will also be clear and unambiguous. In particular, those propositions will have objectively determined truth values: the truth or falsity of such a proposition will be a question of objective fact, not a mere matter of convention or of agreement among experts.

The *principles* of mathematics are its *axioms*, properly so called. They are fundamental propositions that, although true, neither require, nor admit of, proof; and they constitute the ultimate and primary assumptions upon which all mathematical argument finally rests. There is no sense in which the axioms can be construed as giving or determining the meaning of the vocabulary in which they are couched. On the contrary, the meanings of the various items of vocabulary must be given, in advance of the laying down of the axioms, in terms of the elements of the theory, antecedently understood.

1.2 What the foundations of mathematics consist in

The *methods* of mathematics are to be given by laying down the canons of definition and of argument that govern the introduction of new concepts and the construction of proofs. This amounts to specifying the *logic of mathematics*, which we must take care to distinguish from *mathematical logic*: mathematical logic is a particular branch of mathematics, whereas the logic of mathematics governs all mathematical reasoning, including reasoning about the formal languages of mathematical logic and their interpretations. The logic of mathematics cannot be purely formal, since the propositions to which it applies have fixed meanings and the proofs it sanctions are meaningful arguments, not just formal assemblages of signs.

Here it must be said that the need to include an explicit account of logical method is a peculiarity of modern mathematics. Under the Euclidean dispensation, before the advent of set theory as a foundational theory, and when definition played a much more modest role in mathematics, one could, or, in any event, one did, take one's logic more for granted. But with the rise of modern mathematics, in which definition has moved to the centre of the stage, and where mathematicians have gone beyond even Euclid in their quest for accuracy and rigour, it has become necessary to include logical methods among the foundations of the subject. In fact, the central problem here is to explain the logical principles that underlie the modern axiomatic method. This will raise questions of the logic of generality, of the *global* logic of mathematics, that are especially important, and especially delicate, as we shall see[5].

A systematic presentation of the foundations of mathematics thus consists in a presentation of its elements, its principles, and its logical methods. In presenting these things we must strive for *simplicity, clarity, brevity*, and *unity*. These are not mere empty slogans. The requirements for *simplicity* and *clarity* mean, for example, that we cannot take sophisticated mathematical concepts, such as the concept of a category or the concept of a topos, as *foundational* concepts, and that we cannot incorporate "deep" and controversial philosophical theories in our mathematical foundations. Otherwise no one will understand our definitions and no one will be convinced by our proofs.

The ideal of *brevity*, surely, speaks for itself. *Unity* has always been a central goal: unity in principles, unity in logical technique, unity in standards of rigour. With the stupendous expansion that has taken place in mathematics since the middle of the nineteenth century the

---

[5] I shall discuss this point in Sections 3.4 and 3.5.

need to strive for unity in foundations is even more pressing than ever: mathematics must not be allowed to degenerate into a motley of mutually incomprehensible subdisciplines.

This, then, is what an exposition of the foundations of mathematics must contain, and these are the ideals that must inform such an exposition. But the task of *expounding* the foundations of mathematics must be kept separate from the task of *justifying* them: this is required by the logical role that those foundations are called upon to play. A little reflection will disclose, indeed it is obvious, that there can be no question of a *rigorous* justification of proposed foundations: if such a justification were given, then the elements, principles, and logical methods presupposed by that justification would themselves become the foundations of mathematics, properly so called.

Thus the clarity of basic concepts (if they really are basic) and the truth of first principles (if they really are *first* principles) cannot be established by rigorous argument of the sort that mathematicians are accustomed to. Insofar as these things are evident they must be *self*-evident. But that is not to say they are beyond justification; it is only to say that the justification must proceed by persuasion rather than by demonstration: it must be dialectical rather than apodeictic.

In any case, self-evidence, unlike truth, admits of degrees, and, as we shall see, the set-theoretical axioms that sustain modern mathematics are self-evident in differing degrees. One of them – indeed, the most important of them, namely Cantor's Axiom, the so-called Axiom of Infinity – has scarcely any claim to self-evidence at all, and it is one of my principal aims to investigate the possibility, and the consequences, of rejecting it. But what is essential here is this: when we lay down a proposition *as an axiom* what we are thereby claiming directly is that it is *true*; the claim that it is self-evident is, at most, only implicit, and, in any case, is *logically* irrelevant.

1.3 What the foundations of mathematics need not include

It is obvious to anyone who teaches mathematics that means must be devised for presenting its foundations simply, yet rigorously and thoroughly, to apprentice mathematicians: they must be told about sets, about ordered pairs and Cartesian products, about functions and relations; they must be made to grasp the idea of mathematical structure, and of a morphology-preserving map between such structures; more generally, they must be taught the techniques of rigorous proof and rigorous

1.3 What the foundations of mathematics need not include

definition, and, especially, must be led to understand the ideas and strategies that inform the method of *axiomatic definition* – the central technical idea underlying modern mathematics. Much of this is confusing, none of it is easy, and all of it is necessary: those who do not master these foundations will find the road to modern mathematics barred to them.

These practical necessities remind us that in laying down the foundations of mathematics we are actually engaged in mathematics proper. Those foundations are an integral and essential part of mathematics itself. Of course, when we reflect deeply on such fundamental matters, we are bound to encounter profound questions of a general philosophical character. Sometimes we may be forced to face up to them. *But we should make every effort to avoid incorporating purely speculative philosophical ideas into mathematical foundations, properly so called.* That this is necessary from the standpoint of mathematics should be obvious. Mathematicians, like infantrymen, must march off to battle carrying only such equipment as is absolutely essential to their task. But philosophers, too, will benefit if the foundations of mathematics are kept free of philosophical controversy insofar as that is possible. For it is useful to them to know what are the minimal philosophical presuppositions upon which mathematics can rest.

Put in these general terms, all this may seem rather obvious and unexceptionable. But the matter may take on an entirely different colour when I draw what I see as the necessary consequences of these observations. In particular, I take the view that the foundations of mathematics do not require, and therefore should not include, a general theory of the meaning of mathematical propositions, or a general theory of mathematical truth, or a general theory of how mathematical knowledge is acquired. In mathematics it is sufficient if our propositions *have* clear meanings; it is not our business, as mathematicians, to account for what having a clear meaning consists in. Our theorems must be true and our proofs valid; but we are not required to *say* what a proposition's being true or an argument's being valid amounts to. Mathematicians must strive to acquire mathematical knowledge; but they do not need a theory of what the acquisition of such knowledge consists in merely in order to acquire it. Such a theory would belong to psychology, not to mathematics.

The mathematician studies mathematical structures, such as groups or topological spaces, just as the entomologist studies insects or the palaeontologist fossils. It would be an absurd impertinence to demand of an entomologist that he supplement his descriptions of the behaviour and physiology of insects with an account of how it is that human beings can acquire knowledge of this sort, or communicate it to one another once they have acquired it. I say that the same should apply to the mathematician: we may insist that his definitions be precise, his theorems be true, and his proofs be valid. But that is all we can sensibly require, or have any reason to expect.

Of course some may argue that, unlike insects or fossils, mathematical structures are "abstract" or "ideal" entities, that they exist, if at all, only in the minds of mathematicians, and that, in consequence of having these gossamer and insubstantial things for its subject matter, mathematics gives rise to ontological and epistemological difficulties unprecedented in the other sciences, difficulties which must be addressed when the foundations of mathematical science are laid down. I am convinced, however, that this is a mistake. It is perfectly true that in the past mathematics was thought to have certain characteristic abstract or ideal "mathematical objects" for its subject matter. But such views are outmoded, for mathematicians can now use the modern axiomatic method to replace reference to those peculiar "objects" with discourse about mathematical structures. Mathematical *structures*, however, are not the "abstract" and "ideal" entities that the mathematical *objects* of tradition were thought to be, and do not give rise to the ontological and epistemological difficulties inherent in that tradition[6].

The great philosophical questions of meaning, truth and knowledge are no doubt of considerable interest in themselves; but it is not necessary to solve them before getting on with the business of proving, say, that a continuous function on a closed interval assumes a maximum, or that every integer is uniquely factorable into a product of primes. This is indeed fortunate, since *definitive* answers to these philosophical questions are nowhere in sight. Certainly there is not the remotest prospect of *universal agreement* on such answers. But there must be universal, or near universal agreement on what constitutes a valid proof or definition in mathematics – and, indeed, there is. If there were not, the subject would be in chaos.

We must also keep separate from foundations those general questions that, though not really philosophical, are what mathematicians themselves

---

[6] This is an important, difficult, and, I must confess, on the face of it, *controversial* point. What is essential to the claim I am making is the distinction between the "abstract" or "ideal" character of traditional mathematical *objects* and the arithmetical (set-theoretical) character of modern mathematical *structures*. Here I have simply stated, without argument, what I take to be the case. The argument is given, and at some considerable length, in what follows, principally in Chapters 2 and 6.

1.3 What the foundations of mathematics need not include

might call "philosophical": questions of the significance (or otherwise) of theories, of the suitability of mathematical definitions, strategic questions about the importance of problems, or about the most useful ways to tackle them, questions about the overall organisation of mathematics, questions about the relative importance of its various branches, ... These are questions about which all mathematicians, even those with no real philosophical interests, are called upon to think from time to time. We must not underestimate their importance, for it is often decisive, though usually only the very best mathematicians make significant contributions here. But these questions, which call for sound judgement and large experience, cannot be taken to be part of the foundations of mathematics, properly so called, although they are inextricably bound up with its practice. What belongs to mathematics proper – and that includes its foundations – cannot be speculative, or evaluative, or controversial. Indeed, the very word "mathematics" comes from the Greek *"mathêma"* which means simply "what can be taught and learnt", in other words "what is cut-and-dried".

In mathematics our aim is to start from what is simple and obvious, our basic concepts and axioms, and to proceed by obvious steps, our definitions and our inferences, to obtain what is often complex and difficult, our general concepts and our theorems. If this sort of thing is to work, we must strive to make both the starting points and the individual steps as transparent and as obvious as we can make them.

Accordingly, it is no reproach to an account of the foundations of mathematics that its basic concepts and its axioms are remote from the actual practice, and the immediate concerns, of most mathematicians. On the contrary, that very remoteness is rather a measure of the logical depth of our definitions and theorems, and, as such, is probably the best indication we have that our basic concepts and axioms are, in fact, suitable. Such a reproach has frequently been levelled at modern set-theoretical foundations. But that is to misconceive the purpose they are called upon to serve. Whatever the shortcomings of those foundations, remoteness from practice is *not* among them.

To be sure, the basic concepts and axioms of set theory are, indeed, remote from practice. One cannot gain insight into group theory or functional analysis or algebraic geometry by contemplating them. But that fact, though incontrovertible, is utterly irrelevant. The *only* question relevant here is whether those concepts and axioms do, in fact, logically sustain such disciplines. And *that* they unquestionably do.

1.4 Platonism

On the face of it mathematics is full of references to special mathematical objects, "abstract" or "ideal" things that we cannot touch or see. The mathematician's triangles, for example, are not to be identified with those he draws on the blackboard, or with the architect's or the land surveyor's. But even mathematical triangles seem relatively "concrete" when compared to other things that mathematicians regularly talk about: natural numbers, real numbers, functions, "spaces" and "structures" of various kinds .... What are these things? Do they really exist? And if so, how, and in what sense? These questions are as old as mathematics itself. Moreover, they have been a central preoccupation of philosophers in the European tradition since before the time of Plato.

Indeed, Plato himself has been invoked in the present day debate on these matters, for it is now the fashion to describe as "Platonism" the naive idea that the peculiar objects mathematicians talk about exist in their own special way – that they are what they are, so to speak – and that they really have the properties and relations that mathematicians say they have. "Platonism", understood in this sense, is often used as a term of abuse or, alternatively, adopted as a badge of defiance.

But the relation of modern "Platonism" to the opinions on mathematics actually held by Plato and his disciples is not at all what it is commonly supposed to be. It is true that Plato posited a special category of eternal and unchanging objects, the "Mathematicals" or "Intermediates", occupying a place in the realm of being midway between ordinary objects of sense and the Platonic Ideas or Forms. But it is clear that Plato did not regard what we should call "sets" and he called "numbers" (*arithmoi*) as necessarily belonging to the class of Intermediates. On the contrary, only sets of a certain special kind were classed as mathematical objects in his special sense. This is a matter of some significance, as I shall make clear when I come to discuss the question of set existence in the next chapter.

In its modern usage the term "Platonism" does less than justice to the historical facts. Moreover, that usage rests upon a classification of objects into the "abstract" and the "concrete" that is so crude, so simple-minded, and so undiscriminating as to be useless. It is best abandoned. In any case, as I shall show in the next chapter, modern set theory, which in the present day estimation is the very quintessence of "Platonism", is fundamentally Aristotelian, not Platonic, in spirit.

Nevertheless, one of Plato's principal doctrines in the philosophy of

1.4 Platonism

mathematics is an essential component of modern set-theoretical foundations. I mean his anti-operationalism:

> ... no one who has even a slight acquaintance with geometry will deny that the nature of this science is in flat contradiction with the absurd language used by mathematicians, for want of better terms. They constantly talk of "operations" like "squaring", "applying", "adding", and so on, as if the object were to *do* something, whereas the true purpose of the whole subject is knowledge – knowledge, moreover, of what eternally exists, not of anything that comes to be this or that at some time and ceases to be[7].

Plato seems to be invoking his doctrine of Intermediates here, but we should ask whether his doing so is really necessary to his central point. Given that the truths of mathematics are timeless truths, does it then follow that they must, of necessity, be truths about timeless entities like Plato's Intermediates? Such a supposition is natural enough, I suppose, but is it really essential? The question is a deep and difficult one, and deserves a more than merely cursory examination. But that question aside, Plato is surely right in holding that mathematics is not primarily a matter of *doing*, but rather of *knowing*.

I take operationalism in mathematics to be the doctrine that the foundations of mathematics are to be discovered in the activities (actual or idealised) of mathematicians when they count, calculate, write down proofs, invent symbols, draw diagrams, and so on. No doubt we ought to be chary of following Plato in positing "mathematical objects", and, indeed, modern mathematics provides us with the conceptual tools which make this possible; but we ought all to account ourselves "Platonists" in this sense: *considerations of human activities and capacities, actual or idealised, have no place in the foundations of mathematics, and we must therefore make every effort to exclude them from the elements, principles, and methods, upon which we intend to base our mathematics.*

This is no easy matter, for the art of mathematics consists, in large part, in finding suitable symbolic expression for our concepts and propositions with a view to replacing complicated conceptual thought with mere symbolic manipulation – letting our notation do our thinking for us, so to speak[8].

---

[7] *Republic* 527a.
[8] No clearer illustration of this can be given than by contrasting the ancient theory of ratio and proportion given in Book V of the *Elements* with the modern, symbolic handling of cognate material in the algebra of real numbers. Euclid's treatment is complicated and cumbrous, and is carried out purely conceptually, so that the exposition is almost entirely verbal. The modern theory, by contrast, is entirely algebraic, that is to say, is largely a matter of manipulating symbols, so that complicated arguments in Euclid

When we are engaged in mathematics our attention is constantly shifting between the notation we employ and the subject matter. Our concern is now with the symbols themselves as syntactico-combinatorial objects, now with the things for which they are the signs; now those symbols are the objects of investigation, now the medium of expression. Often it proves necessary to find suitable *objective correlatives* for our symbols, that is to say, non-linguistic, non-symbolic objects corresponding to certain symbols or symbol combinations, or to processes of algebraic or numerical calculation. In this way we render our discourse objective by purging it of its "human, all too human" elements; and it is the great strength of modern mathematics that it provides us with powerful techniques for accomplishing this necessary purge.

The need to exclude operationalism, in all of its guises, from the foundations of mathematics is not something that can be established in a few paragraphs of argument: it is the central lesson of the whole modern movement in mathematics, a lesson which mathematicians absorb almost unconsciously in learning their trade, and practise without even reflecting on it. It is built into the conventions of expository style that every mathematician must master. But many of the most widely used and fundamental concepts in mathematics have an operationalist air about them in consequence of their origins in the contingencies of mathematical practice as that practice has developed historically: the concepts of "natural number", "ordered pair", "function", and "relation" are all of this character. To objectify these concepts, so to speak, is, inevitably, to introduce some appearance of arbitrariness and artificiality into our mathematical discourse: we must face up to this as we cannot avoid it. But the bedrock, the concepts in terms of which these concepts are defined, must be free of any operationalist taint.

---

correspond in the modern theory to simple syntactic operations. Of course there is a price to pay for such facility, and whereas it is clear *ab initio* just what Euclid is talking about, mathematicians had to wait until the end of the nineteenth century before an adequate account of the facts that justify the symbolic manipulations of real algebra was given.

2
Simple Arithmetic

2.1 The origin of the natural numbers

The natural numbers $0, 1, 2, \ldots$, as we now understand them, are not simply given to us as part of the "raw data" of mathematics. On the contrary, these numbers were invented, indeed invented fairly recently, along with rational, irrational and negative numbers. There is, in fact, something distinctly unnatural about our "natural" numbers.

This is so important a matter that I want to make doubly sure that no one misunderstands me: when I say that these numbers were invented I am making a particular, historical point, not a general, philosophical one. It is not my intention to resurrect the philosophical claim that mathematics is invention not discovery (surely it is both), nor the more particular claim that the natural numbers are "mental constructions" or anything of that sort. On the contrary, what I am talking about is an actual, historical process of invention that began sometime in the late middle ages and culminated in the late seventeenth century, by which time mathematicians had arrived at what is essentially our modern conception of real number. In the course of this process, the concept of number was drastically altered – no, that is not strong enough: in the course of this process the word "number" was stripped of its customary and traditional meaning to be assigned an entirely new meaning, one which had scarcely anything in common with the original[1].

Not only the fact, but also something of the actual nature of the change in the meaning of "number", can be deduced from the definition given by Isaac Newton in his *Universal Arithmetic*:

By a *Number* we understand not so much a Multitude of Unities, as the abstracted

---

[1] Naturally I do not mean to suggest that this change of meaning was a phenomenon confined to English alone. A similar change occurred in other European languages.

Ratio of any Quantity to another Quantity of the same kind, which we take for Unity.

From this brief passage we can glean several important facts. It is clear that Newton recognised "quantities" of various kinds, and these quantities were not "numbers" in Newton's sense; for the latter are said to be abstracted *ratios*[2] of two quantities of the same kind. Moreover, the notion of "number" that he mentions only to reject ("By a *Number* we mean not so much" – that is to say, not at all – "a Multitude of Unities but ... ") must, by the logic of the sentence, have been the notion of number that some of his readers might have expected or, at least, have been aware of. In fact, it recalls the definition given by Euclid in Book VII of the *Elements*:

> A number (*arithmos*) is a multitude composed of units.

So Newton actually tells us that he doesn't mean by "number" what Euclid meant[3]. What Newton means by "number" is what we should mean by "real number", or, at least, very like what we should mean: for we must not lose sight of the fact that it was important for Newton, as it would not be for a modern mathematician, to ground his theory in the ancient science of quantity, the theory of ratio and proportion expounded in Books V and VI of the *Elements*. Newton's "Quantity" is Euclid's "*megethos*"; such a quantity is, for example, a line, or a surface, or a solid, or a time.

These things may have been regarded as *idealisations* of physical lines, surfaces, solids, and durations, but are clearly not *abstractions* in the sense that Newton's "numbers", or indeed ours, are abstractions. It was, of course, this ancient science of quantity that the new science of (real) number replaced.

How crucially important it is in mathematics to choose the right terminology! The use of "number" for this new concept was especially unfortunate; and there were perfectly good alternatives ready to hand: "(abstract) ratio" and "(abstract) quantity" immediately suggest them-

---

[2] That is to say, relationships in respect of size. (See Euclid's *Elements*, Book V, Definition 3.)

[3] In fact there may already have been a shift in meaning from Euclid's "multitude *composed* of units", which can only mean what we should call a *set* of units, to Newton's "Multitude of Unities" which may refer, not to a set of units, but to its (abstract) cardinality. (See Klein's *Greek Mathematical Thought and the Origin of Algebra*, Chapter 12, especially pp. 201–202). The question of how and why the new notion of number arose is a fascinating one, but for our purposes here the fact, and the nature, of the change in the meaning of "number" are all that are directly relevant.

selves. By choosing this oldest of mathematical words to name what was then the newest of mathematical concepts, the mathematicians of the seventeenth century virtually ensured that their successors would eventually lose sight of the very concept of number as it was understood in antiquity.

In *Frege: Philosophy of Mathematics*, for example, Michael Dummett writes

That the number of objects of a given kind is the set of those objects is sufficiently absurd to need no refuting. (p. 82)

No doubt this view *is* absurd as an account of *our* notion of number. But this "absurd" view was held by Plato, Aristotle, Euclid, Aquinas, and Ockham, and was, as we have seen, acknowledged by Newton, who, however, failed to remark upon its "absurdity", even while he was in the course of explicitly rejecting it.

In fact the ancient concept of number provides, as I intend to show, a simpler, more straightforward, and more natural account of the facts that underlie simple arithmetic than does the modern notion of "natural" number; and if one adopts it one is not burdened like Frege, and, indeed, Dummett, with the task of explaining what *things* those "natural" numbers are.

The original notion of number is so important and so fundamental that it could not remain suppressed. It had eventually to reappear, even if only under another name: what our ancestors knew as "numbers" we now call "sets".

2.2 The abstractness of the natural numbers

I think it unlikely that any modern mathematician would be drawn to Newton's account of number. Of course that account could not be taken as a *definition* of "number", since it does not meet the modern requirements of rigour. But that apart, it is not so much its vagueness as its particularity that seems unsatisfactory. When Newton speaks of "abstracted ratios of quantities" he has something too definite in mind, something quite alien to the modern mathematical sensibility. His abstractions are somehow too concrete for our taste, if I may put it in that somewhat paradoxical way.

This comes of Newton's desire, which I alluded to earlier, to base the new science of number on the old science of concrete quantities. We moderns are, in any case, chary of mixing our natural numbers quite so thoroughly with our reals; and we want all our numbers kept logically independent of geometry.

It seems, then, that Newton's account of the abstractness of numbers won't do. And yet we all agree, do we not, that the modern notion of number, in general, and of natural number, in particular, is highly abstract? The question I want to address now is: "abstract" in what sense?

The "abstractness" of our modern natural numbers is something much simpler, much more insubstantial, than the abstractness of Newton's numbers. Indeed, the abstractness of our numbers is a *fact* about the way we view them, not a *theory* about their natures. It manifests itself in the naive idea that number words and numerals are names, or signs, for particular objects. This idea imposes itself on us, *inter alia*, by our use of certain familiar expressions (e.g. "the number five") and by the way that we understand simple numerical equations (e.g. $128 + 279 = 407$).

The interpretation of "$128 + 279 = 407$" that perhaps most naturally suggests itself is this: if we perform the operation of addition on the natural numbers 128 and 279 (in that order) then we obtain the natural number 407. This way of understanding such equations is, I submit, suggested to us both by the syntactic form of the equations themselves, and by our rules and methods for calculating sums. For the abstract operation of addition here corresponds to the actual procedure of calculation (hence "operation", with its suggestion that something is to be *done*), and the abstract numbers to which that operation applies correspond to the numerals employed in such calculations.

Natural numbers thus present themselves to us as those things, whatever they are, that correspond to the numerals and letters we use in symbolic calculation. They are generated by our notation, and by the syntactic and algorithmic rules that govern its employment. This, no doubt, accounts for their peculiarly thin and insubstantial character, even as "abstract objects".

Abstractions of this *symbol generated* sort, though unknown to the Greeks, are quite common in modern mathematics. Some of them play indispensable technical roles: ordered pairs are an obvious example[4]. But wherever such symbol generated abstractions occur, they are a potential source of perplexity and confusion. For it is never obvious that

---

[4] Functions and relations are also symbol generated abstractions of this sort. I shall discuss the logical status of these key notions in Chapter 4.

there really is anything to which they naturally correspond, outside our symbols themselves that is. They are epiphenomena of our notation.

The naive idea of the natural numbers that I have described here – the idea that they are the particular abstract things named by our number words and numerals – this idea scarcely constitutes a theory, although it is sometimes rhetorically inflated into one. It is really rather a starting point for theories, philosophical or mathematical: it is what those theories have to explain or to explain away. It poses the following dilemma: if there are, in fact, objects of which our number words and numerals are the names, what are those objects? If there are no such objects, what is arithmetic about?

2.3 The original conception of number

Let us consider the idea of number that our modern idea of "natural" number has supplanted, the classical Greek concept of *arithmos*[5]. On that conception, a number (*arithmos*) is a finite plurality (multitude, multiplicity) composed of units, where a unit is whatever counts(!) as one thing in the number under consideration. Thus Trigger, Champion, and Red Rum constitute a number of horses, and each unit in this number is a horse; red, yellow, blue, and green constitute a number of colours, and each unit is a colour. This original meaning of "number" still survives in English, as when we say, "Lieutenant Lightoller was included among the *number of survivors* in the wreck of the Titanic".

In the two examples I have given the units are homogeneous: all of them are horses or all of them are colours. Such numbers provide the most straightforward and unproblematic examples of numbers understood in this ancient and original sense. When the units are all of the same kind, then what it is to be a particular kind of number, for example, a triple or a quadruple, of the kind of thing they are, is determined by what it is to be one thing of that kind, and by what it is to be, say, three things of any kind whatsoever. If you know what a horse is, and you know what a triple is, then you know what a triple of horses is; and if you know which particular horses Trigger, Champion, and Red Rum are, then you know which particular number of horses they compose.

Of course, no one has to know this particular number of horses, or even the horses that make it up, in order for it to *be* a number of horses.

---

[5] See Jacob Klein's *Greek Mathematical Thought and the Origin of Algebra*, especially Chapter 6, Paul Pritchard's *Plato's Philosophy of Mathematics*, Chapters 1–3, and Myles Burnyeat's "Plato on why mathematics is good for the soul".

Simply by being, severally and individually, the particular horses that they are, and by being, collectively, finite (in fact, three) in multitude, Trigger, Champion, and Red Rum make up the particular number of horses that they do. They do not have to be collected together, either in reality or in conception, in order to compose that number: there is nothing that anyone has to think or to do in order to bring it into being. Indeed, this is obvious, on reflection, for a herd of twenty-five horses contains two thousand three hundred such horse triples, most of which, of course, no one would ever separate out, or even think of, not even someone well acquainted with the horses, both individually and as a herd. A number of horses is no more a creature of the mind than are the individual horses that compose it. Since we can count such numbers, it is natural that we "count" them as things.

Getting this point right is important for everything that follows. In the example just considered, one might be tempted to say that what are being counted are not numbers – *arithmoi* – of horses, but, for example, the possible ways of selecting three horses from a herd of twenty-five. That is, indeed, a common way of speaking about such matters, and, moreover, it has a reassuringly "concrete" air about it: one can easily imagine cowboys cutting horses out of herds. But such imaginings are irrelevant, and such confidence in the "concrete", understood in this sense, is misplaced. For it is the existence of the *arithmoi* – the triples – that grounds the possibilities of selection, and not the possibilities of selection that ground the existence of the *arithmoi*. It is impossible to think those triples away: they are simply "there" to be counted as units in the *arithmos* of 2,300 horse triples that they compose.

But to what extent must the units of a number be homogeneous? Are we allowed to count disparate, even incongruous things as together constituting a number? Indeed, are we not forced to acknowledge numbers composed of heterogeneous units? Are they not simply "there" by virtue of their units being "there" in finite multitude? Frege notes, with approval, that

Leibniz rejects the view of the schoolmen that number is not applicable to immaterial things, and calls number a sort of immaterial figure, which results from the union of things of any sorts whatsoever, for example, of God, an angel, a man, and motion, which together are four[6].

If we were to follow Leibniz and Frege and allow the widest possible latitude in the choice of units, then we should have to acknowledge

---

[6] *Die Grundlagen der Arithmetik*, p.31.

2.3 The original conception of number

that finitely many things (e.g. two hundred and ninety-seven) of any kinds whatsoever, however various and heterogeneous, simply by being, individually and severally, the particular and definite things that they are – horses, or men, or ideas, or characters in fiction, or numbers (in the sense discussed here) – and by being, collectively, finitely many things (in the circumstances posited, two hundred and ninety-seven), are the units of a unique number that together they all constitute.

But what kinds of things are suitable to serve as units in a number? Surely, some kinds of thing are too vague, or too indistinct, or too poorly differentiated to count as units. Clouds, ripples on the surface of a liquid, psychological states – such things are usually too indefinite to count. How many psychological states did you experience yesterday? How many clouds are there now overhead in the sky? It's not that these questions have answers that we don't know; it's rather that, in general, they don't have answers – objectively determined answers – at all.

But sometimes we can, in fact, count such things. There are occasions on which we can, for example, say that there are three clouds overhead. And, after all, do we not speak of four Noble Truths, seven types of ambiguity, three theological virtues, thirteen ways of looking at a blackbird,...? What are the numbers that these sorts of things compose? Do Faith, Hope, and Charity form a triple in the way that Trigger, Champion, and Red Rum do? That is rather like the question whether Faith is a *thing* in the way that Trigger or Champion is.

The bafflement and uncertainty we experience when we confront such questions remind us that the ancient conception of number under consideration was not an exact and artificial scientific concept but a concept in common use. Natural concepts in ordinary use characteristically exhibit a fluidity and suppleness that makes them unsuitable for exact, scientific discourse in their raw state, so to speak. The domain of applicability of such a concept typically is sharply and clearly delineated at its centre, but fades into vagueness at its periphery. In the case we are considering, the vagueness that infects the notion of *number* at its boundary is the same vagueness that infects the notion of *thing*.

It is a characteristic of language that it allows us to form substantives by combining expressions in complex ways, and to use them as if they were ordinary nouns in forming sentences. When we form a sentence in this way it seems as though we were predicating something of a thing. In this way we pepper our discourse with references to "possibilities", "ways", "likelihoods", "facts", "circumstances", and so on. Thus arises the illusion (if illusion it be) that there are "things" corresponding to, or denoted by, substantival expressions of the sort we are discussing. Sometimes we want to count such "things", that is to say, to combine them into numbers. But we should not take such "numbers" any more literally than we do the "things" that constitute their "units". Of course it is unlikely that anyone would undertake to do sums with the number of theological virtues or the number of Noble Truths, for example. But the boundary separating these absurd cases from legitimate ones is not easy to draw by means of a simple formula.

If we wish to treat numbers exactly, to attribute definite sizes to them or to apply arithmetic to them, their units must be properly identifiable and distinguishable, at least in principle. They may be ephemeral or evanescent; but they must be definite and distinct. We can count mayflies, light flashes, or peals of a bell, and can conceive numbers composed of such things. But we cannot, under ordinary circumstances, count clouds, or shades of red in a portrait, or sounds made by a crowd.

If a number is to be the subject of an exact mathematical proposition, its units must be subject to the standard laws of identity and difference. As we shall see, the mathematician, whose business it is to treat of numbers in the most general way, achieves the exactness his science requires by ignoring the natures of the units that compose his numbers, considering those units only insofar as they are subject to those standard laws.

2.4 Number words and ascriptions of number

When "number" is understood in the original sense of *arithmos*, number words, such as "five", "twelve" and "six hundred and fifty-nine", should not be taken as names for particular abstract objects; rather they should be taken to stand for *species* of numbers (i.e. of finite pluralities composed of units), just as "horse" and "man" stand for species of animals.

Thus if we were to employ the original concept of number, we should speak of "a five of horses" or of "a six hundred and fifty-nine of members of Parliament", and express ascriptions of number, what Frege calls "*Zahlangaben*", in the following ways:

There is a five of horses in the field.
There is a six hundred and fifty-nine of members of Parliament.

or, alternatively

The number of horses in the field is a five.
The number of members of Parliament is a six hundred and fifty-nine.

2.4 Number words and ascriptions of number

These ways of speaking are, of course, at odds with ordinary English usage. But that usage is not, after all, an infallible guide,[7] and, in any case, *is based on a different notion of number*. In these instances ordinary usage would require:

There are five horses in the field.
There are six hundred and fifty-nine members of Parliament.

or, corresponding to the alternative

The number of horses in the field is five.
The number of members of Parliament is six hundred and fifty-nine.

But these ways of speaking, though more familiar, are less expressive of the facts underlying ascriptions of number than are the "deviant" forms I suggested as according better with the ancient meaning of "number".

Consider number ascriptions of the first kind, in which the choice is between, say, "a five of horses" and "five horses". Here the first alternative, though it sounds odd to the modern ear, captures exactly what is intended on the ancient meaning of "number". The second phrase is misleading even on the modern view of number, because it suggests a false parallel between, say, "five horses" and "white horses". This is a difficulty to which Frege has called attention (e.g. in *Grundlagen* §52). Thus on either the ancient or the modern notion of "number" such adjectival use of number words is deceptive. Naturally this does not mean that we need avoid using such locutions: since we realise such parallels are false, we may simply refuse to draw them, and continue to use number words adjectivally as before.

When we come to consider number ascriptions of the second, alternative kind we arrive at the nub of the matter. The conventional formulation:

The number of horses in the field is five.

enforces – perhaps it even imposes – the idea that number words such as "five" or "six hundred and fifty-nine" are names of particular abstract objects. The absence of the indefinite article before the number word "five" virtually forces us to read the "is" as the "is" of identity rather than as the "is" of predication. This leads to the conviction that the expression "the number of horses in the field" and the word "five" both

---

[7] Moshé Machover has pointed out to me that this form of expression is correct in Arabic and in biblical Hebrew. Indeed, in Arabic it is *mandatory* for numbers 3–10. See W. Wright, *A Grammar of the Arabic Language*, Art. 21.

designate the same particular abstract object, namely, the natural number five. But what *is* that object?

This is the critical question; Frege wrote the *Grundlagen* in order to answer it. But *no answer to this question will ever pass muster*, indeed *no answer to this question can be "correct"*. For any *particular* answer will be unsatisfactory *precisely because of its particularity*. It is the nature of symbol generated abstractions, of which the natural numbers are conspicuous examples, that *any* particular specification of what they are "sounds odd" and seems too explicit. This insubstantiality of our modern "numbers" reflects their origin in syntax and their epiphenomenal nature.

To the man engaged in calculation, any exact specification of the natural numbers as objects seems superfluous: how will it help him in reckoning twice two is four to be told that two is the singleton of the singleton of the empty set or that four is the class of all quadruples, or anything of that sort? During a calculation his ciphers and symbols are the sole objects of his interest and attention; whatever anyone says about the abstract numbers that correspond to them will therefore strike him, correctly, as irrelevant to his calculation. And what is irrelevant to calculation cannot be of significance for the concept of number, not, that is, for the modern, symbol generated concept of "natural" number.

It was the necessity to defend the particularity of his definition of natural number that led Frege to formulate what has since become known as his *context principle*:

never to ask for the meaning of a word in isolation, but only in the context of a proposition[8].

The principle is designed to forestall the obvious objection to Frege's definition or, come to that, to any other *particular* definition of number, namely, that it is not, in fact, what anyone actually means by "number". How could it be that Gauss, for example, did not know the meaning of "number"? But to ask, say, whether five is *really* the extension of the concept

(extension of a) concept equinumerous with the concept *finger on the right hand of Gottlob Frege*

as it is under Frege's definition, is to ask for the meaning of the word "five" outside the context of a proposition, and this violates the context principle.

However, it is clear that in Frege's eyes the principle has only limited

---

[8] *Grundlagen*, p.12.

application. For once he has disclosed, or rather, has laid down, what the numbers are, there is nothing to prevent our asking what "five" means outside the context of any proposition whatsoever, nor, more importantly, to prevent our taking it to mean just what Frege tells us it means.

The strategy he pursued in the *Grundlagen* was first to expose the confusion that surrounded the concept of number in the writings of his predecessors and contemporaries, and then to offer his own definition. To justify that definition all he needed to do was to show that everything we really knew about numbers before he told us what they are, everything really essential, remains true on his definition of number. So to judge whether Frege's definition does the job required of it we need only consider the meanings it confers on the propositions of arithmetic, and on the propositions by which we make ascriptions of number: we need only consider the meaning of number words in the contexts of propositions.

Frege wanted to tell us exactly what *things* numbers are, but to do so in a way that preserves the truth values of all the propositions about which we already agree, at least agree in principle. Once we know *exactly* what things the natural numbers are, then we know *exactly* what mathematical propositions about them mean; we know, for example, *exactly* what it means to say that *there exists* a natural number with such and such a property.

The much discussed context principle is thus a kind of stopgap: it is intended to forestall Frege's readers from dismissing his account of number before Frege has had a chance to demonstrate its merits. Of course Frege realised that his definition of number was not a correct account of what people actually meant by "number": indeed, the context principle surely implies that number words, in Frege's view, *have no meanings outside the context of propositions*, at least insofar as they are employed by ordinary speakers. But this is, according to Frege, a defect of the ordinary way of speaking about numbers, for which his theory supplies the necessary corrective.

There is, or at least there appears to be, a radical alternative to Frege's account of natural number. It starts from the premise, which I believe to be true, that our conception of natural number arises out of our procedures of calculation. From there it advances to the further proposition that, in reality, there are no natural numbers, that is to say, there are no special abstract objects corresponding naturally and uniquely to our number names and numerals. This, too, I believe to be the case. Finally, it maintains – and here I must decidedly part company – that there is nothing to arithmetic (and, by extension, to all of mathematics) but calculation. This was the view of Wittgenstein:

> Mathematics consists entirely of calculations. In mathematics, *everything* is algorithm and *nothing* is meaning; even when it doesn't look like that because we seem to be using *words* to talk *about* mathematical things[9].

Such a conception of mathematics is profoundly unsatisfactory, and, I believe, deeply irrational. Frege dismissed such views with contempt:

> It is possible, of course, to operate with figures mechanically, just as it is possible to speak like a parrot; but that hardly deserves the name of thought. It only becomes possible at all after our mathematical notation has, as a result of genuine thought, been so established that it does the thinking for us, so to speak[10].

Frege is surely right here: the problem is to discover the facts that underlie our calculating procedures and that, alone, give them sense. But are the facts in question facts about natural numbers in our modern sense? If so, then it seems to me that we must despair of ever understanding what those facts are facts *about*[11].

If, however, we do not encumber ourselves with the modern, abstract "natural" numbers as vehicles of explanation, then we shall not be burdened with the necessity of explaining what they are. If we return to the original conception of number, in which numbers are not symbol generated abstractions but have a more concrete[12] character, these particular difficulties disappear.

Ascriptions of number, Frege's *Zahlangaben*, then become completely straightforward: each ascription is simply a matter of assigning a number, that is to say, a finite multitude composed of units, to its proper numerical species. Thus the number of fingers on my right hand is assigned to the numerical species *five*, the number of planets and the number of Muses are both assigned to the numerical species *nine*[13], the number of members

---

[9] *Philosophical Grammar*, p. 468.
[10] *Grundlagen*, p. iv (Introduction).
[11] We can, of course, give an axiomatic definition of the natural numbers, but that is not so much to define them as to define them away. See Section 6.2.
[12] Perhaps it would be more accurate to say that these numbers are concrete *relative to the units that compose them*. A number of colours is, surely, abstract in whatever way it is that colours are abstract, but no more so. I shall discuss this point at greater length in the next section.
[13] The number of Muses is the multitude composed of Clio, Euterpe, Thalia, Melpomene, Terpsichore, Erato, Polymnia, Urania, and Calliope. But surely the Muses do not exist: strictly speaking, they are merely imaginary, mythical creatures. Well, in that case there is, strictly speaking, no *number* of Muses, either, in the sense of "number" used here. The problem of referring to that number is akin to the problem of referring to "its" units individually and severally, and is not a problem about the notion of "number" per se.

of Parliament to the species *six hundred and fifty-nine*, and so on. The things that these ascriptions apply to are particular numbers – *arithmoi*; the things they ascribe to those numbers are numerical *species*.

Of course, even if you take particular number words, such as "five", "nine", or "six hundred and fifty-nine", as standing for species of numbers (*arithmoi*), you may still be tempted to take each such species itself to be an abstract object, just as you may be tempted to take species words in the category of substance (e.g. "horse" or "man") as standing for particular abstract objects – "universals", or "Platonic Ideas", or things of that sort. But such ontological extravagance (if it be extravagance) is not forced upon you.

2.5 The existence of numbers

Are there such things as numbers, finite pluralities composed of units? The straightforward answer is yes, *obviously* there are such things. Counting simply consists in assigning such a number to the correct numerical species: to count the number of pupils in the class is to assign the number composed of those pupils as units to the number species twenty-nine, say; and to do this you don't have to count them out one by one: you can simply observe, for example, that there are four rows each occupied by six pupils and one occupied by five.

This is the natural view – it was the view of the Greeks – and one would be tempted to leave the matter there were it not that the question of the existence of numbers is part of the wider, and highly contentious, question of the existence of "mathematical objects" in general. In fact, this question just *is* the question of the existence of mathematical objects in its entire extent; for, as we shall see, in modern mathematics the only "mathematical objects" we need ever consider are these very numbers, or rather their modern descendants, sets[14].

We must therefore face up to the question whether, and in what sense, these numbers can be said to exist. But we must also face up to the fact that there can be no question of proof here: the matters in question are too basic for that. In particular, we cannot use rigorous argument to compel the sceptic to accept the existence of numbers. The most we can

---

[14] It is, I believe, highly significant that for the Greeks the problem was not the existence of numbers composed of ordinary units, such as men or horses or stones – no one ventured to doubt the existence of *those* numbers – but rather the existence of *mathematical* numbers, numbers whose units were conceived to be pure, simple, timeless, undifferentiated "ones". I shall discuss these mathematical numbers in Section 2.6.

hope to do is to persuade him. Nevertheless, as we shall see, the sceptic must bear a considerable portion of the burden of persuasion here. In any case, his determination not to be caught out affirming a falsehood will prevent his grasping the truth, if it is there to be grasped.

There is, however, a practical problem here, one which is not merely of philosophical interest, but which bears directly on technical questions in mathematics itself. It is the problem, not of the *existence* of numbers, but of their *extent*: how far do the number species stretch?

To the mathematician this problem presents itself in technical form: which pluralities are finite and which "operations" preserve finiteness? It is, in short, the question of what is meant by "finite", and it is as deep as any question in mathematics. What is relevant to this question is not *whether* numbers, in the original sense, exist, but *how* they exist: in particular, how can they be defined and how can they be recognised?

Now the question *how* something exists seems, at first sight, puzzling, perhaps even unanswerable. But a little reflection will suggest that, after all, there is no reason to suppose that all the things that exist do so in exactly the same way. On the contrary, as Aristotle tells us

Things are said to be in many ways[15].

This is not a controversial thesis, but an incontrovertible truth: for it is a truth, not about things, but about what we *say* of them, correctly or not. On such a fundamental and difficult matter, however, surely it is sensible to start from the assumption, which we can always modify or qualify under the pressure of argument or in the light of reflection, that what we say corresponds, more or less faithfully, to what is actually the case.

Let us pause, then, and briefly consider some examples that illustrate Aristotle's point. Events are said to be because they *occur*; colours and shapes, or, more generally, properties, because things *have them* or *could have them*; places are said to be because things are, or could be, *in them*; people are said to be because they are *alive* (Abraham Lincoln *ceased to be* on April 14, 1865). Games are said to be because people *play them*; activities, in general, because people *take part in them*; states because other states *recognise* them; ...

Of especial relevance to our discussion of number is the fact that sometimes many things are said to be one thing: many chapters are said to be one book; many cells to be one animal; many bees to be one swarm;

---

[15] This is the very first sentence in *Metaphysics Z*, the first book of the great trilogy Z, H, Θ, which contains the core of Aristotle's doctrine on being. Indeed, one might say that the non-univocality of the word "being" is the first principle of his metaphysics.

2.5 The existence of numbers

many horses to be one herd; many words to be one utterance; many battles to be one campaign. All these plural or composite things are said to be. How can one make sense of denying that there are such things? Surely there are armies, clubs, legislatures, conspiracies, circles of friends, ... ; and, moreover, these things are most naturally regarded as existing over and above the many things that compose them; for, on the face of it, all of these "plural" things have functions, properties, capacities for action, etc., that cannot be construed as functions, properties, or capacities of the things that compose them, taken individually and severally.

The profusion of these examples is discouraging and mocks any ambition we might have to order or to classify them. I have no such ambition here. On the contrary, I want merely to call attention to these facts so that we may reflect on the abyss of difficulties they disclose. It is, after all, no easy matter to determine what does and what does not exist, notwithstanding that the literature on the foundations of mathematics is full of confident pronouncements on such matters[16].

Clearly we must try somehow to prevent the perplexities and uncertainties occasioned by these metaphysical questions from infecting the foundations of mathematics, insofar as that is possible. Of course, as we have already seen, we shall have to face up to the question in what way numbers, as originally conceived, may be said to exist. But we do not need to provide an absolute answer here: we may content ourselves with a merely relative one. For we need only to compare the way in which numbers exist with the way in which the units that compose them exist.

We may not be able to say – we may not even know – in what sense things of the kind Φ exist. But if we suppose the sense in which Φ's exist somehow to be given, we may be able to form a conception of how a

---

[16] For example, Hartry Field, in *Science without Numbers*, writes

> Nominalism is the doctrine that there are no abstract entities. The term "abstract entity" may not be entirely clear, but one thing that does seem clear is that such alleged entities as numbers, functions, and sets are abstract – that is, they would be abstract if they existed. In defending nominalism, therefore, I am denying that numbers, functions, sets, or any similar entities exist. (p.1)

> But Field's scruples about admitting the existence of "abstract entities" do not extend so far as to prevent him from acknowledging an uncountable infinity of space-time points in his "nominalistic" cosmos. Space-time points, it would seem, are satisfyingly "concrete", whereas, for example, the number (set) of books in Field's office is an ineffable "abstraction". Incidentally, what could Field mean when he speaks (p.31) of postulating "uncountably many physical entities"? Ordinarily one would mean that there is an injection but no bijection from the set of natural numbers to the set of "physical entities" in question. (What, by the way, *is* a "physical entity"?) But for Field there are no functions at all, *a fortiori* no functions counting the given set.

three, or a seven, or a two hundred and thirty-seven, or, in general, a number, of Φ's may be said to exist.

Now a number consists in certain things (its units) present in a certain definite multitude. Accordingly, we may say that each particular number presents two fundamental aspects, namely, a *material* aspect, and a *formal* or *arithmetical* aspect. What I am calling the "material aspect" of a number here is determined by *what* its units are, severally and individually; its formal or arithmetical aspect is determined by *how many* they are, conjointly and collectively. These two aspects constitute the *essence* of a number, the *what it is*[17] of that number. What the number is whose units are $x, y, z, \ldots$ is determined by

(1) What x is, what y is, what z is, ... (This is what the number is in its *material* aspect)

and

(2) What it is to be *exactly as many as* $x, y, z, \ldots$ (This is what the number is in its *formal* or *arithmetical* aspect.)

To say that there is such a thing as being exactly as many as $x, y, z, \ldots$ is just another way of saying that $x, y, z, \ldots$ are finite in multitude. And, of course, $x, y, z, \ldots$ constitute the units of a number by *virtue* of being finite in multitude. Thus the number species that we designate by $2, 3, 4, \ldots$ are determined by, and correspond to, the formal or arithmetical aspects of numbers: they represent ways of being finite[18].

The material and arithmetical aspects of the number whose units are $x, y, z, \ldots$ determine not only *what* the number is but also *that* it is (i.e. that it *exists*): for that number to exist it is necessary and sufficient, first, that the units $x, y, z, \ldots$ exist, individually and severally, and second, that they be, conjointly and collectively, finite in multitude. That they are many things depends upon each of them being *some*thing; that these many things together constitute one single thing (viz. a number) depends upon there being only *finitely* many of them altogether. Finitude is thus the principle of unity for a number.

In arithmetic, as traditionally conceived, it is usually only the formal aspects of numbers that are of significance. But even there this is not uniformly the case, and in modern mathematics it is no longer true. For example, since functions are numbers (sets), what their units are is of

---

[17] In scholastic Latin *quod quid est*, corresponding to the Greek *to ti estin*.

[18] The number species start with 2 since a number, as a *plurality* composed of units, must possess at least two units.

2.5 The existence of numbers

essential significance. In any case, on the ancient conception it is the material aspect of a number that largely determines the way in which it may be said to exist. This, I believe, is why Aristotle, and after him Euclid, stressed the concept of unit in defining number[19]: the mode of being of the units determines the mode of being of the number; and if "things are said to be in many ways" then numbers, too, must be said to be in many ways.

Corresponding to every kind of thing, to every particular way of being a definite something, there is a way of being a two of that kind, a way of being a seven of that kind, a way of being a six million four hundred and eighty-two thousand seven hundred and eighty-three of that kind, ... In general, there is a way of being a finite plurality, i.e. a number, of things of that kind.

Moreover, with *homogeneous* numbers all of whose units are the same kind of thing (e.g. all horses or all colours), it seems natural to say that what it is to be a particular kind of number, say a three, of a particular kind, Φ, of thing is, so to speak, a plural (in this case, a triplicate or threefold) version of what it is to be a Φ (i.e. one Φ); and, conversely, what it is to be a Φ – one Φ – is a singular version of what it is to be a triple of Φ's, say. A number of Φ's, on this view, exists in a manner analogous to the manner in which a single Φ exists, except multiply instead of singly, so to speak.

To see things in this way is to see the number species $2, 3, 4, \ldots$ as akin to what the scholastics called "transcendentals": like "being" and "one" they each can be regarded as having analogous, but distinct, meanings in each of the categories of being[20].

The formal or arithmetical aspects of numbers are more transparent to us, more easy to grasp, than their material aspects. Indeed, I suspect that it is easier for us to grasp what it is to be a three than what it is to be a three of some particular kind, e.g. of horses, or of trees, or of stones[21]. This reflects the fact that the general problem of the being

---

[19] Eudoxus, for example, defined "number" as "limited plurality". No mention of units there. The point, for Aristotle, was that the units need not belong to the category of substance, but can be taken from any of the categories of being.

[20] I do not necessarily mean the Aristotelian or scholastic categories here: I mean, rather, whatever categories or kinds of things we ultimately acknowledge. In any case, I am not suggesting that, for example, the scholastics *did* regard number species in this way: I'm suggesting that they *ought* to have regarded them in this way.

[21] Surely, one might think, being a three of stones is a quite straightforward affair. But what counts as a stone? I mean *exactly*. How big must a stone be? How firmly must its parts adhere to one another for it to remain one stone? Etc. There are clear instances of triples of stones, but also puzzling boundary cases. And the difficulty, in the latter

of numbers, in their material aspects, is the general problem of "being" itself. Arithmetic is, after all, a more straightforward matter than general ontology. In arithmetic we simply take the "being" of the units that compose our numbers as given.

The view that emerges from these considerations is that a number composed of things of a certain kind has the same kind of claim to existence as have individual things of that kind – its units, for example. But does this not run counter to our intuitions? Do we not suspect that a number somehow has a weaker grip on existence than the units that make it up? Are we not inclined to say, for example, that a number is more "abstract" than the units that compose it? These intuitions certainly exist; indeed, I myself share them. But will they stand up under serious scrutiny?

Let us consider the question whether numbers, merely as a consequence of their being numbers, are "abstract" in any intrinsic sense. Is a five of horses any more abstract than a single horse? A *herd* of horses does not strike us as being an abstraction; but then a five of horses is not a herd, not even a small one: for a herd of horses ceases to be a herd when the horses that make it up are dispersed, whereas a five of horses remains the same five no matter where the horses that are its units are located.

But this fact does not distinguish a five of horses from a single horse *in point of being abstract*. For, after all, a single horse remains one horse, and the same horse, no matter where *it* is located: it is no part of what it is to be *this* horse that it is now grazing in *this* field. Of course, it is part of what it is to be a horse that at a given time a single horse will be in a single place. But then it is part of what it is to be a five of horses that at a given time a five of horses will be in a five of places.

If numbers are abstractions, what, we may ask, are the concrete things from which they are abstracted? Can we arrive at the concept *number of horses* by abstracting from the concept *herd of horses*, for example? Obviously not. You can no more obtain the concept *number of horses* from the concept *herd of horses* by abstraction than you can obtain the concept *man* from the concept *insurance salesman* by abstraction. For just as an insurance salesman is a man with an additional property, so a herd of horses is a number of horses with an additional property: according to

---

cases, is with the concept of *stone*, not with the concept of *triple*. Similar difficulties arise in the cases of horse- or tree-triples. This reflects the fact that number species, like *three* or *six hundred and fifty-nine* have what the scholastics called *essential definitions*, whereas species in the category of substance (like *horse*, or *tree*) do not (as Darwin has shown us).

2.5 The existence of numbers

Webster, a herd is "a *number of animals* of one kind, kept together under human control", and according to Dr. Johnson a herd is "a *number of beasts* together".

The additional property in the case of a herd of horses is that of *being together*, and this can only be a property of the number of horses that compose the herd. For although all of the horses in the herd are together, no single horse in the herd is "together". The word "together", unlike "bay" or "Arabian" or "spavined", does not apply to a single horse.

The logical point I am making here is quite general. We cannot say, for example, that numbers exist only by virtue of their units being *conceived* as being together. For, again, what does "together" apply to here? Similarly, we cannot obtain the notion of number by abstraction from the notion of heap or aggregate or anything of that sort.

Numbers are not abstract *per se*, although some numbers are abstract: a number of colours is abstract in whatever way it is that colours themselves are abstract. But if you think there are no such things as "abstract colours", then, naturally, you must think there are no numbers composed of "abstract colours" either.

Whatever kinds of things there are, and in whatever sense things of those kinds are said to exist, there are numbers whose units are things of those kinds, and those numbers may be said to exist in a way analogous to the way in which the things that are their units exist.

Is a number of horses, say a five, a physical object then? No, it is a *number*, a *five*, of physical objects. Numbers are *sui generis*; but then so is every other basic kind of thing: things are said to be in many ways. In any case, conceiving a number of physical objects to be a physical object does not render the problem of its existence more tractable. Indeed, the notion of number, although it is more abstract and general, is much clearer than that of a physical object. Perhaps it is clearer *because* it is more abstract and more general[22].

In our discussion so far we have dealt mainly with numbers composed of units that are homogeneous in kind. These are the most straightforward examples of numbers, and by considering them we are able to grasp

---

[22] In saying this I do *not* contradict myself. It is perfectly true, as I have pointed out, that individual numbers, are no more abstract than the units that compose them. However, the *concept* of number is abstract, as, indeed, is the *concept* of physical object. But the former is more abstract, or, at least, more general, than the latter, inasmuch as it applies in *all* the categories of being, not just in that of substance. In the same way, the concept of *unit* is more abstract and general than that of physical object as well. I have put this point in traditional Aristotelian terms, but the essential idea seems independent of this particular formulation of it.

how it is that a number can be said to exist, and how its existence as a number is related to the existences of the units that compose it. But how are we to conceive a number composed of disparate, non-homogeneous units? Things assembled higgledy-piggledy into a number threaten not to form a coherent whole. In particular, it is difficult to conceive of the way in which such a number may be said to exist.

If we say that the colour purple, the U.S. Constitution, the Empire State Building, and the word "unlikely" together form a number (viz. a four), then the way in which it exists, in its material aspect, is determined by the several ways in which its units exist. But to say that the colour purple, the U.S. Constitution, the Empire State Building, and the word "unlikely" all *exist* sounds more than a little odd; and the oddity stems from the fact that such a use of "exist" seems to amount to a syllepsis or zeugma: like saying "He stood for Parliament and for honest government" or "He ran for Congress and for his life". The difficulty encountered here, which arises out of the equivocality of "exists", confirms again the importance of Aristotle's dictum that things are said to be in many ways.

The oddity here is not diminished if we follow Frege and Dummett in regarding number ascriptions as propositions about (Fregean) concepts. The Fregean concept ξ *is the colour purple or* ξ *is the U.S. Constitution or* ξ *is the Empire State Building or* ξ *is the word "unlikely"* is just as odd as the number these things compose, and for the same reason: on Frege's view these things are all *objects*.

Aristotle addresses the problem of numbers with heterogeneous units in book *N* of the *Metaphysics* (1088a14):

> ... unity denotes a measure of some plurality, and a number denotes a measured plurality and a plurality of measures... The measure must be something that applies to all alike; e.g. if the things are horses, the measure is a horse; if they are men, the measure is a man; and if they are a man, a horse, and a god, the measure will be an animate being, and their number a number of animate beings. If the things are a man, white, and walking, there will scarcely be a number of them, because they all belong to a subject which is one and the same in number; still their number will be a number of genera, or some other such term.

Thus, for Aristotle, the units (he calls them "measures" here) of a number must all be of the same kind, however attenuated its definition or vast its extent. No doubt it is significant that Aristotle refers to "measures" instead of "units" here, since "measure", unlike "unit", suggests a standard of comparison.

The problem of non-homogeneous units, in the full generality considered here, is not really of much importance for mathematics. Because

2.5 The existence of numbers

mathematicians deal in generalities, they do not need to specify what kinds of things the units in their numbers (sets) are, provided those units are not themselves numbers (sets). What is of interest to the mathematician, *qua* mathematician, is not that this five of sheep and that five of goats together make a ten of ruminant animals, but rather that *any* two (disjoint) fives make a ten[23].

Moreover, we are no longer as confident as were Aristotle and the scholastics that the notion of *kind* or *species* upon which the relevant conception of homogeneity depends has the significance they assigned to it. After all, the central Aristotelian examples of species, namely, species of animals, are, since Darwin, no longer seen as sharply delineated and immutable in the way Aristotle and his medieval followers saw them.

However, a special case of the problem of numbers with heterogeneous units does become a matter of serious practical concern in modern mathematics, because we need to posit numbers (sets) some, but not all, of whose units are themselves numbers (sets). As an example, consider the four whose units are Trigger, Champion, Red Rum, and the number whose units are Trigger, Champion, and Red Rum. In standard set-theoretical notation this would be designated by

{Trigger, Champion, Red Rum, {Trigger, Champion, Red Rum}}

Three of the units of this number are horses and the fourth is the number composed of those three. Each of the units in this number – the three horses and the triple they compose – is well-defined. Moreover, the horses are mutually distinct, and the triple of horses that together they compose is a triple of horses, not a single horse, and so the triple is distinct from each of the other units. Surely together they compose a four.

Of course it might be said – indeed, I suspect that Aristotle and the scholastics *would* have said – that each of Trigger, Champion, and Red Rum is *one* thing, whereas the number they compose is *many* things. But there is an important distinction to be made, a distinction that is obscured by the use of the word "thing" here: the plurality composed of Trigger, Champion, and Red Rum is, indeed, many *horses*, but it is one *number*. It is as *one number*, rather than as *many horses*, that this triple occurs as a unit of the number in question, so it is reasonable to suppose that we are dealing with a number composed of non-homogeneous units in this example.

But can we say that all four of these units *exist* without equivocation?

---

[23] This is an important point to which I shall return in Section 2.6.

Are we dealing with units too unlike to enable us to conceive what it is to be the number that they putatively compose? If we adopt the view suggested above and take the unit to be a kind of limiting or degenerate case of plurality, then we may argue that the ways in which Trigger, Champion, Red Rum, and the number the three of them compose all exist are sufficiently alike for us to form a clear conception of the material aspect of the four of which they are the units, and therefore of the way in which that four may be said to exist. That, or something like it, must be the line taken in modern set theory. It has been disputed[24]; but here we have reached bedrock, and argument is ultimately indecisive.

This is perhaps the key respect in which we must go beyond the ancient notion of number (*arithmos*) in order to arrive at the modern notion of set[25]. There are two serious matters at stake here. In the first place, we need to form numbers out of numbers together with other objects in order to give an arithmetical (set-theoretical) account of key notions like *ordered pair, function,* and *relation* in the full generality required. As I shall show[26], all of these things can be taken to be numbers (sets) of certain special kinds. But there is also the problem of the *extent* of the number concept, the question how far out the number species stretch.

Aristotle held that the number species in numbers composed of bodies were *potentially infinite* in the following precise sense: given any number, S, of bodies, a number of bodies larger than S exists *in potentia* in as much as each of the unit bodies in S is potentially divisible into two or more bodies. For example, if S is composed of the bodies $a_1, a_2, \ldots, a_n$ and a_1 is potentially divisible into b_1 and b_2, then if S' were composed of $b_1, b_2, a_2, \ldots, a_n$, it would have two units, b_1 and b_2, where S had only the single unit a_1. Notice, however, that since b_1 and b_2 exist only *in potentia* but not actually, S' exists only *in potentia* and not actually as well[27].

Of course, we no longer believe that physical bodies are potentially divisible to infinity in this way – certainly no one could possibly maintain that this is obviously the case – so this source of a potential infinity of numbers is no longer available to us. But we can say, with confidence, that given a number, S, of Φ's (whatever the kind Φ may be), the number,

---

[24] by Nelson Goodman in *The Structure of Appearance*, for example.
[25] There is also the problem of assimilating the empty set and singleton sets. All of this I shall discuss in Sections 3.1–3.3.
[26] In Sections 4.5 and 4.6.
[27] See the *Physics* 207b ff. What Aristotle gives us here is an exact sense in which numbers are "potentially infinite". His use of "potentiality" here does not refer to human capacities, as we do when we say we can always "add one" to any natural number. For the potentiality here is formal and resides in the body a_1 not in us.

S', whose units are all the numbers of Φ's whose units are taken from among the units of S, is larger than S itself, provided S has at least three units[28]. And, of course, the units of S' are homogeneous insofar as those of S are (i.e. insofar as Φ is a legitimate kind of thing).

The modern view that Φ's and numbers of Φ's can occur together as units of a number has the virtue of greater simplicity: we may start with any two, $\{a,b\}$, and discern an unlimited supply of numbers

$$\{a,b\}; \ \{a,b,\{a,b\}\}; \ \{a,b,\{a,b\},\{a,b,\{a,b\}\}\}; \ \ldots$$

each of which contains the units of its predecessors among its own. In this example the units in the numbers indicated are exactly specified. The difficulty, if difficulty it be, can only consist in our having mixed units with numbers of units in forming these pluralities. But we can count the pairs and triples in a twenty-five (of horses, say): together they comprise a $2600\,(=300+2300)$. Why can we not count its units and pairs as well? Together they comprise a $325\,(=25+300)$.

Where are we to draw the line here? Indeed, is it necessary to draw a line at all? Surely it is at least as puzzling to deny as to affirm that there is a pair

$$\{a,\{a,b\}\}$$

Surely the sceptic must bear at least an equal share of the burden of proof here. Does he not, in common sense, bear much the greater part of it?

But here I have come full circle. *Obviously* there are such things as numbers, finite pluralities composed of units; and, equally obviously, such numbers can be taken, together with other things, including their own units, as units in further numbers.

2.6 Mathematical numbers and pure units

We must take care to distinguish the straightforward notion of a unit, which is whatever is taken as one thing for the purposes of numbering, and the altogether more problematical notion of a *pure* unit, which is, so to speak, just one thing and nothing else. Numbers composed of such pure units were called "mathematical numbers" in antiquity, and it was the existence of these *mathematical* numbers that was taken by the ancient philosophers[29] as requiring justification: that there are ordinary

---

[28] If S has n units, S' has $2^n - n - 1$.
[29] Aristotle in particular, e.g. in *Metaphysics M* and *N*. See 1078a1 ff.

numbers, composed of ordinary units (horses, men, colours, etc.) seemed so obvious as not to require defending. It is this conception of numbers as composed of bare, featureless "ones" that Frege attacks with such devastating effect in sections §§29–44 of the *Grundlagen*.

However, it was not the Greeks, but more modern adherents of the doctrine of pure units, who came under Frege's fire. This is significant, because the modern conception, at least as criticised by Frege and, later, by Dummett, differs from the ancient one in a crucially important respect: on the ancient conception there are many mathematical numbers of the same cardinality; on the modern conception, however, only one. Here is how Dummett explains the modern version:

The point of interpreting numbers as sets of units[30] rather than taking the number of objects of a given kind simply to be the set consisting of those objects themselves, is obviously to guarantee that the number will be independent of the particular objects counted, being determined as it ought to be, solely by how many of those objects there are; if, say, there are just as many spoons as forks on the table, the number of spoons on the table will be the very same abstract entity as the number of forks. This requires that the set of units arrived at by abstraction from the set of spoons shall be the very same set of units as that arrived at by abstraction from the forks[31].

But the classical conception of mathematical number was something quite different. Here is Aristotle's account of the notion of mathematical object that he attributes to Plato:

He [i.e. Plato] states that besides sensible things and the Forms there exist an intermediate class, the objects of mathematics (*ta mathêmatika*) which differ from sensible objects in being eternal and unchanging and from the Forms in that there are many similar objects of mathematics, whereas each Form is unique[32].

Thus Plato conceived that there would be, for example, many *different* fives composed of pure units. Clearly the "mathematical" or "intermediate" numbers of the Platonists were invented to serve a different purpose from that served by the "pure unit" numbers of the moderns: they were to form the subject matter of theoretical arithmetic[33], where their "eternal and unchanging" character might be thought to be more appropriate to

---

[30] Dummett means what I am calling "pure units" here.
[31] *Frege: Philosophy of Mathematics*, p.86.
[32] *Metaphysics A* 987b15
[33] Plato spells out the differing requirements of practical and theoretical arithmetic in the *Philebus* (56d). There he emphasises the importance of the units in a mathematical number being exactly alike.

2.6 Mathematical numbers and pure units

a mathematics that deals "not with that which is subject to change and decay but with that which always and eternally is"[34].

In the modern version of the doctrine of pure units criticised by Frege and Dummett, however, it would seem that each number of this sort is unique, and in this respect they more resemble the Forms than the Intermediates. What we have here, in fact, is an unhappy marriage of two very different ideas: the ancient idea of a number as a finite plurality of things, and the modern idea of a "natural" number as an objective correlative to a numeral or a number word. Surely the criticisms of Frege and Dummett are decisive here.

But the ancient notion of mathematical number, though on the face of it open to the same sort of criticism, becomes, upon being reinterpreted by Aristotle, an altogether more intelligible, more subtle, and more useful conception than either its Platonic or its modern rival.

In reading Aristotle one forms the clear impression that he held it to be a general principle that the expert practitioners of a science, especially a highly developed science like mathematics, are most unlikely to be *just wrong* in what they say about their science. Accordingly, he took great pains to ensure that his account of "mathematical number" and the "mathematical intermediates" should be both scientifically rigorous *and* in accord with what the mathematicians of his day *said* (if not, one suspects, always in accord with what they *meant*).

The heart of Aristotle's account of "mathematical objects" is to be found in *Metaphysics M* (1077b17–1078b7), to which I refer the interested reader as it is too long to quote here in full. The gist of the passage, insofar as it concerns arithmetic, seems reasonably clear, however, and amounts to this, namely, that mathematical numbers are not numbers of a special kind, but are just ordinary numbers viewed in the abstract manner appropriate to mathematicians; in particular, a "pure" unit in such a number is just an ordinary unit, e.g. a man, but viewed not, e.g. *qua* man, but *qua* single, countable thing. Therefore, the units of a "mathematical number", that is to say, of a number viewed in the manner appropriate to a mathematician, are "indivisible" and "without place": for *whatever* they are, their "parts" do not *count* (i.e. as units) in the number in question, and *wherever* they are, their locations do not matter to what the number they compose *is* (i.e. as a plurality).

If we have a number of stones, say a twenty-three, in a box, the stones, though divisible *qua* stones, are not divisible *qua* units in that

---

[34] *Republic* 527a.

twenty-three; and if we scatter them, they are no longer in the same place "together" in the box, but they remain the same twenty-three, the same number of stones.

This gives us a sense in which we may say that ordinary numbers, viewed in the abstract way that is appropriate in mathematics, are, like Plato's mathematical numbers, composed of units that are "indivisible" and "without place".

But Plato's mathematical numbers were composed, as Aristotle tells us, of units that are "eternal and unchanging", whereas men, horses, trees, and stones – the kinds of things that serve as units in ordinary numbers – do not have this character. Ordinary things of this sort pass into and out of existence; they are subject to growth, change, decay, and destruction. Where in this Heraclitean flux are we to find the stability and permanence that we instinctively feel must be essential features of the subject matter of mathematics?

Most mathematicians, no doubt, can understand the motives that prompted Plato to posit a special subject matter for arithmetic, a realm of being outside time and space, inhabited by the perfect, changeless, featureless, pure "ones" that serve as the units in those mathematical numbers which, on the Platonic view, are the proper objects of our arithmetical propositions. Some may be prepared to follow Plato in positing such a realm, and, indeed, to populate it with much more complicated things than pure units and the mathematical numbers that they compose: real numbers, complex numbers, geometrical objects like triangles and spheres, mathematical structures like groups, rings, and topological spaces, etc.

This is, in fact, what is often called "Platonism", namely the view that mathematics has a special subject matter consisting of abstract "mathematical objects". Such a view is not contemptible: it arises naturally out of mathematical experience and represents a straightforward, if no doubt naive, attempt to come to terms with that experience. It may, for all I know, be the view of the majority of practising mathematicians. But it is not, in my judgement, the correct view, nor was it deemed to be so in the judgement of Plato's greatest pupil, Aristotle.

For Aristotle, mathematics has its own special methods and its own special way of looking at things: but it does not have its own special subject matter, in the sense of a special realm of mathematical objects of the sort envisaged by Plato. Nevertheless, he did hold that there is a sense in which we may truly say that mathematics lies outside the Heraclitean flux of things, and that its propositions express timeless truths.

2.6 Mathematical numbers and pure units

To discover Aristotle's treatment of this problem we must turn from *Metaphysics M* to the *Posterior Analytics*. There Aristotle tells us that the arithmetician must posit both *what* the unit is and *that* it exists; in fact, he tells us this at least three times, namely, at 71a15, at 76a35 and at 93b25. This, I believe, is highly significant, for this doctrine lies at the very heart of Aristotle's account of mathematics. To posit *what* the unit is to lay down its definition[35]; but to posit *that* the unit exists, is, in effect, to place outside mathematics the choice of what *kind* of thing the unit is. It is to assume that the units are *given*, but without assuming what they are *given as*.

On the Aristotelian view it is as if the mathematician were handed a commission of the following sort: you are charged with the task of providing us with the arithmetical facts about numbers. Each number is a finite plurality composed of units, and the units, the "things called one", are the individual entities that make it up. But you will not be told what *particular* things these units are. You may take them to be definite, properly distinguished things, subject to the laws of Identity and Difference; but any knowledge of what they are beyond that would be superfluous to your task.

This idea, the idea that the mathematician, in his theoretical role, merely *posits* that the units exist, is a subtle and important one. It provides the means by which we may prevent mathematics from becoming entangled in the thickets of metaphysics and general ontology. It is, after all, possible to raise objections, to *any* kind of thing's being suitable to serve as a unit in a number: these proposed units no longer exist; those are too vague, as particular things, to be properly individuated; those, yet again, are too evanescent, or too disparate, or .... But such objections concern what I called in Section 2.5 the *material* aspects of numbers, and it is their *formal* or *arithmetical* aspects that are the proper concern of the mathematician, *qua* mathematician.

By *positing* the unit, in the sense just explained, mathematicians circumvent these difficulties. Since they have not specified their units, their propositions possess a hypothetical, or formal, generality that renders them more transparent and hence more certain.

It is not that they posit their units as having no individuating peculiarities – that would be to posit the "pure" units of the Platonists – it is rather that they posit their units *without positing what their individuating*

---

[35] i.e. what the unit is *qua* unit, as in Euclid's definition of "unit" in *Elements* VII, not what it is *qua* man or horse or whatever it happens to be.

peculiarities are. And their posited units are therefore timeless and not subject to change, by default, so to speak; they are, as posits, "eternal and unchanging" in a sense, though not in the literal sense Plato meant, of course. The mutability or destructibility of the units is irrelevant to the formal aspect of the number they compose, and that is the only aspect of the number that concerns the mathematician *qua* mathematician. The truths of arithmetic must be seen as necessary truths so long as it is granted that there is *any* kind of thing – animal, vegetable, mineral, mental, ideal, abstract, or whatever you like – suitable to serve as a unit in a number.

In Aristotle's conception of mathematical number, we have the best vehicle yet devised to account for the facts of theoretical arithmetic. In arithmetical reasoning the mathematician regards things in the most abstract and general way conceivable, namely, only insofar as they are subject to the laws of identity and difference. That there are things subject to those laws he simply takes for granted. He does not burden himself with the task of specifying the units exactly, and in this way he keeps his science clear of entanglement in metaphysics and ontology. He is not called upon to make general pronouncements on what there is. Indeed, his arithmetic is ontologically neutral, except insofar as it must acknowledge the existence of finite pluralities, that is to say, of numbers, composed of whatever things are acknowledged as suitable units.

The identification of any particular kind of thing as suitable to serve as a unit in a number must always, in the nature of things, be provisional. As I have said, it will always be possible to cavil over a particular choice. But by *positing* the unit, the mathematician brushes such considerations aside. "You tell me", he says, "what *you* count as a suitable unit, and I will give you the facts about numbers composed of such units."

Although the mathematician, in his professional capacity, deals with things abstractly and generally, the things to which his discourse applies are themselves neither abstract nor general, *per se*, though abstractions and generalities, too, insofar as they can count as individual things, are grist for his mill. If he applies his science to numbers composed of slices of gingerbread, or of pebbles, then in that application he is, indeed, engaged in gingerbread- or pebble-arithmetic, as Frege's gibe[36] would have it. But his real concern, as a mathematician, is with the formal or arithmetical aspects of such numbers, not with their material aspects, not with the shapes of the pebbles or with the flavour of the gingerbread.

---

[36] In the introduction to the *Grundlagen*. Frege's sarcasm was directed at J. S. Mill.

For Aristotle, arithmetic was the most exact of the sciences *because the most abstract*. That is not because "abstract objects" are more accessible to the mind's eye than ordinary ones. Indeed, there are no "abstract objects" *sensu stricto* in Aristotle's account of arithmetic. It is rather that by viewing ordinary objects abstractly, that is to say, by ignoring those features that make them the particular objects they are, we ignore all that is most problematic and most puzzling about them. Those features are simply not relevant to the contribution such objects make to the formal aspects of the numbers in which they appear as units. To determine the arithmetical facts, all we need to know is that the units in our numbers are determinate things, properly distinguished from one another, subject to the usual laws of identity and difference.

2.7 Ascriptions of number: Frege or Aristotle?

On the original conception of a number, an ascription of number, what Frege calls a *Zahlangabe*, consists in saying of a particular number, that is to say, of a particular finite multitude composed of units, which *species* of number it belongs to: for example, saying of a three that it is a three, of a thirty-seven that it is a thirty-seven, or of a one million nine hundred seventy-six thousand two hundred and thirty-nine that it is a one million nine hundred seventy-six thousand two hundred and thirty-nine.

This is an utterly straight forward matter: if I am asked "How many are there?", I reply with a *species* name in the category of number (e.g. "a thirty-seven"), just as if I am asked "What is it?", I reply with a *species* name in the category of substance (e.g. "a horse"). Frege's answer was, on the surface of it, quite different:

> *Die Zahlangabe enthält eine Aussage von einem Begriffe.*

In Dummett's careful translation this becomes

The content of an ascription of number consists in predicating something of a concept[37].

Thus, for example, when I say "Jupiter has four Galilean satellites" I predicate something of the concept *Galilean satellite of Jupiter*. The reason for this rather roundabout way of speaking is that, according to Frege, numbers must be objects, so that we are not allowed simply to say that *four* is the second level concept of *being a first level concept under*

---

[37] *Frege: Philosophy of Mathematics*, p. 88.

which exactly four objects fall and that the first level concept *Galilean satellite of Jupiter* falls under it.

But, in the end, the "object" that is the number four, in Frege's account, namely, the extension of the concept *extension of a concept under which exactly four objects fall*[38], is obviously very closely related to that second level concept; in fact, it is not at all far in conception from the numerical species *four* in the classical account. So there is almost an agreement here. Indeed, we may well ask: wherein does the essential difference lie?

Obviously the principal difference in the two accounts is to be found in their conceptions of what it is that numbers are ascribed to in *Zahlangaben*, that is to say, in their conceptions of what it is that possesses cardinality. On the classical view of Aristotle, that which possesses cardinality is a finite plurality composed of units, viz. a *number*, in the original sense of "number"; whereas on the Fregean view it is the extension of a first level concept (in the *Grundgesetze*) or just the concept itself (in the *Grundlagen*)[39].

Thus on Aristotle's account, for example, four is to be assigned to the finite plurality composed of Jupiter's Galilean satellites, that is to say, to the number whose units are Io, Europa, Ganymede, and Callisto, whereas on Frege's account it is ascribed to the extension of the concept *Galilean satellite of Jupiter* (in the *Grundgesetze*), or to the concept itself (in the *Grundlagen*).

But is this difference of any significance? It is, indeed, of great significance; for it draws our attention to Frege's conviction that all concepts have extensions, that those extensions are objects, and that, as objects, they fall within the common domain of definition of all concepts, and, in particular, of those concepts of which they are themselves the extensions. (Recall that in his mature theory concepts are functions defined at all objects and having truth values, *the True* or *the False*, as their values.) Frege is deeply in error here, and his opinions on these matters lead directly to contradictions.

But what do the views of Aristotle have to do with Frege's error? Just this: according to Aristotle, what it is that cardinality is ascribed to is a *finite* plurality composed of units; so *the distinction between*

---

[38] This is the definition given in the *Grundgesetze*; the original definition in the *Grundlagen* is faulty in that it treats concepts as if they were objects.

[39] It may be that despite having altered the definition of *the number that belongs to the concept* Φ in passing from the *Grundlagen* to the *Grundgesetze*, Frege continued to regard the concept itself as the bearer of ascriptions of number. If so, that does not in the least affect the argument I am developing here.

2.7 Ascriptions of number: Frege or Aristotle?

finite pluralities, which have sizes, and infinite pluralities, which do not, is built into the very foundations of Aristotelian arithmetic; but in Fregean arithmetic there is no such distinction, because in Fregean arithmetic *all* pluralities have sizes. And this idea is simply incoherent.

When Frege was writing the *Grundlagen*, Cantor had just developed a radically new theory of cardinality in which definite sizes could be assigned to pluralities that Aristotle, and, indeed, all subsequent thinkers up to the time of Cantor, would have regarded as not even having a size, as being *in*-finite. The natural numbers, the points on a geometric line, the continuous curves in space – each of these pluralities, although infinitely large on the classical view, was assigned a definite size (i.e. cardinality) by Cantor: indeed, as is well known, *different* sizes were assigned to the plurality composed of the natural numbers and that composed of the points on a line. Nevertheless – and this is the crucially important point – nevertheless, Cantor held, with Aristotle and against Frege, that *there are pluralities that are too large to be assigned cardinalities*. Such pluralities, which Cantor called "*genuinely infinite*" or "*absolutely infinite*" pluralities, include, for example, the plurality composed of all sets, and that composed of all ordinal numbers.

Cantor sometimes spoke of assigning "powers" or "numbers" to "infinite" totalities, but that is not the best way of regarding his innovations. On the contrary, to see his revolutionary ideas in their true light, we must see him, not as having brought certain infinite pluralities within the domain of number, but rather as having extended the domain of the finite to include totalities, thitherto universally regarded as infinite, within that domain.

Cantor's key doctrine maintains that a plurality may be of the same size as certain of its proper subpluralities and nevertheless remain finite, in the essential and original sense of "finite", the non-technical sense, viz. "limited (or definite or bounded) in size". In effect, he rejected the classical notion of finiteness, the notion employed by both Aristotle and Euclid, by rejecting Euclid's Common Notion 5 (*Elements*, Book I)

The whole is greater than the part

and in doing so he replaced the traditional *Euclidean* finitism by a new *Cantorian* finitism, the point of view that underlies and informs modern mathematics. The modern conception of set is, essentially, the classical conception of *number* (*arithmos*) that Aristotle employed, namely, the conception of a number as a finite plurality composed of units, but with "finite" understood in its new, Cantorian sense, rather than in the traditional, Euclidean one.

Let me make it clear exactly what I am, and am not, claiming here, since it is very important to all that follows. I am *not* claiming that Cantor himself would have regarded his innovations *exactly* as I have described them here, or would have described his theory as a form of finitism. I am claiming that that is how *we* should should regard those innovations and how *we* should describe that theory. But I am convinced that my terminology, though novel, is in accordance with the essence of Cantor's views[40]. Perhaps he would not have been entirely unhappy with it.

We can now see why Aristotle's account of ascriptions of number is superior to Frege's: Aristotle's account generalises to the correct, set-theoretical account of cardinality, including transfinite cardinality, whereas Frege's is flawed in its essentials. Paradoxically, it is the more ancient theory that is more modern in spirit.

But what about the criticisms Frege himself directs at theories like Aristotle's? In fact, most of what Frege has to say that might be construed as applicable to the classical number concept is aimed at the notion of mathematical number with its "pure units", and I have already discussed these criticisms in section 2.6. There is, however, a brief passage in *Grundlagen* §25 that might appear to be relevant here:

Some writers define number as a set or multitude or plurality. All these views suffer from the drawback that the concept will not then cover the numbers 0 and 1[41]. Moreover, these terms are utterly vague: sometimes they approximate in meaning to "heap" or "group" or "agglomeration", referring to a juxtaposition in space, sometimes they are so used as to be practically equivalent to "Number" only vaguer. No analysis of Number, therefore, is to be found in a definition of this kind.

But whatever the merits of these remarks, they do not apply to the *classical* notion of number (*arithmos*), because the units of such a number, even if they are spatial objects, need not be "juxtaposed" in order to compose that number; and, indeed, they need not be spatial objects at all.

---

[40] Cantor did insist that, for example, the natural numbers form an "actually infinite" plurality, but the "infinity" in question is an "unboundedly increasable" one, and in this repect is like a finte plurality. He therefore distinguishes the "increasable" finite and transfinite, from the "unincreasable" Absolute. There is a good discussion of these issues in Hallett's *Cantorian Set Theory and Limitation of Size* (especially Sections 1.3 and 1.4) to which I refer the interested reader.

[41] This is a perfectly valid point. I shall address it in Section 3.3.

2.7 Ascriptions of number: Frege or Aristotle?

Of course Frege is right in claiming that these things cannot be *natural* numbers; which "set or multitude or plurality" could be the natural number thirty-seven, for example? Dummett, however, suggests that they cannot even be the things to which number is ascribed:

When we regarded it [i.e. a (natural) number] as ascribed to a complex, an aggregate, it seemed that the number to be ascribed depended on our subjective way of regarding it; as one copse, or as five trees; as four companies, or as five hundred men. But there is nothing subjective about it: it is the concept copse or tree, company or man which we invoke in the ascription of number, that determines objectively which number it must be[42].

But a number in the original sense, an *arithmos*, unlike an "aggregate", understood in the sense Dummett is using here, presents itself to us, so to speak, with its units already determined: what it is is determined by what its units are. Insofar as we succeed in specifying such a number at all, we specify its units as well, directly or indirectly, explicitly or implicitly. Dummett goes on to say:

For a number to be ascribed to a given concept, there need be no physical relationship between the objects falling under it, nor do we need to perform any mental operations upon our ideas of those objects: the concept itself performs the only function of gathering them together or of singling them out that is needed.

Exactly so. But then for an *arithmos* to belong to its proper numerical species there need be no physical relationship among its units either, nor need we perform any mental operation on our ideas of them. Indeed, *we need not even bring them together under one concept*: simply by their being, individually and severally, the definite, particular things that they are, and by their being, collectively, finite in multitude, the number they compose exists and belongs to the numerical species that it does belong to.

Is the concept really essential here? Suppose I ask a computer to select $67,843,271$ U.S. citizens at random from the 1990 census returns. Once this selection is made, then the number species $67,843,271$ can be ascribed to the concept:

ξ was one of the U.S. citizens selected by computer on May 3, 1994.

But those very citizens were $67,843,271$ in number *before* the selection was made. What could $67,843,271$ have been ascribed to then? Do I really have to posit a monstrosity like the Fregean concept

---

[42] *Frege: Philosophy of Mathematics*, p. 88.

ξ *is William Jefferson Clinton, or ξ is Willard Van Orman Quine, or* (and so on, through a list of the names of 67,843,268 U.S. citizens) ..., *or ξ is Arnold Schwarzenegger.*

in order for there to be something to which I can ascribe 67,843,271 here?

Moreover, even if I had not ascribed 67,843,271 to this number, is it not a number, and, indeed, a 67,843,271, nonetheless? Surely my naming it, by bringing its units under one concept, does not confer existence on it, any more than does my specifying its size endow it with a size.

This example raises an important question: isn't there a subjective element built into Frege's analysis simply by virtue of his electing to talk about *ascriptions* of number? After all, it is *we* who make such ascriptions. But what about the objective facts to which we call attention in making them, and which determine their truth or falsity? We must not allow ourselves to be distracted by the circumstance that to invoke, or to describe, or to allude to, those facts we must employ concepts. Surely the natural view – it was, after all, the view of the Greeks – is that it is to the *arithmos* itself that (a) number (species) is ascribed, not to any concept that just happens to single it out.

That *these* five horses are a five is surely independent of the fact that they are the horses now grazing in that field, or are the horses recently purchased by me from the Aga Khan. Surely the natural view is that the number ascription belongs not to these concepts primarily, but to the five horses collectively (i.e. to the *arithmos* they compose) primarily and to the concepts only secondarily. They just *happen* to be the horses grazing in that field, and they just *happen* to be the horses recently purchased by me from the Aga Khan; but they are necessarily the five horses that they are, and, of course, they are necessarily a *five* of horses.

The "binding together" accomplished by the concept in Dummett's account seems, on the face of it, to be a function of what Frege came, after the publication of the *Grundlagen*, to call the *sense* of the unsaturated expression that denotes the concept. Surely, in any case, it is that sense that allows us to "bind together", with one mental act, the objects that fall under the concept. But this raises a much more serious objection to the view that concepts, or their extensions, *understood as Frege understood them in the Grundlagen*, should be the bearers of ascriptions of number.

Suppose, for example, we wish to ascribe a number to the concept *extension of a concept under which only natural numbers fall*. Here, to be sure, Aristotle cannot help us; but we can perfectly well deal with the

2.7 Ascriptions of number: Frege or Aristotle?

problem using Cantor's extension of Aristotle's conception of number. In any case, this is the sort of question with which Frege's theory would eventually have had to deal. Considering it forces us to pass from the loose notion of "concept" employed in the *Grundlagen* over to the more precise account given in the *Grundgesetze* according to which concepts are (globally defined) functions whose values are always truth values. We must do this because here we are forced to consider concepts which do not have names in the *Begriffsschrift*, Frege's formal language, and that raises many difficult, not to say embarrassing, questions: what *are* such concepts; what are their extensions; what is the connection, in general, between concepts and formulas in the *Begriffsschrift*, etc., etc.?

It is perhaps not difficult to imagine how each of the concepts

ξ *is an even number*
ξ *is a perfect square*
ξ *is a prime number greater than 10,000*
.
.
.

"gathers together" or "singles out" (to use Dummett's own expressions) natural numbers into numerable extensions. But what performs that office for the $2^{\aleph_0}$ concept extensions of the sort under discussion whose determining concepts do *not* have names in the *Begriffsschrift*?

I do not believe that Frege was even aware of such a difficulty. His silence on the matter is surely strong evidence for this. Moreover, his analysis of the notion of function is based on an analysis of the nature of function *signs* as syntactic objects. Once we have grasped the peculiar, "unsaturated" character of function signs then, according to Frege, we can come to understand, by analogy, the peculiar nature of the functions themselves that are named by such signs. But when we cut functions free of their names – as we must do in modern mathematics on cardinality considerations alone – then such syntactic analogies lose their explanatory force[43].

The classical view was that it is to finite pluralities – to numbers in the ancient meaning of "number" – that size or cardinality is ascribed. This is also the modern view, if we interpret "finite" in its extended, Cantorian sense. Frege's theory rests upon a profound misapprehension. He failed to perceive that the notion of finiteness, whether in its traditional or in its

---

[43] I shall discuss Frege's notion of function at greater length in Section 3.4.

52   *Simple Arithmetic*

modern, Cantorian sense, must be built into the objects of *Zahlangaben*. This was the source of the disaster which befell him[44]. That is why the Aristotelian account of ascriptions of number is superior.

2.8 Simple numerical equations

What account of simple numerical equations can be given from the standpoint of the original conception of number? By "simple numerical equations" I mean equations like

$$7 + 5 = 12$$
$$13 \times 13 = 169$$
$$128 + 279 = 407$$
$$271 \times 362 = 98102$$

and so forth.

Let us begin by considering the decimal numerals employed in such equations. On their own, these numerals, like number words, stand for species of number. They are of two sorts, *simple* and *compound*. A simple numeral (e.g. "5") stands directly for a particular species of number. A compound numeral (e.g. "659") stands for such a species by indicating, in the well-known manner, how particular numbers belonging to that species can be partitioned into multiples, less than ten, of powers of ten (i.e. of units, tens, hundreds, thousands, etc.)[45]. The facts underlying this notation are very familiar to us now and, indeed, they were known to Archimedes, as he demonstrates in his *Sand Reckoner*. But when combined with the place notation and the use of the symbol "0", they give rise to what is perhaps still the most powerful and useful of mathematical inventions.

Now let us turn our attention to the equal sign, "=". Here we must notice that numbers, in the original sense, can be *equal* without being *identical*. The number of fingers on my right hand is the number whose units are the fingers on my right hand, whereas the number of toes on my right foot is the number whose units are the toes on my right foot.

---

[44] Cantor himself pointed this out in his review of Frege's *Grundlagen der Arithmetik*. (See page 440 of Cantor's *Gesammelte Abhandlungen*.)

[45] It is perhaps worth pointing out that the number of such partitions, even for relatively small numbers, is very large: for a twenty-five it is nearly a ten billion. But when we count out a number we thereby specify such a partition. The 287th unit in a 659, for example, is the seventh unit in the ninth ten of the third hundred of the partition, consisting of six hundreds, five tens and nine units, determined by the particular counting out.

2.8 Simple numerical equations

These numbers are equal, both being fives, but not identical, since the one is composed of fingers, the other of toes. This is a simple and quite straightforward point, but one on which it is possible to be confused, if we may judge from the rather long-winded explanation Aristotle gives at the end of Book IV of the *Physics* (224a2–16), which is designed to forestall such confusion.

This meaning of the equal sign must be borne in mind in order to understand even such simple equations and inequalities as

$$67 = 67$$
$$67 \neq 59$$

The first of these does *not* assert that a certain abstract object (viz. "the number 67") is self-identical, but rather that any 67 is equal to any other 67 (more precisely, that any number that can be partitioned into six tens and seven units is equal to any other number that can be so partitioned). By the same token, the inequality does not assert that two abstract objects (viz. "the number 67" and "the number 59") are distinct objects, but rather that no 67 is equal to any 59 (more precisely, that a number partitionable into six tens and seven units is not equal to any number partitionable into five tens and nine units).

Both the facts conveyed by these symbolic expressions are obviously true, but neither is true simply by "convention" or "definition". Strictly speaking, both of them require proof. Of course with the particular (small) number species employed in these examples proof may seem unnecessary, but the more general propositions, of which these are special cases, are not trivial. In particular, the existence and uniqueness of the decimal representations must be formulated precisely and proved[46].

I have referred to *more* general propositions here because both

$$67 = 67$$

and

$$67 \neq 59$$

are *already* general propositions. It is this extra, and customarily unacknowledged, level of generality incorporated in such apparently *particular* propositions that gives rise, in part, to the illusion that we are dealing with particular, but *abstract* objects in these propositions. The generality

---

[46] The problem here is to formulate, exactly, what is meant by a decimal representation. We also require a rigorous definition of equality between numbers.

forces itself upon our attention, so to speak, but appears in the guise of abstraction rather than generality.

Finally, let us consider simple numerical equations involving the signs for addition and multiplication. Again, the propositions expressed by these equations express general facts about numbers (finite pluralities composed of units).

Thus

$$7 + 5 = 12$$

means[47]

> For any numbers x, y, and z, if x is a 7, y is a 5, and z is a 12 (i.e. z has a partition into 1 ten and 2 units), and if x and y have no units in common, then the sum (simple union, $x \cup y$) of x and y is equal to z.

Similarly

$$13 \times 13 = 169$$

means

> For any numbers x and y, if x is a 13 composed of disjoint 13s (i.e. if each unit of x is a 13 and no two such units have any of *their* units in common), and if y is a 169, then the sum (union, $\bigcup x$) of x is equal to y.

These facts, and others like them, are so familiar and so easy for us to take in that one would have thought it otiose actually to spell them out. And yet ... How easily we are hoodwinked by the surface forms of these familiar propositions! How easily we forget the *kind* of fact they convey and hypostatize the "natural numbers" as the objects correlated to the numerals and "abstract operations" on such numbers as correlates of the addition and multiplication signs!

We all first encountered serious mathematics when we were taught the decimal algorithms for addition and multiplication in school. In carrying out calculations in accordance with these algorithms, we were taught to regard numerals, not as meaningful signs, but as mere figures, that is to say, syntactic assemblages or configurations. We lay them out in rows and in columns, proceeding in a purely rote, mechanical fashion, in accordance with the rules, until we obtain the required answer. Here the actual configuration of the ciphers on the page, or on the blackboard, is

---

[47] It goes without saying that there are other, equally valid interpretations along these lines. The facts here are over-determined.

2.8 Simple numerical equations

the chief and essential thing, and most of us can, no doubt, remember being marked down when our work was not neat, with properly aligned rows and columns and legible numerals – and quite properly so, for this is not mere schoolmarmish pedantry: the actual configurations produced are the essence of the matter.

Now these algorithms permit us – indeed, they require us – to regard the numerals, and the configurations composed of them which we write down when implementing the algorithms, as mere figures, and to ignore their meanings as signs. But a calculation carried out in accordance with these algorithms, and displayed in the conventional manner, can be regarded as a schematised *proof* of the general proposition expressed by the corresponding numerical equation.

Let us consider, for example, the calculation of the product 271×362. Following the prescribed rules we are led to lay down the following configuration:

$$
\begin{array}{r}
3\,6\,2 \\
\times\,2\,7\,1 \\
\hline
3\,6\,2 \\
2\,5^4 3^1 4\,0 \\
7^1 2\,4\,0\,0 \\
\hline
9\,8^1 1^1 0\,2
\end{array}
$$

Hence $271 \times 362 = 98,102$. Notice that I have written down the numerals "carried" rather than doing the carrying "in my head". The zeros at the ends of the 5th and 6th rows are, of course, optional. I have written them down in aid of making a point in what follows.

What is the significance of the configuration just laid down? From the standpoint of the algorithm it has no significance other than that it was laid down in accordance with that algorithm. That it was so laid down is a *geometrical* or *combinatorial* fact, and we could easily program a computer to check such facts. But the significant point is that we can also "read" the configuration as an abbreviated and schematised proof of the proposition that $271 \times 362 = 98,102$; and remember, the equation is itself an abbreviation and schematisation of the general proposition that, for all numbers x and y, if x is a 271 composed of disjoint 362s, and y is a 98,102, then the sum, $\bigcup x$, of x is equal to y. Let us look at this in more detail.

Suppose that the number x is a 271 composed of disjoint 362s. Let us call the 362s that are units in x units$_1$ and the units in the 362s, units$_2$. Since x is a 271, it has a partition into 1 unit$_1$, 7 tens$_1$ (each composed of units$_1$, i.e. 362s), and 2 hundreds$_1$, (each, again, composed of units$_1$). The units$_2$ in the sum, $\bigcup x$, of x thus comprise

(1) The units$_2$ in the one unit$_1$, in the given partition of x
(2) The units$_2$ in the sum of the seven tens$_1$, in the given partition of x
(3) The units$_2$ in the sum of the two hundreds$_1$, in the given partition of x

Turning to the configuration produced by the calculation, we see that the numerals occupying the third, fourth, and fifth lines, viz. 362; 25,340; and 72,400; respectively, name the number species which the numbers, composed of the units$_2$ named in (1), (2), and (3), respectively, belong to.

How do we know this? In the case of (1) it is obvious, since each unit$_1$ is a 362 composed of units$_2$. In the case of (2), suppose that y is a number composed of seven tens$_1$, each unit$_1$ of which is a 362 composed of units$_2$ and no two of which tens$_1$ have any units$_1$ in common. Then the number species to which y belongs ...

But a little of this goes a very long way indeed, and I am sure that the point I am making is now clear. Writing down such an argument in detail is tedious in the extreme: it would be much better to *talk* one's way through the argument as one actually carried out the calculation – writing out the required configuration on a blackboard, say, and explaining, as one went along, the significance of what one was writing down.

But why am I belabouring the obvious here? For all these things *are* obvious on a little reflection[48]. It is equally obvious that the inventor, or inventors, of the decimal algorithms must have reasoned things out in much the same sort of way as I have done here. There is, however, a surprisingly widespread view that somehow the decimal algorithms themselves *determine* the truth or falsity of the associated numerical equations, so that the purport of such an equation is that if you follow the decimal rule correctly then you will obtain the "answer" given by the equation. Wittgenstein held such a view:

But on the other hand I say again: "Calculating is right – as it is done." There can be no mistake of calculation in "$12 \times 12 = 144$." Why? The proposition has assumed a place among the rules[49].

How can it be just a *rule* that twelve disjoint twelves together make a

---

[48] Perhaps I should say they are obvious in retrospect. It took a mathematician of genius to *discover* them, and to grasp their significance.
[49] *Remarks on the Foundations of Mathematics*, p.197.

2.8 Simple numerical equations 57

one hundred and forty-four? It is not even a *rule* that seven eights make a fifty-six, although we all learned it as a rule when we were children.

When I was a child, I spake as a child, I understood as a child, I thought as a child; but now I have become a man I have put away childish things.

Wittgenstein's views, inspired, no doubt, by his experiences as a school teacher, stand the facts squarely on their heads[50]. Let me put the matter as simply and as briefly as I can. The decimal algorithms are of use precisely because the assemblages of signs we construct when we follow them express, in schematised and symbolic form, simple reasonings about finite pluralities, that is to say, about *numbers*, as originally conceived: they depend upon underlying facts about such finite pluralities for their significance, though not, of course, for their formulation or for their execution. *It is the meaning of the completed calculation that alone provides the rationale for the syntactic manipulations that produced it and the algorithm that guided those manipulations.*

The interpretation of simple numerical equations that I have given here is quite straightforward and easily grasped. As an explanation of the facts underlying simple arithmetic it is vastly superior to the naive, conventional interpretation in which such equations are seen as assertions of identity between abstract "natural" numbers that have been subjected to the "abstract operations" of addition and multiplication.

But that is not the whole story! If it were, we should be tempted to say that the introduction of the concept of natural number was just a stupendous blunder. But that is the very reverse of the case. On the contrary, the historical process, briefly described in Section 2.1, was the major and essential ingredient in the scientific revolution that has completely transformed our world. As vehicles of explanation the natural numbers are useless; as vehicles of invention and discovery they are indispensable.

If we want to find out the facts about "How many" we turn to the familiar decimal algorithms, in which the numerals are treated as "things" to be manipulated syntactically. As Frege said, we let our symbols do our thinking for us. To dwell on what our symbols really mean would seriously hinder us in performing those manipulations: it would defeat the whole purpose of having introduced the symbols in the first place. And, of course, the symbolism of arithmetic gives rise to the more abstract and general symbolism of algebra.

---

[50] For a concise statement of these views see the passage from Wittgenstein's writings I quoted in Section 2.4.

The power of the use of symbols in mathematics is that it enables us to render our abstractions concrete and our generalities particular, so to speak: it permits us to replace thought by mere symbolic manipulation. The Greeks, for all their mathematical acuity, never suspected the power of symbols, and in this respect, which is the crucial respect, we have advanced far beyond them.

But this very advance brings with it foundational problems of a complexity much greater than anything the Greeks had to face. For we have to explain what our symbol ridden and symbol inspired propositions mean. If we cannot do this then the whole enterprise of proving theorems, that is to say, the whole enterprise of mathematics, becomes futile.

The problem of natural numbers is representative of this class of problems. For us, arithmetic *is* natural number arithmetic. And yet the natural numbers, which form the subject matter of our science of arithmetic, are symbol generated abstractions: they are what Hegelians or Marxists might call "reified and alienated" products of our counting and calculating activities: they have no claim to objective existence outside those activities. To seek to discover what they "really" are is to pursue a will-o'-the-wisp. That is why they have no explanatory power.

But can we not simply identify the natural numbers with numerals? Well, the numerals themselves are real enough, no doubt, the ones we actually write down, that is. But we need an unlimited supply of natural numbers, and there are, as far as we know, only a limited number of numeral *tokens* potentially available to us.

So the identification must be with numeral *types*. But that notion is not at all clear. One says that "in principle" or "in theory" there is an unlimited supply of numerals; but that just raises the questions: What principle? What theory? If the theory is theoretical physics, there may be an upper bound on the length of possible tokens.

A numeral is a finite string of digits. But what does "finite" mean here? And doesn't attaching "finite" to "string" simply compound the problem? Are we not just confronting the problem of *finiteness* again, the problem that the "natural" numbers are supposed to solve, but this time with the additional necessity of explaining what a *string* of symbols is?

But the number *species* are objective. Why not identify *them* with the natural numbers? True, the number species are *objective*, but it is problematic whether we can simply treat them as *objects*. For each of these species is *essentially* infinite, and to treat these infinite things as ordinary objects is not possible in any straightforward way. Any attempt to do so is fraught with difficulties: at worst we encounter contradictions; at best we blind ourselves to subtle but essential logical distinctions[51].

Numbers, in the original sense, are not symbol generated abstractions. They were the objects of rigorous mathematical investigation long before mathematicians had mastered the art of thinking with symbols. They are not even abstractions, *per se*. This concept of number is the oldest scientific concept that we possess. But the things that fall under that concept, the numbers themselves, exist quite independently of human beings, their concepts, and their mental activities. These are the things that the facts of arithmetic, the most certain scientific facts we possess, are facts *about*. Long before there was an eye to see them or a mind to count them the planets circling our sun composed a nine; and that nine had partitions composed of three threes; and the number of those partitions was a one thousand six hundred and eighty; ...

2.9 *Arithmetica universalis*

Mathematicians today are the beneficiaries of two revolutions in the foundations of their science, one in the seventeenth century, the other in the nineteenth. The second of these was a revolution in the original sense of that much abused term, namely, a return to earlier standards and ideals, in this case to Greek standards of rigour in argument, and to Greek ideals of concreteness in concepts and economy in assumptions. Indeed, the mathematicians of the nineteenth century went beyond the Greeks in adhering to those standards and in realising those ideals.

Numbers – natural, rational, real, and complex – are leftovers from the earlier revolution. But we have now forgotten the terms of the seventeenth century debate which accompanied the introduction of these novelties, with the consequence that all these numbers appear to us cloaked in the guise of "raw data", that is to say, as things just "given" – the "abstract objects" that mathematical discourse is discourse *about*. Naively, we take

---

[51] We can give a perfectly rigorous *axiomatic* formulation of the theory of natural numbers as particular objects. That is the orthodox solution of the difficulties I have been discussing here, and I shall present such a formulation in Section 5.1. But that theory, like all axiomatic theories, logically presupposes the original notion of number (in its modern guise as the notion of set). Moreover this axiomatic theory rests on powerful existential assumptions, and, if we do not make those assumptions, the orthodox axiomatic approach to natural number arithmetic becomes unfeasible. In those circumstances the difficulties I have alluded to here make themselves manifest. I shall discuss these matters at length in Part Four.

them to be simply what our symbols stand for, even though officially, so to speak, we feel obliged to give rigorous, axiomatic accounts of them[52].

For us, therefore, these "given" numbers are not creatures of our own conscious invention, as they were for our seventeenth century forebears, but rather illusions engendered by the way we employ symbols in arithmetic and algebra; and, as such, they have no claim to independent existence. To attempt to determine what they really are is, in consequence, utterly futile. There simply are no objects named by our number words and numerals, and to suppose otherwise is to fall into confusion and error.

Numbers in the original sense, however – *arithmoi* – pluralities composed of units – these things are not, like "natural numbers", mere fabrications of the mind, but on the contrary, are authentic inhabitants of the world, independent of human beings and their mental activity; they are things that we are obliged to acknowledge if we are to make any sense at all of our mathematical experience.

Nevertheless, the difficulties which must necessarily attend any attempt to give a "natural" account of "natural" number, difficulties with which Frege struggled so heroically, are extremely instructive, and I offer no apology for discussing them at such length. For natural numbers are included among what I am calling "symbol generated abstractions", and it is a central technical problem for the foundations of mathematics to give an account of such things. Ordered pairs, functions, relations – all these things are symbol generated abstractions and, moreover, unlike natural numbers, are indispensable to the formulation of our most basic mathematical concepts and techniques.

We can, after all, give an axiomatic definition of the natural numbers. To be sure, such a definition does not so much define them as define them away. But, in any case, no such option is available to us in respect of those other notions. For unless the notions of ordered pair, function and relation are already in place, we cannot give an account of the axiomatic method itself.

This, in short, is the problem: since ordered pairs, functions, and re-

---

[52] An axiomatic account does *not* consist merely in laying down basic axioms in the theory being axiomatised with a view, say, to to making explicit the assumptions which govern the proofs of propositions about the "mathematical objects" of our theory, even though such a view of the axiomatic method is surprisingly widespread. The whole point of laying down axioms is to *avoid* reference to such supposed objects – "natural" numbers, for example – and, strictly and logically speaking, the axioms are elements in the definition of a species of mathematical structures, namely those in which the axioms hold true. I shall discuss this point in detail in Section 6.2.

lations are symbol generated abstractions, alienated and reified products of our symbolic procedures of calculation and definition, there is nothing in heaven or earth – nothing outside those procedures – that naturally answers to the name "ordered pair", or "function", or "relation".

Accordingly, unless we wish our existence claims in mathematics to amount, ultimately, to claims that human beings have, actually or "in theory", certain capacities, real or imagined, for manipulating symbols – and such claims must, by their very nature, either be uselessly vague or be false, *sensu stricto* – then we must find objective correlatives to those symbol generated abstractions, that is to say, we must find genuine objects that have the essential properties we attribute to those imaginary ones. In modern mathematics, the objective correlatives of all these things are defined to be numbers, but numbers as conceived in the original sense, that is to say, *arithmoi*, or, as we now prefer to call them, *sets*.

In the development of set theory as a foundational theory, modern mathematics has attained the goal, first envisaged by Descartes and Leibniz, of a *mathesis universalis*, a universal framework for the exact part of human thought. But perhaps it is more accurately described as an *arithmetica universalis*: it is *arithmetic* in that it is based on the classical notion of number, refined and generalised; and it is *universal* in that all mathematical concepts are defined, ultimately, in terms of its basic concepts, and all mathematical proofs rest, ultimately, upon its basic truths.

The fundamental assumption underlying set theory is this: *whatever things there are, finite pluralities composed of those things are also themselves things, and, as such, serve as units in still further pluralities*. The technical problem in this is to spell out, more precisely, what is meant by "finite", and here we have a critical choice to make. We can follow Euclid and adopt his Common Notion 5 (*Elements*, Book I)

> The whole is greater than the part

This leads to what I shall call *Euclidean* set theory, which takes as a fundamental assumption the following axiom:

Euclidean Finiteness *If S is a set and $f : S \to S$ is an injective (one to one) function from S to S, then f is also surjective (onto).*

This axiom simply reformulates Common Notion 5 in modern, set-theoretical terminology.

The alternative is to follow Cantor in rejecting Euclid's Common Notion 5. As I have already explained, Cantor held that a plurality can *have* a size, and still be of the *same* size as certain of its proper subpluralities; and to have a size, in this sense, is to be *finite* in the root sense of "finite", i.e. is to be an *arithmos* or, in modern terminology, to be a set. The resulting *Cantorian* set theory takes as a fundamental assumption the following axiom:

Cantorian Finiteness *The species of (von Neumann) natural numbers*[53] *is finite, and therefore forms a set, ω.*

This axiom is incompatible with the Euclidean axiom, as it is easily shown that the correspondence $x \to x \cup \{x\}$ defines a function $f : \omega \to \omega$ which is injective but not surjective.

This reinterpretation of "finite" is the central idea that leads to the modern way of doing mathematics. On either view arithmetic (i.e. set theory) is finitary; it is a matter of opting for *Cantorian finitism* or the traditional *Euclidean finitism*. To be sure, Cantor's innovation was bold and radical – his contemporaries saw it as such. But over one hundred years of mathematical practice have not revealed any contradiction lurking in it. If it is mistaken, then the mistake is a deep and subtle mistake, not a gross and obvious one.

The possibility of an *arithmetica universalis* arose with the development of the axiomatic method, understood in its modern sense of axiomatic *definition*. In an axiomatic definition what is defined is a *kind* or *species* of mathematical structure, where a particular such structure typically consists of an *underlying domain* or *set* equipped with a *morphology*[54]. As the axiomatic method took over mathematics – and, in retrospect, the take-over seems to have been inevitable – set theory, without which that method cannot even be formulated, moved to the centre of things. Sets provide the underlying domains of mathematical structures and, via the "arithmetised" versions of ordered pair, function, and relation, the morphologies of such structures as well.

But set theory, as it is essentially, and in itself, is the theory of finite pluralities, and questions of mathematical structures and of the logic of axiomatic definition simply do not arise when the fundamental concepts and principles of set theory are under consideration. We can discover, and evaluate, those concepts and principles quite independently of any

---

[53] In von Neumann's definition, each natural number is identified with the set of all its predecessors, so that $0 = \emptyset$, $1 = \{0\} = \{\emptyset\}$, $2 = \{0,1\} = \{\emptyset,\{\emptyset\}\}$, ...; in general, $n+1 = n \cup \{n\}$. I shall give an exact account of these matters in Section 5.2.

[54] In a group this is provided by a binary operation on the underlying set, in a ring by two such operations, in a partial ordering by a binary relation, etc.

2.9 Arithmetica universalis

consideration of set theory's role as a foundational theory. The issues involved in establishing the theory of sets are simple[55] and could have been understood by Aristotle or Euclid. This is the great strength of set theory as a foundational theory.

We can now say, with confidence, that all mathematical proofs rest, ultimately, on set-theoretical principles, and all mathematical definitions rest, ultimately, on set-theoretical concepts. But that is by no means to say that set theory is the whole of mathematics. If I am acquainted only with the basic principles and concepts of set theory, I have, to be sure, made a start in mathematics; but only a start. Virtually the whole of the subject remains unknown to me. I know no real analysis, no complex analysis, no group theory, no linear algebra, no differential geometry, ... But I have, in fact, acquired the logical tools required to master these various branches of the subject.

It is important that we see these matters in their true light. Each branch of mathematics has its own ethos, its own special techniques of proof and of definition. These techniques can, of course, be traced back to their roots in set theory, where they find their ultimate logical justification; but once they have been firmly and rigorously established, it would be unnecessary and distracting continually to refer back to the details of that establishment.

The art of mathematics consists in encapsulating precise and intricate conceptual analysis in concise and rigorous definitions and proofs, expressed in simple and elegant symbols. With each advance in abstraction and generality, the previous level of abstraction and generality attained takes on the appearance of concrete, "given" reality. Of course from the standpoint of logical analysis such appearances are illusory; but these "illusions" are beneficial, for they free the mind from unnecessary preoccupation with inessential and irrelevant details.

However, all these generalities will no longer suffice: we must now turn to a proper mathematical investigation. Even our discussion of simple arithmetic, which has taken up most of the present chapter, cannot advance unless we give precise definitions of the concepts we employ (the concept of "equality" between numbers, for example). Our immediate task, however, is to lay the foundations of general set theory, and this will be addressed in Part Two; Parts Three and Four will deal with the Cantorian and the Euclidean conceptions of finiteness, respectively.

---

[55] Naturally, by "simple" I do not mean "easy". Simple notions are often among the most difficult to grasp.

Part Two
Basic Set Theory

It seems to me that the tree I have planted must lift a colossal weight of stone in order to gain space and light.

Frege

3
Semantics, Ontology, and Logic

3.1 Objects and identity

Arithmetic has taken over the whole of mathematics: not the arithmetic of natural numbers, but the generalised and refined version of classical, Greek arithmetic that we now call set theory. But if it is present everywhere, its presence usually goes unnoticed. Indeed, it is most successful when it is least conspicuous, for its essential role is a logical one, so that it appears to the mathematician practising his trade simply as a collection of tools which enable him to get on with his real job – algebraic geometry, functional analysis, or whatever it happens to be.

Its most important function is to provide the logical foundations for the axiomatic method, which is the central method of modern mathematics. When we lay down a system of axioms (e.g. the axioms for groups, or for real numbers), we are defining a class or species of mathematical structures (the species of all groups, or the species of all complete ordered fields). We must therefore say, exactly, what these "structures" are, and that is where set theory comes in. For the raw materials out of which such structures are manufactured are sets.

In fact, since the axiomatic method only allows us to define our structures up to isomorphism, the particular objects that go to make up the universe of discourse of a structure make no contribution to the structure's mathematical properties – only the size (cardinality) of that universe is mathematically relevant. Thus the traditional "mathematical objects" have been rendered logically superfluous by the axiomatic method: the *ideal* objects of geometry and rational kinematics – line segments, circles, triangles, curves, and the continuous motions of points, of lines, of surfaces and of solids – and the *abstract* objects of traditional algebra –

natural numbers, fractions, surds, imaginary and complex numbers – all these things no longer have any *logical* role to play in mathematics.

We now use the axiomatic method to define these things, or rather, the structures in which they occur, up to isomorphism. We no longer see them as simply "given" to us; their essential natures do not seem transparent to us as they did to our ancestors, and we have come to feel that talk about these "things" requires justification.

From these familiar facts it follows that all mathematical propositions rest, ultimately, on propositions about sets and their members; in particular, all existence claims in mathematics rest, ultimately, on claims to the effect that certain sets, or certain kinds of sets, exist. Moreover, we cannot use the axiomatic method to define the notion of *set* itself: that would involve us in the most gross and obvious sort of vicious circle, for set theory must be developed before we can *establish* the axiomatic method[1]. It follows that if we are to understand modern mathematics we must understand the axiomatic method; and if we are to understand the axiomatic method we must have a clear idea of what sets are, of how they can be defined, and of what basic relations, operations, and constructions apply to them.

Mathematical propositions must be precise, definite, and unambiguous. This means that they must have precise, definite, and unambiguous truth values, *true* or *false*: *tertium non datur*[2]. It follows that simple propositions such as "*a* is a member of the set *S*", "*S* is a subset of the set *T*", "the set *S* has exactly three members", etc., must be of this character. This, in turn, imposes the requirement that sets be composed of *definite* elements, things that are determinately individuated so that they are sharply and objectively distinguished from one another. But what does that requirement entail?

We can distinguish two separate problems here. The first of these problems is this: what conditions must we place upon the things referred to in mathematical discourse, what features must we suppose them to possess, in order for mathematical definitions and proofs to work; and how can we formulate those conditions in a mathematically usable way?

This is the problem of the *semantics* of mathematical discourse, and it leads naturally to the second problem, namely: what kinds of things, if any, satisfy those conditions? This is the problem of the *ontology* underlying mathematics. It is most important, indeed it is vital, that we

---

[1] This is an important point which I shall discuss at length in Chapter 6.
[2] "No third (possibility) is given".

3.1 Objects and identity

keep these problems separate. For the semantic problem can be solved, but the ontological problem is open ended.

Let us first turn our attention, then, to the semantics of mathematical discourse. How are we to put the requirement that the elements of a set must be *definite* into mathematically usable form? To fix the terminology I shall call the definite, determinately individuated things that can be members of sets *objects*, employing that expression, borrowed from Frege, as a piece of technical jargon.

The essential requirement that we must place on what I am proposing to call "objects" is that *propositions asserting the identity of objects must always have determinate truth values*. By this I mean that if s and t are names for objects then the proposition expressed by

s *is identical with* t

where "*is identical with*" is to be understood as "*is one and the same object as*", either is determinately true or is determinately false; and whichever is, in fact, the case, it is so quite independently of anyone's *knowing* that it is the case.

Conversely, propositions of identity, if they are taken literally, ought to be construed as identifying *objects*. Where doubt arises that the truth value of a putative identity proposition is objectively determined, what is called into question is the implicit claim that the supposed objects being identified really are *bona fide* objects.

This, then, is the practical, semantic content of the requirement that objects be *definite* things: they must be subject to the familiar laws of identity and difference. They are what mathematical discourse is about, and in that discourse we can refer to them, identify them, and distinguish them. This semantic requirement is the minimum necessary if our discourse is to have the rigour and precision needed for definition and proof.

But what things, or what kinds of things, if any, satisfy this requirement? That is the ontological problem, and it is most unlikely that we shall ever find a definitive solution to it. Fortunately, we do not need a definitive solution.

The modern concept of set derives from the ancient concept of number (*arithmos*). A set is a plurality, finite in multitude, composed of determinate, definite things, that is to say, of *objects*, in the technical, Fregean sense of "object" I have adopted here. The simplest, and most straightforward examples of sets are those composed of two or more *individuals*, that is to say, objects that are not themselves sets[3]. In such sets, the individuals play the role of Euclid's units (*Elements* VII, Definition 1).

But Euclid makes no attempt to catalogue the various kinds of things that make suitable units, nor should we attempt to catalogue the kinds of things that are suitable to play the role of individuals. For the purposes of *mathematics* this is not necessary. Indeed to do so would be most ill-advised, for mathematics ought to be ontologically neutral, insofar as that is possible. The mathematician has absolutely nothing to gain, in his capacity as mathematician, by committing himself to the existence of individuals of certain particular kinds. As Aristotle says, he need only *posit* the unit, or the individual[4].

"Positing" doesn't mean "inventing" or "conjuring out of thin air" here: it means rather "accepting as given without further specification or analysis". The mathematician should regard individuals, not as unspecifi*able*, or abstract in the manner of the "pure units" that make up Plato's "mathematical numbers", but as unspecifi*ed*. Mathematical discourse, therefore, is abstract in the Aristotelian sense of being formal and general, rather than in the Platonic sense of being about abstract "mathematical objects". The mathematician says, in effect, "if you specify what things you count as individuals, then my formal and general propositions can be given specific content, and can convey to you the facts about the sets built up from those individuals."

There is one ontological issue, however, that set theory cannot avoid. Whatever objects there are, sets composed of those objects must count as further objects which can themselves appear as members of still further sets. This entails, in particular, that we must accept that a set has just as legitimate a claim to existence as have the objects that compose it. That claim is not hypothetical but absolute: given that a and b are objects, the set, $\{a,b\}$, composed of a and b is also an object. In the next two sections I shall look more closely at the nature of this ontological commitment.

3.2 *Arithmoi* and their units

In Chapter 2 I pointed out that in Greek arithmetic a number (*arithmos*) is what we should nowadays call a set and its units are what we should now call its members. But not everything we call a set is an *arithmos*.

---

[3] Zermelo called these objects "primitive elements" (*urelementen*).
[4] I have discussed this Aristotelian doctrine in Section 2.6.

3.2 Arithmoi *and their units*

For an *arithmos*, as a *plurality* composed of units, must necessarily have at least two units.

This was certainly the way the Greeks regarded *arithmoi*, the fact that lies behind the widely maintained, but misleading, view that they did not recognise zero and one as numbers. It is misleading because they did not recognise *anything* as a number, in our sense of "number". It would be more correct to say that they didn't recognise empty or one-element pluralities, for the good and sufficient reason that "plurality" implies more than one thing. Thus singleton sets (sets with exactly one member) and the empty set are not *arithmoi*.

Surely the Greek view here is the natural one. There is something very baffling, indeed dubious, about an empty *plurality* (it seems, as Frege said, to "disappear along with its members") or about a plurality consisting of exactly one element (how are we to distinguish such a "plurality" from its sole member?). Thus although "set" is better than "number" as a translation of *"arithmos"* it is not an exact translation.

This calls attention to a difficulty with the notion of set as it is conventionally understood at present. For by acknowledging empty and one-element "pluralities" in modern set theory we purchase convenience at the price of intelligibility.

Indeed, on the most widespread account the entire universe of sets is built up from the empty set in a kind of creation *ex nihilo*, so that the whole of set theory, and with it the whole of mathematics, is conceived as balancing, not on the head of a pin like the angels of the scholastics, but on absolutely nothing at all!

Such a notion, taken literally, is surely unintelligible. To put it forward as the sole foundation for mathematics is surely implausible in the extreme.

From the standpoint of ontology, the difficulties we encounter in making sense of the empty set and singleton sets are not trivial. There was a serious debate on the status of the unit in antiquity which reflects our problem with the status of singletons[5].

To avoid these difficulties I propose that we take, not the notion of set, but the notion of *arithmos*, as fundamental, and then define the notion of set in terms of it. If we do this then the fundamental assumption upon which set theory, and with it all of mathematics, rests comes to this: *whatever objects there are, there are also finite pluralities composed*

---

[5] See Heath's commentary on *Elements* VII, Definition 1, in his translation of Euclid.

of those objects, namely, arithmoi; moreover, these arithmoi are themselves objects in their turn and, as such, can serve as units in further arithmoi.

Of course, to assert that *arithmoi* are objects is to assert that there really are such *things* as finite pluralities, and that they have the same kind of claim to real existence as definite, independent entities as have the objects that make them up.

But this then becomes the sum total of the mathematician's "ontological commitment". The only "mathematical objects" he need acknowledge are *arithmoi*. The whole of his science is subsumed under arithmetic, the generalised and refined version of classical arithmetic he knows as set theory, and he may take as his motto the Pythagorean dictum "Everything is number."

I have already discussed the existence of *arithmoi* at length in Chapter 2 (especially in Section 2.5). But it will be useful to supplement that discussion by considering why *arithmoi* can be accounted *objects*, in the light of what was said about the notion of object in the previous section. Briefly, to count as an object an *arithmos* must be a single, definite thing: it must possess unity (so that it is *one* thing) and be clearly individuated (so that it can be sharply distinguished from every other thing).

That *arithmoi* can be properly individuated is quite a straightforward matter. For the *principle of extensionality*, by virtue of which the *arithmos* S is identical with the *arithmos* T if, and only if, S and T are composed of exactly the same units, provides a clear and unambiguous criterion of identity for *arithmoi*. This guarantees that *arithmoi* can function as objects insofar as they are the subjects of propositions of identity.

But what is the source of unity in an *arithmos*? What is it that makes an *arithmos* a single thing, given that it is composed of, and, indeed, in some sense just is, many things? Of course, common sense suggests that many objects are composite: they are composed of many other objects, just as my body is composed of limbs and organs, those of cells, cells of molecules, molecules of atoms, ... (But perhaps some of these things are not objects at all; perhaps they are *indefinite* things like clouds ...)

Surely there cannot be any objection to composite objects *per se*. Surely the mere fact of a thing's being composed of many other things cannot tell decisively, *per se*, against its unity, that is to say, against its being one single, definite thing, i.e. an object, in our technical sense.

But even if we grant that there are composite objects, how does it follow that *arithmoi* are such objects? In all the examples cited, the unity

3.2 Arithmoi *and their units*

of the composite seems to arise essentially from what we might call the mode of its composition, that is to say, the particular way in which it is composed out of its elements: for example, in the case of a human body's being composed of cells, the spatio-temporal relations among those cells, their chemical interaction, their common origin in embryonic cells, etc.

But if this mode of composition is destroyed (e.g. if a human body is completely dismembered and then dissolved into its component molecules) the composite object ceases to exist. What is it, then, that makes an *arithmos*, which is a *bare* plurality without such a characteristic mode of composition, a single definite thing, i.e. an object? What is the principle of unity for an *arithmos*?

The answer, which I have already given, is that the principle of unity for an *arithmos* is *finitude*: simply by being objects, and therefore, by hypothesis, having, severally, claim to real, independent existence, *and* by being, conjointly, finite in multitude, the units of an *arithmos* together and collectively constitute a single well-defined thing, viz. an object.

On this explanation, the notion of finiteness is constitutive of the very concept of *arithmos*. It is present in that concept from the beginning, so to speak, so that without it we cannot form any conception of *arithmos* at all. But this raises the difficulty that we do not have a sharp notion of finiteness.

The difficulty here does not consist principally in there being rival notions of finiteness – the Euclidean and the Cantorian – competing for our approval. For even if we decide between them, finiteness remains a concept with clear instances but vague in extent. We can give clear examples of finite pluralities, and clear examples of infinite ones; but we cannot establish an unambiguous criterion that separates them. Indeed, it is true almost by definition that the notion of finiteness cannot be delimited or circumscribed: it cannot be "bounded" or "limited" since finiteness is the concept which itself determines the very notions of "bound" or "limit" that are in question here.

But this is not the last word on the matter. On the contrary, what we know about finiteness (or, indeed, what we believe or conjecture) can be expressed in the form of general laws or axioms – the axioms of set theory, in fact. It is the role of those laws to convey, partially and indirectly, just what the concept of finiteness amounts to.

It remains for me to define the notion of *set* in terms of the notion of *arithmos*. This I shall do in the next section.

3.3 Sets

Just as natural number arithmetic runs more smoothly with zero and one, so set theory runs more smoothly with their correlatives, the empty set and singleton sets (sets with exactly one element). These sets present themselves as the "limiting cases" of finite plurality, so to speak: naively, the empty set is a plurality with no members, and a singleton set is a plurality with just one member.

It is not difficult to see why postulating the empty set and singletons makes set theory run more smoothly; for if we suppose there to be such sets, then we may lay down the assumption that if S is a set and Φ any definite property that applies, truly or falsely, to every member of S, then there is a unique subset, $\{x \in S : \Phi(x)\}$, of S consisting precisely of those elements of S (if any) that satisfy the property Φ. This assumption, which is, in fact, the Axiom of Comprehension, provides us with the means to achieve a purely extensional, set-theoretical treatment of "properties" or "concepts". On the other hand, without the empty set and singletons, we should have to make special provision for the cases in which Φ is satisfied by no elements of S, or by exactly one element.

But with the empty set and singletons we are confronted with the question: how are we to conceive these things as *objects*? How, in particular, can we conceive them as *pluralities*? A finite plurality of two or more objects is "there" in virtue of its members being "there" in finite multitude. But it seems nonsensical to say that the empty set is "there" in virtue of its members *not* being "there". The empty set doesn't seem to be anything at all. The set $\{a,b\}$ is composed of a and b; but what is the empty set composed of? Nothing? And how are singletons to be conceived as *pluralities*? How are such supposed "pluralities" to be distinguished from their sole members?

An apparent way out of these difficulties now seems to suggest itself. Although we cannot conceive of a mere plurality composed of *no* units, we can easily think of concepts under which no object falls. Similarly, we are in no danger of confusing a concept under which exactly one object falls with the unique object that falls under it. Might we not, therefore, follow Frege in taking the empty set and singletons, and, by extrapolation, all sets, to be "extensions of concepts"?

This seemingly simple and natural suggestion raises a matter of absolutely fundamental importance; for what is at issue is the question of which of two ancient and rival notions is to be taken as the fundamental notion underlying mathematics, namely, the *mathematical* notion

3.3 Sets

of number (*arithmos*) or the *logical* notion of species or class (*eidos* or *genos*).

At stake here are deep and difficult issues, and I shall discuss them at some length in Section 3.5 below. But even on the face of it one question does seem to depend on whether we take sets to be arbitrary pluralities finite in size (*arithmoi*) or pluralities which can be defined as the extension of some concept. I mean the question whether the notion of set is *objective*, whether sets exist in reality, and not merely in conception or in thought. Indeed, the very word "concept" in this context seems to suggests a creature of the mind.

There is, to be sure, an old and respectable philosophical tradition in which concepts or ideas are held to possess an objective existence independent of any particular mind that apprehends them. This view is now often described as "Platonism" or "metaphysical realism".

But such a view, though it is certainly not contemptible, and perhaps, for all we know, may even be true, is hardly suitable as a basis for mathematics. For, as I have already explained[6], mathematics must be seen to rest upon non-controversial ideas, insofar as that is possible; and this kind of metaphysical realism, justified or unjustified, legitimate or illegitimate, true or false, is, beyond question, *controversial*. Surely it is not desirable that a mathematician be required to subscribe to a disputed philosophical doctrine merely in order to get on with the job of defining and proving things.

In any case, the notion of concept that would be required here, at least for the Cantorian version of the theory, would have to encompass "concepts" that are inexpressible, even in principle, in any actually employable language, natural or artificial. And moreover, even if we were to accept the notion of concepts which we cannot formulate in language or entertain in thought as legitimate, there is still the further question what the *extensions* of such concepts might be. For we know that not all such extensions can be accounted objects[7].

Arithmoi, by way of contrast, do not give rise to such difficulties. That there are finite pluralities composed of whatever kinds of things there are is, as I have explained at length, the common sense view. That is why we must base our notion of set, and with it the whole of our mathematics,

---

[6] In Section 1.3.
[7] This is the import of the so-called "paradoxes" of set theory, which I shall discuss in Section 3.5.

on the *mathematical* notion of *arithmos* rather than on the *logical* notion of the extension of a concept.

But the notion of *arithmos* does not extend to include the empty set and singletons. Since it makes no sense to speak of pluralities that do not comprise at least two elements, it makes no sense to speak of such sets as pluralities, and we must therefore regard them merely as useful fictions, things we pretend exist in order to simplify mathematical discourse. Zermelo himself took this view of the empty set, though he was willing to accept singletons[8]. Others[9] have had serious doubts about singletons as well.

Surely this view, the view of the Greeks on this matter, is the natural one: nothing that really exists, no object, answers to the standard descriptions of the empty set or of singleton sets. Talk about such "things" is a mere *façon de parler*; although we could not dispense with such talk without considerable inconvenience. The question that confronts us here is how to handle such talk technically, that is to say, what technical conventions to adopt concerning "reference" to these non-existent entities.

Here it is important to remember that since set theory is not a formal axiomatic theory, *we are not at liberty simply to "postulate" the empty set and singleton sets by laying down suitable axioms.*

In a formal axiomatic theory the logical role of the axioms is to define the class or species of all models of those axioms, and what is mathematically significant in any particular model is what is common to all models in its isomorphism class: in a particular model of Euclidean geometry it does not matter at all what kinds of things play the role of lines or of points in that model, so long as, for example, two distinct points (whatever "points" may be) both stand in the relation of "lying on" (whatever that may consist in) to exactly one line (whatever "lines" are taken to be). The "particularity" of any one model of Euclidean geometry will be factored out, so to speak, when we pass from consideration of that model to consideration of its isomorphism class.

But the universe of set theory is not an arbitrarily chosen model of set-theoretical axioms – the very idea that it could be is logically and foundationally incoherent. Thus it is necessary that we know *exactly* what we are talking about when we speak of "the empty set" or "the singleton of the element a". It is, however, a matter here of adopting

---

[8] "Über Grenzzahlen und Mengenbereiche".
[9] e.g. David Lewis in *Parts of Classes*.

3.3 Sets

suitable technical conventions. Several possibilities come immediately to mind, and I do not claim that the conventions I shall adopt are in any way privileged or optimal. What is required is that we adopt *some* conventions to stipulate what reference to the empty set and to singletons really means.

I shall describe a convention in accordance with which the concept of set, or strictly speaking, many equivalent concepts of set, can be defined in terms of the more fundamental concept of *arithmos*. Sets so defined will include among their number the empty set and singleton sets, or rather, objects that go proxy for them.

To fix a notion of set, choose $\bot$ and $\top$ to be any two distinct individuals (so that neither $\bot$ nor $\top$ is an *arithmos*) and define a *set*, relative to this particular choice of $\bot$ and $\top$, to be an arithmos that includes both $\bot$ and $\top$ among its units[10]. The *members* of a set are then defined to be the units of the *arithmos* which *is* that set *other than* $\bot$ *and* $\top$.

Thus the empty set is the *arithmos* $\{\bot, \top\}$, whose only units are $\bot$ and $\top$, the singleton set of the object a, i.e. the set whose only member is a, is the *arithmos* $\{a, \bot, \top\}$, whose only units are $a, \bot$, and $\top$, and so on. We must further stipulate that in the theory of sets determined by a particular choice of $\bot$ and $\top$, *neither* $\bot$ *nor* $\top$ *is to count as an object*. That theory may incorporate individuals, but the distinguished individuals $\bot$ and $\top$ are not to be included among them; their only role is to serve as markers for the *arithmoi* that are to play the role of sets: as units they are inert, for they do not count as *members* of sets, or, indeed, as objects of any sort in the set theory based of that particular choice of $\bot$ and $\top$.

As this brief explanation already suggests, it will prove useful to modify our basic terminology in set-theoretical contexts. Accordingly, let us adopt the following conventions relative to a particular choice of $\bot$ and $\top$:

---

[10] Someone may object to my positing as many as two individuals, on the grounds that the existence of two distinct things is not a self-evident truth, or need not hold in all possible worlds, or whatever. (Even the *cogito* may establish the existence of only one solipsist.) I am unimpressed by such arguments, and indeed can see no harm in positing any definite number of individuals, so long as they are posited severally and individually, and are given proper christenings, so to speak. What is unwarranted is the positing of *arithmoi* composed of individuals without positing those individuals severally and separately: we cannot, for example, simply posit a Euclidean infinite set of individuals as an assumption underlying all mathematics in the way that Russell and Whitehead did with their Axiom of Infinity, although there is no harm in entertaining such suppositions "locally", so to speak, e.g. in the hypotheses of theorems.

(1) By an *individual* we now mean any individual (in the original sense) except for the two individuals ⊥ and ⊤.
(2) By an *object* we now mean either an individual (in the sense determined by (i)) or a set (in the sense determined by our choice of ⊥ and ⊤).
(3) By a *set* we mean an *arithmos* which includes both ⊥ and ⊤ among its units and whose units (if any) other than ⊥ or ⊤ are all objects (in the sense given by (ii)).
(4) The *members* of a set are the units of the *arithmos* that *is* that set, apart from ⊥ and ⊤.

By means of these conventions and changes in terminology we ensure that our ordinary set-theoretical discourse always has objective content, even when we refer to singletons or the empty set. The conventions (i)–(iv) cannot, of course, be construed as a definition of the concept *set*, for the conditions laid down in (ii) and (iii) are *prima facie* circular. But there is also an implicit circularity in the characterisation of an *arithmos* as a finite plurality composed of objects, given that *arithmoi* are themselves objects. The difficulty is not an insurmountable one, but it is not convenient to deal with it here. The solution is to be found in the Principle of Regularity from which the well-foundedness of the membership relation can be deduced[11].

But the essential point about the conventions laid down here is that we can now, in effect, forget them. In what follows I shall present set theory in a straightforward and unselfconscious way, freely referring to the empty set and to singletons, even employing the fiction that they are zero- and one-element "pluralities". Any other way of proceeding would be confusing and tedious. In any case, it is only when we are discussing the very deepest layer of the ontology underlying set theory that we need to take these conventions into account. On that level all reference in our discourse is either to individuals (understood in the original sense), which, by hypothesis, really exist, or to finite pluralities of objects, *arithmoi*, whose real existence is the fundamental assumption on which the whole theory rests.

3.4 Global functions

Among the elements of set theory we must include, in addition to sets, the *operations* that apply to sets and the *relations* that obtain among them.

---

[11] See Section 4.9.

3.4 Global functions

The former I shall call *global functions*, the latter *global propositional functions*, for reasons that will soon become apparent.

A first level global function, σ, of one argument can be defined by specifying, for each object a, the *value*, $\sigma(a)$, that σ takes for a as *argument*. The value must always be specified unambiguously in terms of the argument a. Such a function is *global* because it is defined at *all* objects; it is of first level because its arguments are *objects*. Second level global functions having first level global functions as arguments can also be defined, as can first level global functions of two or more arguments.

The *identity function*, which assigns to each object a as argument the object a as value, is an example, no doubt the simplest example, of a first level global function of one argument. The familiar Boolean operation of *union*, which assigns to the arguments a and b the set, $a \cup b$, whose elements consist of all those objects (if any) that are members of a or of b or of both, is a first level global function of two arguments.

In addition to ordinary global functions we shall have to consider global *propositional* functions as well. A first level global propositional function, Φ, of one argument can be defined by specifying for each object a, the truth value, *true* or *false* that Φ takes for a as argument. Global propositional functions of two or more arguments and of second level are also required.

My terminology is taken from Frege[12], and I shall follow him in maintaining a sharp distinction between global functions and objects, although my reasons for doing so do not coincide with his. For Frege the essence of the notion of *function* was to be discovered in the nature of the signs that denote them. He held that the "incompleteness" or "unsaturatedness" of the function sign contained the clue to the nature of the function itself, which, on his view, is what such a sign *names* or *means* (*bedeutet*).

We can judge the importance that Frege attached to this doctrine by the fact that it is expounded in §1 of his *Grundgesetze*, which is entitled "*The function is unsaturated*". He begins this section by asking in what way an algebraic expression, such as

$$(2 + 3x^2)x$$

---

[12] Frege does not attach the adjective "global" to the word "function". For him *all* functions are global in the sense intended here. I shall, however, have occasion to introduce a notion of *local* function by way of contrast. Thereby hangs a tale, as we shall see.

determines a function, and after pointing out that the expression itself cannot be the function, goes on to say

For example, if for "x" in the expression "$(2 + 3x^2)x$" we substitute the sequence of numerals "0", "1", "2", "3", we obtain as corresponding meanings for the resulting expressions the numbers $0, 5, 28, 87$. None of the numbers thus meant can claim to be our function. The essence of the function manifests itself rather in the correspondence it sets up between the numbers whose signs we substitute for "x" and the numbers which then appear as what our expression means – a correspondence which can be represented in intuition by the trace of the curve whose equation in rectilinear co-ordinates is

$$"y = (2 + 3x^2)x"$$

Accordingly, the essence of the function lies in that part of the expression that is there over and above the "x". The expression for a function is *in need of completion, unsaturated*. The letter "x" serves only to hold open places for a numeral to be put in, and thereby makes recognisable the characteristic sort of need for completion that constitutes the specific nature of the function designated[13].

Thus to grasp what a function is, we must grasp what function signs are, and how they work. The essence of the function, the thing signified, is disclosed by the syntactic character of its sign:

The peculiarity of function signs, which we have called "unsaturatedness" here, naturally has something answering to it in the functions themselves. They too may be called "unsaturated" and in this way we mark them out as different from numbers[14].

Here the true origin of Frege's conception of function is disclosed by the way in which he explains the ontological facts by means of a parallel with the syntactic ones. Again he says:

The function is completed by the argument: I call what it becomes on completion the *value* of the function for the argument[15].

But this pulls us up short. How, we may ask, can a function, which is presumably a timeless abstraction, *become* anything at all?

The argument is not to be counted in with the *function* but serves to complete the function which, in itself, is unsaturated[16].

How does the argument accomplish its task here? What does it or do we have to *do* in order to "complete" the function?

---

[13] *Die Grundgesetze der Arithmetik*, p.5.
[14] *Was ist eine Function?*
[15] *Die Grundgesetze der Arithmetic*, p.6.
[16] ibid., p.6.

3.4 Global functions

Perhaps I can be accused of being unfair to Frege in taking what he says so literally. After all, he himself, no doubt anticipating such objections, has entered a caveat:

> Of course this [i.e. his account of function] is no definition; but likewise none is here possible. I must confine myself to hinting at what I have in mind by means of metaphorical expressions, and here I rely on my reader's agreeing to meet me half-way[17].

Perhaps I have not met Frege half-way here. But half-way to where? Surely it is significant that his central metaphor is *syntactic*: the function is *like* the function sign in respect of the latter's being "unsaturated". And just as we must draw a strong distinction *in syntax* between *function names*, which are unsaturated, and *object names* which are not, so must we draw a parallel distinction *in ontology* between *functions* which are unsaturated and *objects* which are not.

But, for us, the critical question here is whether what Frege calls "function names" really are *names* standing for *things*, albeit for "things" of a different sort or logical type from objects. That is certainly what Frege himself maintained. But ought we to follow him in this? What *explanatory* purpose does hypostatising such entities serve[18]?

In fact, Frege's conception of functions as the abstract entities named by function signs is of no help in explaining the logical and syntactic roles of those signs; on the contrary, his explanation travels in the opposite direction, using the syntactic properties of the signs to describe, by means of the metaphor of saturation, the ontological properties of the supposed *nominata* of those signs, the functions themselves. But Frege's careful syntactic analysis shows, and his attempted ontological explanation confirms, that *the notion of function is syntactic in origin* and that, accordingly, *functions, understood in this most fundamental sense, are symbol generated abstractions*.

I shall not, therefore, follow Frege in hypostatising global functions as the entities named by function signs. On the contrary, I hold that where we come up against such a symbol generated concept, we must either replace it by a genuine mathematical concept, rigorously defined in terms of fundamental concepts that are not of this symbol generated sort, or we must treat the supposed "entities" that fall under such concepts as

---

[17] *Was ist eine Funktion?*, p.665.
[18] For Frege the meaning (*Bedeutung*) of an ordinary (object) name is just the object that it names. No doubt this doctrine predisposed him to postulate "entities" to serve as the meanings of function names.

fictions – what Quine calls "virtual objects" – and interpret all talk about "them" in a suitably Pickwickian sense.

In fact, both approaches are required here and, accordingly, I make a sharp distinction between *global* functions, which are mere *virtual* objects in this sense, and *local* functions, which are genuine objects – in fact, sets. Global functions, like Frege's functions, are defined at all objects; a local function has a particular set as its domain of definition.

We need *local* functions as objects, cut free of any dependence on symbols that name them, in order to carry out standard mathematical constructions, such as forming the space $C[0, 1]$, whose "points" comprise all continuous, real-valued functions on the closed interval $[0, 1]$, for example. There are, as is well known, too many points in that space for all of them to be named. But these "points", and, indeed, the space they compose, must be genuine objects. I shall explain these matters in detail in due course[19].

The question now arises: what first level global functions (of both sorts) are there, or, more precisely, what first level function names can we employ? It is essential that this question be answered in careful detail, because first level functions can occur as arguments of second level functions, or, to put the matter more accurately and without employing the fiction that functions of either level are real entities, there are general techniques which enable us to form complex terms denoting objects by inserting a first level function name into the appropriate argument places of a second level function name.

But it is even misleading to call these expressions "names", since we are "naming" only *virtual* objects in these cases. These first and second level function "names" are really *complex syncategorematic signs*, in the sense of "syncategorematic" given by Ockham in his *Summa Logicae*[20]. There is nothing which such expressions name or denote, but the occurrence of such an expression as part of a complex name or sentence helps to determine, in its own exact and characteristic fashion, the object named by the complex name, or the proposition expressed by the sentence.

This last point is a subtle and difficult one: but it is of absolutely fundamental importance. Systematically to avoid speaking of global functions in general, or of particular global functions ("the power set function", "the Boolean union function", etc.) would render our discourse prolix and tedious in the extreme. I will not impose such a self-denying ordi-

---

[19] In Section 4.6.
[20] Part I, Section 4.

3.4 Global functions

nance on myself. But it is essential to remember that in the final analysis, and all talk about global functions of the various sorts to the contrary notwithstanding, *there are no such "things" as global functions*; and when we speak of such functions we are ultimately talking about our own notational conventions for referring to sets, although, to be sure, those conventions reflect our convictions (or conjectures) about what sets there are and therefore about what it is to be finite.

I now need to explain the conventions I shall follow in forming those syncategorematic expressions that I call "first level function names" (of both sorts). Here I shall be speaking largely in generalities. In the next chapter I shall introduce particular symbols as names for the primitive first level global functions that correspond to the basic "operations" of set theory by means of which we can "construct" sets[21] – the pair set, union (or sum set), and power set functions – and symbols for the second level global functions of comprehension and replacement. The comprehension function takes first level global *propositional* functions as arguments, and that is why we need to define that sort of global function in generality.

First let us consider the basic global propositional functions of *identity*, *membership*, and *inclusion*. I shall use the standard symbols "=", "∈", and "⊆", respectively, as primitive "names" for these functions.

The identity and membership signs have their usual meanings; for the inclusion sign I shall adopt the convention that **s** ⊆ **t** is true precisely when every member (if any) of the object named by **s** is also a member of the object named by **t**. This means that if **s** denotes an individual (which therefore has no members) then **s** ⊆ **t** is true no matter what **t** denotes, and **t** ⊆ **s** is true only when **t** denotes the empty set or an individual.

Clearly, any symbolic expression of one of the forms **s** = **t**, **s** ∈ **t**, or **s** ⊆ **t**, where **s** and **t** stand for particular objects, corresponds to a definite mathematical proposition which is determinately true or determintely false. I shall call such expressions *formal (atomic) sentences* since they can be used in place of ordinary sentences to express mathematical propositions.

Moreover, we must also be allowed to form *compound* formal sentences from previously given formal sentences using the familiar *propositional connectives*, which I shall simply designate by the English words *"and"*, *"or"*, *"not"*, *"implies"*, and *"iff"* (*"if, and only if"*) in the interests of

---

[21] Needless to say, talk about "operations" on sets or "construction" of sets is just metaphorical.

readability. Again, all these compound formal sentences correspond, at least in principle, to ordinary sentences expressing mathematical propositions with unambiguous truth values as determined by the familiar truth tables associated with these connectives.

The object-naming expressions that are components of formal sentences I shall call *formal terms*. They are formed by combining simple global function names with previously given formal terms in the usual manner (e.g. $P(\mathbf{s} \cup \mathbf{t})$, where "$P$" names the power set function, "$\cup$" the Boolean union function, and **s** and **t** are previously given formal terms). Of course the business of combining has to start somewhere, so I shall include certain primitive and defined *individual constants* among the formal terms, in particular the symbol "$\emptyset$", which names the empty set, and, in the Cantorian version of the theory only, the symbol "ω", which names the set of von Neumann natural numbers.

We need the means for conveying or expressing generality in our formal notation; this can be done by using suitable letters as *parameters* or *free variables* in the familiar way. In fact we need such parameters not only for objects but for first level global functions of both sorts[22]. Following Frege we may say that a formal expression containing such parameters does not name an object or express a proposition but "indefinitely indicates" an object or "indefinitely expresses" a proposition.

As Frege's analysis of syntax reveals, we can always regard a complex sentence or a complex object name as the result of applying a global function name to suitable arguments. Moreover, in the general case we can do this in more than one way. Global function names of both sorts are thus contained implicitly in formal terms and sentences, and therefore what global functions we must recognise is determined by what formal terms and sentences we permit ourselves to form.

We need a notational convention like Frege's use of Greek letters that will permit us to "refer" to global functions outside the context of formal expressions. I shall use the lambda notation of Church to form complex first level function names of both sorts.

Thus suppose that **E** is a complex formal expression – either a complex term or a sentence – and let **a** be an object parameter, which may, but need not, occur in **E** (**E** may contain other parameters of other types as well). I shall then form the function name $\lambda x\, \mathbf{E}[x/\mathbf{a}]$, where $\mathbf{E}[x/\mathbf{a}]$

---

[22] I shall usually employ italic letters (a, b, S, T, etc. – with or without subscripts) as free variables for objects, small Greek letters (σ, τ, etc. – with or without subscripts) as free variables for ordinary first level functions, and capital Greek letters (Φ, Ψ, etc. – with or without subscripts) as free variables for first level propositional functions.

is the expression that results from **E** when all occurrences (if any) of the parameter **a** are replaced by occurrences of the *binding letter* **x**. If **E** is a term, the resulting lambda expression names, or indefinitely indicates (if **E** contains parameters other than **a**), an ordinary first level global function of one argument, namely, the function whose value at the argument x would be the object named by the expression **E**, if **a** were taken as naming x. If **E** is a sentence, the lambda expression names, or indefinitely indicates, a first level global propositional function of one argument, in an analogous manner.

Names for global functions of two or more arguments can also be formed

$$\lambda xy\ E[x/a, y/b]$$

provided the parameters **a** and **b** are distinct, as are **x** and **y**.

The matter is of sufficient importance to bear repeating: these complex first level function names formed using the lambda notation are not really and literally *names* of anything whatsoever; they are syncategorematic signs like "=", "∈", and "⊆". Global functions of either sort are merely *virtual* objects, that is to say, fictions, and talk about such functions can always be recast in such a way that apparent reference to "them" disappears.

I hope that this brief account will make the notational conventions that I shall employ sufficiently clear. In fact, in laying down these conventions I have been forced to sketch what is, in effect, a conceptual notation or *Begriffsschrift* which is sufficiently rich to express, with the help of suitable auxiliary definitions, all the propositions of mathematics.

But if I have had to follow Frege thus far, I shall not attempt follow his example in carrying out a strictly formal exposition of set theory using this formal notation. This point requires further comment and, indeed, justification, and I shall return to discuss it in Sections 3.6 and 3.7 below. First however I want to treat of matters which will make it clear why it is necessary to exclude global functions of both sorts from the realm of objects.

3.5 Species

The notion of *class*, or, as I shall say, for reasons that will soon become apparent, the notion of *species*, is one that naturally arises when set theory and the foundations of mathematics come under discussion. Indeed, the most difficult problem that confronted the founders of modern set theory was precisely that of disentangling the notion of set from that of species (class). There is still considerable confusion surrounding this matter, confusion compounded by the widespread failure on the part of mathematicians to acknowledge the dependence of mathematical logic and the axiomatic method upon set theory.

I shall not burden the reader with a catalogue of errors, nor with an account of the history of the notion of "class". Rather I shall try to explain, as simply and as briefly as I can, what the notion of species is, what its relation to the notion of set is, what its logical status is, and what its role in the exposition of the axioms of set theory ought to be.

Consider any well-defined property of objects – the property of *being a set with three or more elements* will serve as an example. The totality of all objects that have such a property is what I shall call a *species*. Thus we may speak of the *species of all sets with three or more elements*. A species always has an associated *defining property*, and the objects that have that property are called the *elements* or *members* of the species. It may be that the elements of a species are limited or definite, that is to say *finite*, in number. In that case we say that the species *forms* a set[23]. If a species does not form a set, that is because it is either *infinite* or *indefinite* in multitude.

A species is *intensional* in the sense that it is indissolubly attached to its defining property. In this respect a species stands in strong contrast to a set which, since it is *extensional*, is just a plurality, and is not in any way tied to any property that happens to be shared exclusively by its members. For this reason it is necessary to distinguish between a set, s, and the species of all its members: the latter might be said to have *being a member of s* as its defining property.

The difference between sets and species, as pluralities, can be roughly expressed in the following way: a set is, in general, *heterogeneous in kind* (its elements need not have any significant property in common) but *definite in number*; a species, on the other hand, is *homogeneous in kind* (its elements all share its defining property) but, in general, *infinite or indefinite in number*. Put more briefly, the distinction is between pluralities that are *quantitatively definite*, namely sets, and those that are *qualitatively definite*, namely species.

But this characterisation, though it will, I hope, help to bring out the

---

[23] As we shall soon see (Section 3.6) there is a *logical* complication in asserting that a species forms a set (the proposition is Σ_2). Where I make such claims implicitly (e.g. in the presentation of the axioms of set theory), I have taken care not to assert them directly.

3.5 Species

distinction I have in mind, will not suffice as a final word on the matter: it is too imprecise for that, although here, to be sure, we are dealing with matters so fundamental that we cannot hope to obtain precision, certainly not the kind of precision required of a mathematical definition.

The essential difference between sets and species, however, is that sets exist whereas species do not. By this I mean that species are not objects: they are *fictions* or *virtual objects*, like global functions. Thus although we may, in carefully circumscribed circumstances, speak of species *as if* they were objects, that is to say, as if there were such *things* as species, any proposition that appears to refer to species can be rephrased in such a way that this apparent reference disappears.

For example, the proposition

The object a belongs to the species of all objects that have the property Φ

can be replaced by

The object a has the property Φ

or, even more simply, by

$$\Phi(a)$$

There is a quite straightforward logical reason why species, in general, could not all be regarded as objects. For suppose that all species were objects, and hence the kind of thing that can belong to species. Now consider the species, R, whose defining property is *being a species that does not belong to itself*. Then R would belong to itself if, and only if, it did not belong to itself – a flat contradiction. This is the famous Russell Paradox.

Thus, in any case, the question would arise: which species are (or form) objects? The answer given in modern set theory is: those that have a determinate size (in the sense of cardinality), the *finite* ones. The objects that share the defining property of a species do not collectively constitute a *single* object unless they are finite in number. *Finitude* is the principle of unity for pluralities: it is both a necessary and a sufficient condition for a plurality to constitute an object, namely, a *set*.

But if a species forms a set *solely* in virtue of being finite, the existence of that set as an object is independent of the fact that its members happen to share the defining property of the species. That is why sets, unlike species, even finite ones, are *extensional* in character. It is perfectly true that in all cases in which the number of elements in a set is too large to list we invariably single out the set by giving a property that its members and only its members share: we get at sets through species, so to speak. But for genuine objects, the fact that we can only refer to them in particular ways is neither here nor there: how we refer to things may *reflect*, but it does not *determine* what they are – or, indeed, *whether* they are.

But there is another reason why species cannot be accounted objects, and this goes to the heart of a central logical problem in the foundations of mathematics. Species, since they can be *infinite* or *indefinite* pluralities, cannot be clearly individuated. There is no general criterion of identity for species in virtue of which every proposition asserting the identity of species could be assigned a definite truth value. And where identity cannot be applied, we cannot be dealing with objects.

Here we are entering deep waters. In all candour I must confess that there is, finally, no way of *establishing* the point I have just made, or, indeed, the general point of view I am about to recommend, by means of argument.

The difficulty is that what is at issue is precisely – logic. I don't mean *formal* or *mathematical* logic, but logic itself, the principles that must govern our actual reasoning. Here we reach bedrock, and we can only *express* our logical principles, we cannot argue, in any straightforward way, for their correctness.

In the last analysis the justification for those principles can be only this, that they make sense of our intellectual experience; that when we adopt them we can gain an overview of mathematics as a science which is compatible with the ideals that animate it and with the claims to clarity and rigour that are made on its behalf.

The problem at issue here is the logic of infinite, or indefinite, totalities. In particular, it is the problem of quantification over such totalities: what does such quantification mean and what logical laws govern its employment?

To get a practical grip on these questions, let us return to the problem of individuating species. At first glance it appears that in order to individuate species we should first have to individuate their defining properties; and this seems to confront us with formidable difficulties. When are two properties the same? Are *being a man* and *being a featherless biped* the same property, supposing, of course, that they are true of exactly the same things. A tempting solution to these difficulties now presents itself: why not separate species from their defining properties and identify them when they have the same *extensions*, that is to say, when they have exactly

3.5 Species

the same members? In this way we abandon the notion of species, as I have described it, and arrive at the traditional, logical notion of *class* as the extension of a property.

Of course now the logical basis for the distinction between sets and classes becomes less intelligible because it is merely a matter of size: sets are finite – limited in size; classes are infinite in some absolute sense; but both are extensional collections, properly individuated by the principle of extensionality. Let us, however, put that difficulty aside to ask how the passage from intensional species to extensional classes is to be made.

Let us first consider how we would *say* that two species S and T have the same extension.

> *For every object x, x belongs to the species S*
> *if, and only if, x belongs to the species T.*

Thus to *define* identity for classes (extensional identity for species) we must employ a *global* quantifier whose domain of quantification is an infinite, or indefinite, species; in this case the domain of quantification is the infinite species of all objects (for every object x...).

Conversely, given identity for classes we can define global quantification over infinite species. Thus

> For every member, x, of the species $S : ...x...$

can be rendered

> The species S is identical with the species of all members
> x of S such that $...x...$

Hence the problem of individuating species just *is* the problem of global quantification over the elements of a species.

What are the logical principles that should govern global quantification? This is a difficult question and I am not confident that I can answer it in full. But I propose to adopt a partial answer, namely

Brouwer's Principle (i) *Conventional (i.e. what Brouwer calls "classical") logic is the logic of finite domains. In particular the conventional laws of quantification apply only when the domains of quantification are finite.*

(ii) *Propositions that require global quantification for their expression cannot be assigned conventional truth values, true or false. They can only be classified as justified or unjustified.*

Thus, for example, suppose that S is an infinite species and that Φ is a well-defined property of objects so that, for each object a, $\Phi(a)$ has a determinate truth value. Then in accordance with Brouwer's Principle, the assertion

$$\text{For all objects } x \text{ in } S\colon \Phi(x)$$

is not a conventional ("classical") proposition with a determinate truth value. It is not *true* or *false*, but *justified* or *unjustified*.

To say that such a proposition is justified is to say that we have grounds for asserting any proposition of the form

$$\Phi(\mathbf{t})$$

where $\mathbf{t}$ is any expression that denotes, or could denote[24] an object.

To say that the assertion is unjustified, on the other hand, is merely to say that we do not have such grounds; and that is not the same thing as saying that we have grounds for denying it. Indeed, it is not clear *how* to deny it, since, in accordance with Brouwer's Principle, such assertions do not have conventional negations: the meaning of *not* is determined by its truth table, and is therefore defined only in cases in which it is applied to a conventional proposition that possesses a determinate truth value. The same observation applies, of course, to all the other propositional connectives: we cannot apply them to propositions expressible only by means of global quantifiers.

By adopting Brouwer's Principle I have excluded the possibility of introducing a notion of identity for species, subject to the conventional laws of identify. For were we to do so, we could, as I have already pointed out, define

$$\text{For all } x \text{ in the species } S : \ldots x \ldots$$

to mean

$$S = T$$

where T is the species of all members, x, of S that possess the property $\ldots x \ldots$. But if we suppose that

$$S = T$$

is a conventional proposition with a truth value, and hence combinable with other such propositions into complex propositions using the

---

[24] For example, $\mathbf{t}$ might be a term containing one or more parameters. If they were assigned values, then $\mathbf{t}$ would denote a particular object.

3.5 Species

machinery of propositional connectives, then the original globally quantified sentence would have a conventional negation, and, indeed, all the standard laws of quantification could easily be derived for such global quantifiers.

Brouwer's Principle thus gives real logical and mathematical content to the notion of infinity. It provides a clear rationale for maintaining a sharp distinction between sets and species, and for excluding species from the realm of objects.

For Brouwer, conventional logic is the logic of the finite. But for him, as for Euclid before him, an essential feature of finiteness is that a finite species, that is to say, a species that *has* a size, cannot be of the *same* size as any of its proper subspecies. As Euclid put it,

> The whole is greater than the part

It follows from this, for example, that the natural numbers, however defined, must compose an infinite species and that, consequently, the logic of general propositions about natural numbers cannot, in accordance with Brouwer's Principle, be conventional logic.

Modern mathematics, however, following Cantor, *rejects this additional assumption about finiteness*. In modern mathematics, a set, which *by definition* is finite, that is to say, has a definite size, can have the *same* size as certain of its proper subsets. In particular, the species of natural numbers (suitably defined) is taken to be finite in the root sense of "finite". That, indeed, is the fundamental assumption upon which the modern approach to mathematics rests.

Of course it sounds odd to say "The infinite set $\{0, 1, 2, \ldots\}$ is actually finite"; but this is analogous to saying "The imaginary number $\sqrt{-1}$ is actually real". There is no mystery in saying the latter. Every mathematician realises that the "oddity" in the latter proposition rests on an equivocation: it comes from taking "imaginary" in the technical sense and "real" in the ordinary. Something analogous is happening in the first example as well, for there "infinite" is taken in the technical sense and "finite" in the ordinary.

Both "infinite" and "imaginary" are used as technical terms in mathematics, and when they are used in that way they are cut loose from their ordinary meanings. They are terminological fossils that preserve dead ideas, as in amber.

But perhaps I go too far: the idea that $\sqrt{-1}$ is just imaginary is surely as dead as any idea can be; but it may be too early to perform obsequies over the traditional, pre-Cantorian idea that the natural numbers constitute an infinite species.

Be that as it may, however, we should not allow an accident of usage to disguise from us the real nature of Cantor's revolution in mathematics: he did not tame the infinite; rather he extended the boundaries of the finite. The infinite, what Cantor called the "absolute" or "genuine" infinite, continues to play a role, if largely a peripheral one, in modern mathematics – peripheral, but not insignificant.

Why do I make so much of what is, after all, a matter of terminology? Because in really fundamental matters an injudicious choice of terminology can lead to confusion of thought: if we do not call things by their proper names we may not give them their proper due.

Even the terms "real" and "imaginary" continue to confuse students, although, of course, they no longer serve to mark a really fundamental distinction, that is to say, a foundational one. But the co-opting of the terms "finite" and "infinite" to mark a technical distinction within mathematics is an altogether more serious matter, much more productive of confusion of thought on fundamental issues.

We can see the truth of this in connection with Brouwer's Principle. It is probably true to say that the small band of contemporary mathematicians who reject the Cantorian dispensation accept Brouwer's Principle in some form or other. But the vast majority of mathematicians have either never heard of it, or, if they have heard of it, regard it as a statement of Brouwer's anti-Cantorianism.

But there is good evidence that Cantor himself held something very close to Brouwer's view on this matter:

> I have no doubt whatsoever that in this way [i.e. in the generation of the various transfinite number classes] we can extend ever further, never reaching a barrier that cannot be surpassed, but also never approaching even an approximate comprehension of the Absolute. The Absolute can only be acknowledged (*anerkannt*) never known (*erkannt*), not even by approximation[25].

and again

> The *transfinite* [i.e. that part of Cantorian finite that lies beyond the Euclidean finite] with its abundance of configurations and forms points the way, of necessity, to an Absolute, a "genuine infinite" whose magnitude admits neither of increase nor of decrease, and which is therefore to be regarded quantitatively as an absolute maximum. The latter transcends the human capacity for comprehension, so to speak, and, in particular, lies beyond mathematical determination[26].

---

[25] *Gesammelte Abhandlungen*, p.205 ("Über unendliche lineare Punktmannigfaltigkeiten").
[26] ibid., p.405 ("Mitteilungen zur Lehre vom Transfinitum").

3.5 Species

Brouwer's quarrel with Cantor, correctly understood, is not so much over the logical properties of the infinite, the "absolute" as Cantor called it, as over the question where to draw the boundary between the finite and the infinite, if I may put it in that somewhat paradoxical way. Indeed, although Brouwer refuses to follow Cantor in extending the boundaries of the finite, he goes beyond Cantor in claiming that the infinite, properly so called, does admit of "mathematical determination". But the logic of infinite totalities is not, in Brouwer's view, the familiar "classical" logic.

Brouwer's Principle allows us to make sense of the distinction between sets and "proper classes" (i.e. infinite species). If we suppose, contrary to that principle, that quantification over the members of an infinite species/proper class is governed by the "classical" laws of conventional logic, then, as we have seen, we can define a notion of identity for species/classes which itself is governed by those conventional laws. What then is the rationale for not treating these "things" as objects? Since we can individuate them can we not count them? Do they not form pluralities, "super classes", and, if so, do not these, in turn, form still further pluralities, "super-super classes" perhaps? Do the operations that apply to sets apply to these things? If not, why not? Where does the process of constructing higher and higher level collections stop? And why?

Brouwer's Principle, together with our fundamental underlying assumption that sets are *finite* pluralities, tells us that sets are radically different from non-finite pluralities, i.e. infinite species; it tells us that infinite species cannot be accounted objects because, as a matter of *logic*, they cannot be properly individuated; it tells us that the "universe" of set theory cannot be the underlying domain of a conventional mathematical structure. This last is especially significant for the light it sheds on difficulties that have arisen in the foundations of the theory of categories and functors.

The requirements of precision and intellectual economy dictate that we avoid reference to species, and indeed to finiteness, in the *statements* of our axioms, and, accordingly, the official formulations of the axioms will take a different form[27]. But these notions will have a role to play, indeed, an important role in the general exposition. For one way of regarding

---

[27] If that form somewhat obscures the insights and motivations that underlie those axioms, it has the compensating virtue of presenting them so that what they assert is absolutely transparent in meaning. That, after all, is the principal requirement: we must understand what our propositions say before we can use them in mathematical discourse. This is a significant point to which I shall return in the next chapter.

the central axioms of set theory, and perhaps it is the most illuminating way, is as *finiteness assumptions*, that is, as assertions to the effect that such and such a species is finite (of course, in the appropriate sense of "finite") and therefore forms a set.

Viewed in that light, the axioms take on an aura of self-evidence, (though unsurprisingly, some are more self-evident than others). Indeed one of them, the central assumption of *Cantorian* set theory that the species of natural numbers (suitably defined) is finite, and therefore forms a set, is not self-evident at all. But then it is not self-evidently false either, and it is, after all, the key assumption that makes possible the whole modern approach to mathematics. In any case, it is perfectly clear what we are committing ourselves to when we adopt it: we enter the world of modern mathematics with our eyes open, and without illusions.

Brouwer's Principle also tells us that global functions cannot be accounted objects. Like species such functions cannot be individuated: we cannot define an identity relation for such "things"– that would require global quantification over the species of all objects; and where there is no identity there can be no objects. Accordingly, simple or complex "names" for such functions are not *names* at all: they are, as I have explained, *syncategorematic* signs.

3.6 Formalisation

When I introduced the notion of global function in Section 3.4 I was forced, in effect, to set up a formal notation – what Frege would have called a *Begriffsschrift* – in order to give a precise explanation of what those global functions are and how they can be defined. But as I have already explained I shall use this notation only when doing so makes it easier to express what I have in mind. This calls for some explanation.

There is, of course, a tradition that presentations of the foundations of mathematics should be couched in formalised theories. The tradition is a recent one: originating with Frege, continued by Russell and Whitehead, it has been observed, to a greater or lesser degree, by authors of treatises on set theory and the foundations of mathematics ever since.

We know what impelled Frege to formalise his discourse, because he said in the Preface to the *Begriffschrift* that since his aim was to establish that natural number arithmetic can be developed from logic alone, he was forced to develop the conceptual notation in order to show that "nothing intuitive (*anschaulich*) could penetrate unnoticed" into his proofs. The purpose of that notation, he tells us, was "to provide us with

3.6 Formalisation

the most reliable test of a chain of inferences, and to point out every presupposition that tries to sneak in unnoticed".

In short, Frege formalised his discourse in the interests of ensuring the cogency of his proofs by making all their presuppositions manifest. Similar considerations lay behind the use by Russell and Whitehead of a formalised language in their *Principia Mathematica*[28].

Partly as a result of these great pioneering works, however, mathematicians in general, and mathematical logicians in particular, are no longer convinced of the utility of formalisation as a prophylactic against logical error, and the motives that dictate the formalisation of more recent presentations of mathematical foundations are quite different.

Nowadays the theory of sets is usually presented as a formal first order theory. This has very definite advantages: first order logic has been the principal preoccupation of mathematical logicians, and all their marvellous technical discoveries can be brought to bear on any first order theory.

But however useful such formalisations of set theory might prove to be, they cannot be regarded as providing a foundation for mathematics in any straightforward sense. For strictly speaking, and taking the notion of axiomatic theory in its precise, mathematical sense, the idea that *any* formal axiomatic theory could, as such, constitute a foundation for mathematics is logically absurd. As I have already said, we cannot make sense of formal axiomatics unless the theory of sets is already in place. And the idea that a *first order* axiomatic theory could serve in such a capacity is doubly absurd. For we cannot axiomatise even natural number arithmetic in a first order theory.

Let me not be misunderstood here: the fact that a formal axiomatic theory cannot, *qua* formal axiomatic theory, constitute a foundation for mathematics does *not* mean that there is no theoretical interest in formalising our foundational theory. On the contrary, such formalisation can be of the utmost theoretical interest. But to understand precisely what that interest is we must carefully distinguish between *laying down* a foundation for mathematics, on the one hand, and subjecting that foundation to a *mathematical investigation*, on the other.

For example, although it makes no sense simply to *identify* mathematical proof with formal proof in a particular formal proof theory, nevertheless we must set up a system of formal proof sufficient to formalise our actual proofs if we wish to give a mathematically precise

---

[28] See the introduction to the first edition.

formulation of the question whether a given proposition is provable from the basic principles of our foundational theory. And such questions of provability are of central importance when we are *investigating*, as opposed to merely *employing*, that theory.

The famous Incompleteness Theorems of Gödel and, in set theory itself, the consistency and independence results of Gödel and Cohen depend essentially on formalisation. But formalised proofs, properly understood, are mathematical *objects* not mathematical *arguments*. That is why they can be studied mathematically; but that is also why they cannot be used directly to establish truth, but must first be translated (if that is the word for it) into genuine mathematical discourse.

It follows that the celebrated incompleteness and independence results of Gödel and Cohen depend, for their foundational *significance*, upon the fact that the formalised proofs of the systems to which they apply correspond to genuine proofs – proofs properly so called that employ genuine propositions which express meanings that are independent of, and antecedent to, the proofs in which they occur. And even that is not enough.

For these results to be significant it is not sufficient that each formalised proof correspond to a proof in the ordinary sense: it is crucial that every ordinary proof have a formal counterpart. Otherwise, the fact that this or that proposition is *not* formally provable is of little theoretical significance.

Now it is relatively easy to judge whether the formal proofs of a given formal system convert – at least in principle – into genuine proofs; for that is a matter simply of checking that the formal axioms correspond to genuine first principles and that the rules of inference preserve truth. But the converse judgement – which is the critical one for incompleteness and independence proofs – is not at all easy to make. In the case of theories formalised as conventional first order axiomatic theories, general propeties of first order logic – for example, its completeness – will be relevant to those judgements. And those general properties cannot be established unless the principles of set theory are already in place.

In any case, it is essential to take note of the direction in which explanation and justification move here. It is the ordinary notion of proof that explains and justifies the system of formalised proofs, not the formalised proofs that justify their ordinary counterparts. Indeed, a formalised proof, insofar as it is strictly formalised, is not really a proof, and, insofar as it is really a proof, is not merely formal.

This last point calls attention to a further difficulty. For we cannot

3.6 Formalisation

discuss these matters sensibly until we determine what *kind* of thing formalised proofs are. Are they arithmetical (i.e. set-theoretical) or geometrical in character? And if geometrical, are they geometrical in the old-fashioned sense of *anschaulich* geometry, or in the modern sense in which geometry is defined axiomatically? What is the status of the configurations we actually write down? Are they themselves formal proofs? Or are they rather analogous to the drawn triangles, circles, and squares of *anschaulich* geometry, from which the triangles, circles, and squares of geometrical theory are obtained by idealisation? Clearly, there is nothing here corresponding to the *exactness* of ideal geometrical figures.

These matters are so important, and the confusion surrounding them is so widespread, that I shall be forced to return to discuss them at length (in Section 6.3). Meanwhile, I have included these brief remarks on formalisation mainly by way of explaining why I shall *not* be presenting set theory as a fully formalised theory.

Nevertheless, a certain measure of formalisation is necessary for two specific purposes: first, as I have already explained, in order to give an account of the key notions of *global function*, and *global propositional function*, and second, in order to give an account of the *global logic* of the theory, and, in particular, of the *global quantifiers* which play the central role in that logic.

Of course the logic of Cantorian finite domains – what is sometimes called "classical" logic – is now very familiar to us, and we do not need to have recourse to formalisation merely in order to keep our bearings there. But these matters of global logic are so imperfectly understood, and so little studied, that we find ourselves facing the kind of difficulty that faced Frege or Russell and Whitehead, and here formalisation will help us, as it helped them, to formulate our ideas with greater clarity and precision.

I have already called attention to the logical difficulties raised by the employment of global quantifiers, which range over the entire universe of sets and individuals, and have proposed the adoption of Brouwer's Principle as a partial solution to those difficulties. As we have already seen, Brouwer's Principle raises issues which concern the fundamental distinction between the Cantorian finite and the absolutely infinite which have been a source of perplexity in the foundations of set theory since the time of Cantor.

In fact, for most practical purposes we shall need to consider only two sorts of *global propositions*, namely, Π_1 and Σ_1 propositions, which are most conveniently described using the formalism outlined in Section 3.4.

Π_1 propositions are those that can be expressed symbolically in the form

$$\forall x_1 \ldots \forall x_n A[x_1/a_1, \ldots, x_n/a_n]$$

which is read "For all $x_1, \ldots$, for all x_n, $A[x_1/a_1, \ldots, x_n/a_n]$". Here **A** is a "local" formula, free of global quantifiers, and $a_1, \ldots, a_n$ are distinct free varables.

Similarly Σ_1 propositions are those that can be expressed symbolically in the form

$$\exists x_1 \ldots \exists x_n A[x_1/a_1, \ldots, x_n/a_n]$$

which is read "There exists $x_1, \ldots$, there exists x_n such that $A[x_1/a_1, \ldots, x_n/a_n]$". Again **A** is local and $a_1, \ldots, a_n$ are distinct.

Brouwer's Principle requires that Π_1 and Σ_1 propositions cannot be accounted *true* or *false*, but must rather be described as *justified* or *unjustified*. This means that they cannot be combined with each other, or with ordinary propositions, using conventional propositional connectives to form complex propositions. In this sense they are logically inert. The principal use to which I shall put such propositions is in definitions and in the use of Π_1 propositions as hypotheses in theorems. A formal treatment of global propositions is to be found in Appendix 1. For the most part we can deal with the logical complications occasioned by the employment of such propositions as they arise[29].

Because set theory is not, and cannot logically be, a formal axiomatic theory, the Π_1 propositions that I shall lay down as its axioms must be justified, and this justification can only take the form of observing that all the ordinary propositions that are their instances are true. But I haven't given, and won't give, a general, mathematical definition of truth for such propositions. And I shall get on with the business of defining and proving things in ordinary prose (supplemented by mathematical notation), without giving precise, mathematical definitions of what defining and proving are. There may be those who wonder whether such a way of proceeding is legitimate. I shall address their worries in the next section.

---

[29] It is worth pointing out, however, that I shall take it as a basic principle of the theory that non-empty sets are *inhabited*, so that from the premise

$$\mathbf{t} \text{ is a set and } \mathbf{t} \neq \emptyset$$

we may infer

$$\exists x[x \in \mathbf{t}]$$

3.7 Truth and proof in mathematics

There is a surprisingly widespread view that the very notion of truth in mathematics is somehow suspect. Insofar as this view is not a mere reflection of the "post modern" cult of the irrational, it stems from modern logic's discovery of the semantic paradoxes of self-reference. Perhaps I ought to speak of rediscovery here, for the most important of these paradoxes, the Liar, was known and widely discussed both in antiquity and in the middle ages. But modern logic has produced new paradoxes – that of Richard, for example, which turns on the notion of *definability* – and has formulated all of these paradoxes, old and new, with a sharpness and precision previously unknown.

Mathematicians, aware of such paradoxes and of the difficulties encountered in resolving them, have become chary of appealing to the traditional concept of truth explicitly and directly in definitions and proofs. But this understandable, and indeed sensible, circumspection has, in some, swollen to the conviction that this concept, in its ordinary, non-technical sense, is devoid of significance. Here, for example is Bourbaki on the subject of mathematical truth[30].

Mathematicians have always been convinced that what they prove is "true". It is clear that such a conviction can only be of a sentimental or metaphysical order, and cannot be justified, or even ascribed a meaning which is not tautological, within the domain of mathematics.

On the face of it this is just nonsense. How could mathematicians not claim as true what they claim to have proved? How could such claims be dismissed as "sentimental" or "metaphysical"? Am I not justified in laying claim to truth *by virtue* of having produced a proof? Or is Bourbaki telling us that no mathematical proof has demonstrative force? If so, why has he devoted his life – or rather his lives – to producing such proofs? Why has he filled the many pages of his *Elements* with them? Surely he would have been well advised to postpone the disclosure of this scandal until a later volume – if only in the interest of retaining his readers.

But perhaps we should suppress our natural inclination to laugh at this folly. For behind the mask of magisterial irony that Bourbaki has assumed in propounding it, we can discern the features of a profound and bitter disappointment, of bright hopes dashed and of worthy ambitions thwarted.

---

[30] *Elements of Mathematics*, vol. 1, p.306.

In fact, large and important issues are at stake here, issues that go beyond the problem of providing sound foundations for mathematics. For Bourbaki has evidently succumbed to that "Pyrrhonic and Academic scepticism" which Cantor so deplored in his own day, and which has grown even more fashionable in ours. Indeed, even among those who account themselves scholars or scientists are to be found many who believe that to deny the possibility of objective truth is the mark of an advanced thinker, and who never write down the word "true" without archly enclosing it in quotation marks. In these circumstances Bourbaki's example is all the more deplorable, given his unrivalled intellectual eminence and scientific prestige.

Have things really come to such a pass that we can no longer make use of the concepts of truth and falsehood? The semantic paradoxes disclose that if we are not careful in our use of these concepts we may stray into error or nonsense. But we must not allow ourselves to become hamstrung by excessive caution in the face of these inconvenient facts. In any case, we need not take counsel of our fears, for we know exactly where the danger lies: all the semantic paradoxes turn on the *reflexive* use of these concepts. In ordinary, non-reflexive use they are as straightforward and unproblematic in their application as is the discourse to which they are applied: no more, no less. If I know, antecedently, what it means to say that Φ, then I know what it means to say that Φ is true or that Φ is false. And that simple observation is all that is required to vindicate the use of these important concepts in ordinary, non-reflexive contexts.

Unfortunately, however, matters are not so easily put right. We cannot simply dismiss Bourbaki's views on truth out of hand, even though they come stamped with the hallmark of absurdity. The issue is too important, the confusion is too great, and those caught up in it are too numerous. We must trace such views to their origins, and make a serious and sympathetic effort to understand what motivates them. *Then* we may dismiss them. But where to begin?

The etymologies of the words "true" and "false" suggest that they, or the words from which they have descended, were originally applied to speakers, as descriptions of their intentions or of their trustworthiness: it was the *speaker* who was true (i.e. faithful) or false (i.e. deceitful). One could say that his words were true or false, by a species of metonymy, just as one could say they were angry or vengeful or just. But a faithful man may sometimes say what is not the case, and a deceitful man may sometimes say what is – indeed, he must do so if he is to be successful in practising his deceit. It is not difficult to imagine, therefore, how the

3.7 Truth and proof in mathematics 101

words "true" and "false" came to be attached to the speech itself rather than to the speaker: true speech is what a true speaker would say if he were in possession of the facts, and false speech is what a false speaker practising his deceit would say were he in possession of the facts.

Perhaps these etymological speculations cannot be sustained. That doesn't matter; for, in any case, the example of modern formal logic encourages us attribute truth and falsehood, not to speakers, nor to meaningful utterances, spoken or written, but to mere syntactic assemblages: successions of sounds, or configurations of inscriptions. This may be an important source of our difficulty with these concepts, for questions of truth and falsity, as these terms are ordinarily used outside formal logic, arise only when a succession of sounds is understood as articulate speech, uttered with the intention of conveying information, or when a configuration of inscriptions is understood as written language similarly intended[31]. That is what the words *true* and *false* have come to mean, whatever their provenance in our remote past.

Therein lies the difficulty. To make judgements of truth and falsity, taking those terms in their ordinary senses, we must first grasp the meaning of the utterances or inscriptions to be so judged. Indeed, whoever grasps the meaning of an utterance or inscription grasps, *ipso facto*, what it is for the proposition it expresses to be true or to be false. But what is it to grasp the meaning of an utterance or of an inscription? What is it to understand language? That is where the mystery of language lies: and it is a deep mystery.

But the mystery in this case is the mystery of the familiar. There is clearly a sense in which grasping meanings is something utterly familiar and utterly obvious. We have all had the experience of understanding both spoken and written language; surely to doubt that would be incoherent. Of course, we do not understand everything we hear or read; some people speak or write unclearly or unintelligibly. Moreover, there is no sharp line dividing the meaningful from the meaningless; if there were, language would lack the flexibility it requires.

But even when we do grasp the meaning of an utterance or inscription we cannot *say* what it is that constitutes our doing so. Should this be

---

[31] If I say "What I am now saying is false", then it cannot be with the intention of conveying information. Some readers may be tempted to object that I have not defined what I mean by "intention" or "information" here. Those readers are caught up in the very confusion that I am now attempting to dispel, namely, the false assumption that such words must be give mathematically rigorous definitions before they can be meaningfully employed.

taken, then, as casting doubt on our capacity for grasping meanings? Certainly not. After all, not every experience, not even every common experience, can be explained or put into words that do more than merely name it. We no more have to abandon the conviction that we can grasp meanings, on the grounds that we cannot explain how it is done, than we need abandon the conviction that we can see colours, although we cannot explain how that is done either. And when we can grasp the meaning of a proposition, when we can understand it, we can grasp what it is for that proposition to be true or false.

What is required, then, of the mathematician when he is laying down the foundations of his subject? All scientists, of whatever stripe, must try to discover the truth about the matters with which their sciences are concerned. But no one feels that it is incumbent upon geologists, chemists or botanists to explain what truth is, *per se*, before they set about the business of pursuing it. Why, then, should this be required of mathematicians? Perhaps it was the thought that he might be expected to supply such an explanation, combined with the certain knowledge that he could not do so, that led Bourbaki to propound the nonsensical views I have quoted above.

The demand that mathematicians should provide *ab initio* a rigorous, mathematical definition of mathematical truth is a senseless one. But it is not difficult to see why they, alone among scientists, should be confronted by it, or should impose it upon themselves. Mathematicians, after all, have grown used to the necessity for providing rigorous definitions of the concepts they investigate. If it is necessary to give a rigorous definition of *continuity* in order to establish that certain functions are continuous, then surely it is equally necessary to give a rigorous definition of *truth* in order to establish that certain propositions are true. How can we carry out a mathematical investigation of a concept which has not been properly defined?

But the analogy between continuity and truth is a false one. Mathematicians do not *investigate* truth, they *pursue* it. If we fail to make this crucial distinction, we are led to draw the false analogy in question. *Truth* is just not a concept of the same logical type as *continuity*, or, indeed, as any other concept that stands in need of rigorous, mathematical definition. Predications of truth are *sui generis*, and in *particular* cases are logically superfluous. Thus the claim that Φ is true is just the claim that Φ, so to make that claim in that particular case the notion of truth need not be explicitly invoked. But the concept of continuity, and similar concepts that mathematicians *investigate*, cannot be dispensed with in

3.7 Truth and proof in mathematics

particular cases, and in this respect are sharply to be distinguished from the concept of truth.

But don't mathematicians investigate, and indeed (following Tarski) even *define* truth, when studying interpretations of their formal languages? Yes, in a sense – but then these formal languages, the structures that interpret them, and Tarski's definition of truth in those structures are all *essentially* set-theoretical in character. We cannot stand outside mathematics and provide a similar definition of truth for set theory itself.

In any case, Tarski's definition cannot be taken as an *explanation* of what is meant by truth in the way, say, that the definition of continuity is an explanation of what is meant by continuity. For the principle of duality in logic means that a formal definition of *truth* in a structure is formally indistingushable from a formal definition of *falsity*, and, indeed, the various possible formal definitions of *proof* are indistinguishable from the corresponding dual definitions of *refutation*. Unless we know what the formal logical symbols (propositional connectives and quantifiers) are intended to *mean*, that is to say, unless we can translate them into ordinary mathematical discourse, in which, of course, the notion of truth – the real one, not Tarski's formal simulacrum – is implicitly given, we cannot see, from its purely *formal* structure, that it is *truth*, rather than *falsity* that Tarski's definition captures.

Thus the significance of these formal definitions of "truth" and "proof" for formalised languages does not consist in their providing *explications* of those notions: it rather lies in the fact that they allow us to avoid explicit appeal to truth and proof, in their full "pre-formal" meanings, in laying down mathematical definitions and carrying out mathematical proofs. In short, the formal definitions make it possible for us to avoid the paradox-engendering *reflexive* uses of those notions in our mathematical investigations of logic.

Since mathematicians deal in proof, their truth claims have an absolute character that distinguishes those claims from the truth claims put forward by other scientists. It makes perfect sense to say "This law *may* be true for it applies correctly, to within experimental error, to the known cases"; but it makes no sense to say "This theorem *may* be true, for I have proved it"; one must say "This *is* true for I have proved it". Is a proof then a guarantee of a proposition's truth? Of course it is, in so far as it is valid, that is to say, really a proof. That is just what the word "proof" *means*.

But if that is what "proof" means, perhaps it represents an unattainable ideal. Perhaps we ought to consider the possibility that no one has ever succeeded in constructing a *genuine* proof.

Is such a possibility really unthinkable? In the introduction to *Continuity and Irrational Numbers*, Dedekind expressed the opinion that up to that time no one had succeeded in producing a valid proof of the simple proposition that $\sqrt{2} \times \sqrt{3} = \sqrt{6}$. Most mathematicians would now agree that he was right. Naturally Dedekind's scepticism did not extend to *all* previous proofs of *all* theorems; but this particular example raises the general point. We can legitimately criticise our predecessors in respect of the rigour of their arguments. Is there any reason to suppose that this sort of thing will no longer be possible in the future? May not our descendants find similar fault with us, and theirs with them, and so on, as long as men retain an interest in mathematics?

Thus it might seem that by reflecting on the history of mathematics we could conclude that the ideal of a completely rigorous proof, which settles a question once and for all subsequent time, may not be attainable. Such doubts might be reinforced by our reflecting on the difficulty of achieving certainty, of dispelling all doubt as to the truth of a proposition.

But the concept of certainty bristles with difficulties. Certainty is not, strictly speaking, a property that attaches to propositions or to proofs: it is rather a state of mind which may or may not be induced in someone who takes in the argument that a proof presents. In the case of short, simple proofs, most of us achieve it with relative ease; if a proof is long and complex, it may be only after severe effort and struggle that we come to understand the argument and to experience that quality of conviction that we call certainty. But many, perhaps the vast majority of mankind, are incapable of following any but the simplest of arguments; and even gifted mathematicians are sometimes fooled by an argument into accepting a falsehood as true: they feel a certainty that is completely unwarranted.

When one is presenting a proof of a proposition, one's aim is to produce the conviction, in oneself and in others, that the proposition *must* be true; one wants to establish in oneself and to induce in others, a sense of *certainty* that this is the case. But that sense of certainty, the characteristic feeling that accompanies conviction, is not, in itself, a guarantee of truth. A proof guarantees the truth of the proposition proved by virtue of the fact – and we must suppose it to be a fact, since we are speaking of *proof* here – that its starting points, the propositions it lays down as true without argument, are true, and its inferences are valid; these are the facts the grasping of which produces the conviction of

3.7 Truth and proof in mathematics

certainty. In any event, it *ought* to produce such conviction. If someone were to say "Yes, the premises of the proof are true, and, yes, all the inferences by which the argument advances to its conclusion are valid; nevertheless, that conclusion is false", and if he were sincere in saying this, then one could only say, with Frege, "We have here a hitherto unknown form of madness".

In practice, systematic scepticism about mathematical proof can be made plausible only by talking in generalities. "How can it be", asks the sceptic, "that mathematical proofs *force* us to accept their conclusions? What is the nature of the *compulsion* that operates here?" To which the proper reply is "Which *particular* proof did you have in mind?" For the logical force of a particular proof is particular to that proof.

The sceptic reminds us of the fallibility of the human mind; he recalls examples of clever men, even men of genius, who have been taken in by fallacious arguments; he points out to us how men have taken mere shibboleths for necessary truths; he shows us how the very concept of rigour in proof has changed over the centuries; he asks us to explain how the capacity to recognise abstract truths has arisen in the course of human evolution, and how, given that origin, we can have any confidence in that capacity; he challenges us to explain the causal mechanism by which the "abstract objects" of mathematics affect the physical circuitry of our brains ...

But doubts engendered by such general considerations melt away when we turn to a *particular* proof, following it line by line, checking and, if necessary, rechecking the truth of its premises and the validity of its inferences. If we cannot be convinced by *this* how can we be convinced by those infinitely vaguer arguments from history, sociology, psychology, anthropology, and biology that the sceptic has adduced? When we descend from the clouds of generality and abstraction and feel the solid earth of detailed argument beneath our feet, then the sceptic's challenges do not, indeed, disappear, but rather assume their proper proportions and priority.

But can we be *certain* that the basic concepts upon which mathematics rests are precisely determined, and that its basic principles are valid? It's not easy to answer these questions because it's not really clear what that certainty would consist in or entail. Does being certain require that we be unwilling, or unable, to entertain criticism of our concepts and principles?

Surely that would not be desirable. But one thing is clear: from the standpoint of logic, it is not necessary that our principles be *certain*, only that they be *true*. And *general* scepticism about the possibility of *certainty* does not cast *particular* doubt on the *truth* of our axioms. After all, it is quite possible to be certain of the truth of specific, individual propositions without being able to give a general account of what certainty or truth consists in. Just because we are uncertain about certainty, or about truth *in general*, it does not follow that we must be uncertain about *everything*.

But we mustn't make a bugbear of the problem of certainty. In the sixteenth and seventeenth centuries our ancestors were preoccupied with it, and we have inherited their preoccupation. But what for us is an interesting philosophical puzzle, though, no doubt, an intellectually respectable one, for them was a burning issue with profound moral and political consequences. After all, if you are going to deprive a man of his civic rights, perhaps even burn him at the stake, for holding opinions that differ from your own, you had better be *certain* that your opinions are right.

Those earlier thinkers, therefore, had deep and pressing motives for pushing scepticism as far as they possibly could: if you can't even be certain of the accepted facts of history, or of the existence of an external world or of other people, or about the truths of mathematics, how can you justify burning a man just because he holds erroneous views concerning the Real Presence or the governance of the church?

But we have no such overriding reasons for pushing scepticism to its extreme limits in this way. Indeed, quite the reverse. For where our ultimate object is understanding, as well as justification, such a stubborn and thoroughgoing scepticism may simply prevent us from arriving at the truth without providing us with any compensating benefit.

However, I fear that in pursuing these general speculations on certainty, I may be succumbing to the very vice that I am trying to expose. Our task is not to reflect on the notions of first principles and certainty in general, but to judge the truth of a mere handful of *particular* first principles, namely the axioms of set theory. And the question each of us must ask of each of these axioms is just "Is this true?" not "Am I certain that this is true?"

For if one concentrates on the latter question instead of the former, one turns from what is simple and accessible – mathematics – to what is murky and uncertain – introspective psychology – and one is forced eventually to confront the deepest and most intractable of mysteries, namely, what is it to grasp the meaning of a proposition, what is it to be convinced of its truth, what, indeed, is it to think, in short, what is it to be a creature of the sort that each of us is? This is a sure recipe

3.7 Truth and proof in mathematics

for dissolving all our convictions and for turning every certainty into a doubt.

But we have not yet reached the nub of the matter. For the source and origin of all confusion here is the belief that the notions of truth and proof in mathematics can be legitimate only if they can be given rigorous *mathematical* definitions. It is this belief, I am convinced, that caused Bourbaki to produce the series of mistaken, and indeed absurd, pronouncements on the concepts of truth and proof, one example of which I have already given, that so disfigure his volume on set theory.

The basis for Bourbaki's mistake was the false hope held out by Hilbert's program. Bourbaki, like so many of his most gifted contemporaries[32], was inspired and excited by Hilbert's claim that the traditional notion of a mathematical proof could be replaced by a purely formal notion of proof, defined with mathematical precision, and that, moreover, once this replacement had been effected, the *completeness* of the formal systems of proof for the essential theories (arithmetic, analysis, set theory, etc.) would mean that the notion of truth itself for those theories could be given a purely formal definition. In Hilbert's vision, all of mathematics could be reduced to the elementary combinatorial geometry required for formal proof theory. How satisfying – how reassuring – it must have seemed that the perplexing and disturbing philosophical problems traditionally associated with the foundations of mathematics could finally, and with mathematical certainty, be laid to rest!

Of course we now know that Hilbert's program cannot be carried out in the way he conceived it: it founders on the Incompleteness Theorems of Gödel. But Hilbert's idea was not, *prima facie*, absurd; on the contrary, his program was one of the most fruitful foundational proposals ever put forward, and it required Gödel's profound theorems to dash the hopes that it raised. To understand the depth of Bourbaki's disappointment we must bear in mind the general strategy that underlay Hilbert's program, and the prospect that it held out of achieving a radical solution to the problem of mathematical foundations.

But even if the Incompleteness Theorems have put paid to the hope that a formal definition of proof would yield a formal definition of truth itself, nonetheless do we not have several notions of formal proof, any one of which allows us to convert all of our ordinary proofs into formal ones which can be checked for correctness in an algorithmic, purely

---

[32] Bourbaki, of course, *was* many of his most gifted contemporaries. We must assume, however, that in his foundational views he was single-minded.

syntactic way? And should we not, in the interests of rigour, carry out our proofs in this formal manner? Or, if that is a counsel of perfection that we cannot realise in practice, should we not regard our ordinary proofs as arguments to the effect that such formal proofs could be carried out, at least in principle? This latter is, in fact, the line taken by Bourbaki himself.

In practice, the mathematician who wishes to satisfy himself of the perfect correctness or "rigour" of a proof or a theory hardly ever has recourse to one or another of the complete formalisations available nowadays, nor even usually to the incomplete and partial formalisations provided by algebraic and other calculi. In general he is content to bring the exposition to a point where his experience and flair tell him that translation into formal languages would be no more than an exercise of patience (though doubtless a very tedious one)[33].

Of course this does not describe the way most mathematicians *actually* regard their proofs, but we may take it to be Bourbaki's opinion of how they *ought* to regard them. However, even then it is misconceived.

As the notion of proof is ordinarily, and correctly, understood, the purpose of a proof is, after all, to explain why its conclusion *must* be true in such a way as to compel assent to the truth of that conclusion in anyone who understands that explanation. Every consideration in deciding how to present a proof must be subordinate to that purpose. The ideal, perhaps, is to present a proof in the form of a string of propositions connected by inferences every individual one of which can be seen to be obviously correct. Pushed to the ultimate limit, such a method of presentation might eventually approach what a logician would call a fully formalised proof.

But, in practice, when we subject the individual trees to minute inspection of this kind we are apt to lose sight of the forest. And, in any case, even if we are employing a formalised conceptual notation of the sort described in Appendix 1, we are speaking here of a proof, every line of which is understood and acknowledged true, and every constituent inference in which is followed and its apodeictic force experienced.

Following such a proof is not a purely combinatorial or syntactic process of counting brackets and taking note of similarities in the shapes of signs: in short, such a proof, though it may be thoroughly formalised, is not a mere assemblage of meaningless signs arranged in accordance with purely formal rules. It may very well be, if the rules are formulated with sufficient niceness and exactitude, that the question whether the

---

[33] *Elements*, vol. I, Introduction (p.8).

3.7 Truth and proof in mathematics

inference from **A, B, C,** ... to **D** is in accordance with those rules, can be settled by purely combinatorial or syntactic means. But conviction can be compelled only if the logical force of the inference

$$\mathbf{A}, \mathbf{B}, \mathbf{C}, \ldots : \text{ therefore } \mathbf{D}$$

is felt, and that is not a matter of purely combinatorial insight[34].

A proof is, among other things, an attempt to convey ideas, and in presenting a proof one must always be sensitive to the problem of rhetoric. Naturally by "rhetoric" here I do not mean tricks of eloquence or persuasion, but the art of choosing one's vocabulary (including one's symbolic vocabulary) and arranging one's discourse so as to make one's ideas as clear and as accessible as possible. If you include too much routine detail in your argument, you may succeed only in obscuring it. I suppose it is theoretically possible that disagreement over the efficacy of a purported proof might reach such a pitch that it could only be resolved by resort to a full formalisation. But that possibility is so remote that we may surely discount it. Such circumstances have never arisen in practice, as far as I am aware, unless the question in dispute has actually been the existence of a formal proof; and there, of course, it is not the conclusion of the proof that is verified but the existence of a formal proof to that conclusion.

A proof, then should not be conceived as a mere assemblage of signs; nor should it be seen as an argument to show that a strictly formal proof could ("in principle", of course) be constructed. My decision not to carry out a formal exposition of set theory does not, therefore, in itself constitute a derogation from rigour. For rigour, like good manners, is not simply a matter of conforming to rigid, formal rules. And, like good manners, it is easy to recognise but difficult to define.

In any case, my primary claim is that such theorems as I shall present are *true*. The proofs that I shall offer will be in support of that claim. But the truth of a mathematical proposition does not depend on the correctness, formal or otherwise, of any purported proof of it. Indeed, it does not depend, in the general case, even on the *possibility* of a proof, although in some cases it may *entail* that possibility. If anyone feels that

---

[34] We must distinguish "formal" from "combinatorial" insight. It may well be that we grasp the validity of the inference from **A, B, C,** ... to **D** by virtue of the forms of **A, B, C,** ..., **D**, e.g. we may infer **B** from **A** *implies* **B** and **A** without thinking about the meanings of **A** and **B**; but we must think of the meaning of "*implies*" to grasp the validity of the inference. Combinatorial checking of the purely syntactic correctness of an inference, however, can be carried out by a computer, which cannot think of any meanings at all.

the proofs I offer do not support my claims, on the grounds that they are not fully formalised, let him supply his own, fully formalised proofs using the system of formal proof given in Appendix 1, or in some equivalent system of his own choosing.

4
The Principal Axioms and Definitions of Set Theory

4.1 The Axiom of Comprehension and Russell's Theorem

In this chapter I shall present the principal axioms of set theory. As I have already explained (Section 3.5), all these axioms can be construed as embodying claims to the effect that certain species are finite (in the appropriate sense), and therefore form sets: in brief, these axioms can be construed as *principles of finiteness*.

The finiteness principles that are embodied in these axioms can all be formulated in the following general way:

Given certain objects and/or first level global functions (propositional or ordinary), such-and-such a species of objects (the membership condition for which is specified unamb iguously in terms of the given objects and global functions) is *finite*, and therefore forms a set.

But this is not the form the axioms proper actually take. In each case the set whose existence is claimed is regarded as the value of a certain global function for the given objects and global functions as arguments. The defining condition for the species whose finitude is asserted by the axiom, albeit indirectly, provides necessary and sufficient conditions for membership in the set that is the value of the global function for those arguments. In laying down these axioms, I shall, in each case, introduce a new symbol, a *global function constant*, of the appropriate level, to represent the global function whose values are specified by the axiom.

The first of these principal axioms that I shall present is the *Axiom of Comprehension* (Zermelo's *Aussonderungsaxiom*). Thus let Φ be any first level global propositional function of one argument and let S be any set. Then the species of all objects x such that x is a member of S and Φ holds true at x is finite, and therefore forms a set, which we may denote in the conventional way by

$$\{x \in S : \Phi(x)\} \text{ (read "the set of all } x \text{ in } S \text{ such that } \Phi(x)\text{")}$$

In general, the correspondence

$$\Phi, a \mapsto \{x \in a : \Phi(x)\}$$

determines a global function of second level having as arguments the first level global propositional function, Φ, of one argument, and one object, a[1].

The Axiom of Comprehension can thus be formulated precisely as follows.

4.1.1 Axiom *(The Axiom of Comprehension) Let Φ be any first level global propositional function of one argument, and let a be any set or individual. Then $\{x \in a : \Phi(x)\}$ is the set of all objects x (if any) that are members of a and satisfy Φ. In symbolic notation*

(i) $\forall y[\{x \in y : \Phi(x)\} \text{ is a set}]$[2]
(ii) $\forall y \forall z [z \in \{x \in y : \Phi(x)\} \ .\textit{iff.} \ z \in y \text{ and } \Phi(z)]$

Syntactically, the symbol for the second level *comprehension function* characterised by this axiom is a *variable binding operator* which allows us to form the *formal comprehension term*

$$\{x \in \mathbf{t} : \mathbf{A}[x/\mathbf{a}]\}[3]$$

which is obtained from $\{x \in a : \Phi(x)\}$ by replacing a by $\mathbf{t}$ and Φ by $\lambda x \mathbf{A}[x/\mathbf{a}]$.

I have already pointed out that the Axiom of Comprehension does not actually *express* the principle of finiteness that it *embodies*. Surely that finiteness principle itself must be accounted a self-evident truth: on what conceivable grounds could anyone refuse to acknowledge the finiteness of a decidable subspecies of a finite species, that is to say, a subspecies

---

[1] When a is an individual (and hence has no members) we take $\{x \in a : \Phi(x)\}$ to be the empty set. Hence we must lay down the stipulation that $\{x \in a : \Phi(x)\}$ is always a set as part of the Axiom of Comprehension.

[2] By virtue of the conventions in 3.4 governing the meaning of "$\subseteq$", "S is a set" can be expressed in symbolic notation by "$S \subseteq \emptyset$ implies $S = \emptyset$".

[3] In this formal expression, "x" is a bound variable whose leftmost occurrence is a *binding occurrence* and whose *scope* is the expression $\mathbf{A}[x/\mathbf{a}]$ together with the formal term $\mathbf{t}$. $\mathbf{A}[x/\mathbf{a}]$ is obtained from the formal sentence $\mathbf{A}$ by replacing all occurrences of the free variable $\mathbf{a}$ with occurrences of the bound variable x. As usual we must stipulate that none of the occurrences of $\mathbf{a}$ in $\mathbf{A}$ fall within the scope of binding occurrence of x in $\mathbf{A}$, i.e. that x is *free for* $\mathbf{a}$ *in* $\mathbf{A}$.

4.1 The Axiom of Comprehension and Russell's Theorem

whose defining property definitely and objectively holds or definitely and objectively fails to hold, at each member of the given species[4].

But however evident it may be that the elements of a set that satisfy a given propositional function must themselves form a set, that is not what the Axiom of Comprehension actually *says*. It merely stipulates that the object named by a certain term is a set, and gives necessary and sufficient conditions for membership in that set. Because sets are extensional, these conditions actually *define* it. In this way the axiom itself actually fixes the meanings of comprehension terms. It thus reads more like a definition or stipulation of meaning than an axiom. This is a point of some significance, and I shall return to it in Section 4.4.

What the Axiom of Comprehension says is absolutely clear; but our reasons for believing it to be true are to be found in our intuitions about the concept of finiteness, a concept that informs, but is not directly incorporated into, our mathematical discourse.

Despite its "obvious" character, the Axiom of Comprehension permits us to deduce an important consequence which, if we may judge from the circumstances surrounding its discovery, it not obvious, despite its short, straightforward proof.

4.1.2 Theorem *(Russell's Theorem)* *Let S be any set. Then S does not contain $\{x \in S : x \notin x\}$ as a member.*

The proof here is just an observation: if $\{x \in S : x \notin x\}$ were a member of S, then the Axiom of Comprehension, interpreting Φ as $\lambda x[x \notin x]$, would guarantee that $\{x \in S : x \notin x\} \in \{x \in S : x \notin x\}$ if, and only if, $\{x \in S : x \notin x\} \notin \{x \in S : x \notin x\}$, which is a contradiction. This is a deep result with a trivial proof. Since each set fails to contain its "Russell subset" as a member, it follows that no set contains all sets, *a fortiori*, no set contains all objects, i.e. there is no universal set.

The Axiom of Comprehension also permits us to define *local* quantifiers, quantifiers with *finite* domains of discourse, whichever meaning we eventually decide to attach to the word "finite"; and a finite domain of discourse is just a *set*.

---

[4] It is difficult to imagine how *any* subspecies of a finite species could be *infinite*. But one can perhaps imagine an *indefinite* subspecies of a finite species that could not, by virtue of its indefiniteness, be counted as a set. Is there a *set*, that is to say a *definite number*, of bald men currently serving in the U.S. Senate? The example is frivolous, but there are mathematically significant properties that share with "baldness" the characteristic of indefiniteness – Π_1 and Σ_1 properties, for example. Such properties are not decidable in the sense intended, for it is not, in general, objectively determined that such a property holds true, or fails to hold true, of a given object.

4.1.3 Definition *Let Φ be any global propositional function of one argument and a any object. Then by definition*

(i) $(\forall x \in a)\Phi(x)$ *.iff.* $a = \{x \in a : \Phi(x)\}$
(ii) $(\exists x \in a)\Phi(x)$ *.iff.* $not[a = \{x \in a : not\Phi(x)\}]$

It is a straightforward matter to show that, where S is a set, the quantifiers $(\forall x \in S)$ and $(\exists x \in S)$ satisfy all the usual properties, and obey all the usual rules of inference, provided it is taken into account that S can be empty. Of course if the domain of discourse a of a quantifier $(\forall x \in a)$ or $(\exists x \in a)$ is an individual, then that quantifier behaves as if a were the empty set.

4.2 Singleton selection and description

Now that the local quantifiers have been defined we can settle a technical problem: our symbolic notation must permit us to name the sole member of a set that has exactly one member. If that sole member is itself a set then the set-theoretical operations to be introduced later in this chapter will permit us to do this. But if the sole member of a set is an individual that method will not work. Since the operation of *singleton selection* must be put to constant use, I shall make it explicit by introducing a new first level global function constant, ι, of two arguments and adopting the following axiom.

4.2.1 Axiom *(The Axiom of Singleton Selection) Let S and a be sets or individuals. Then if S has exactly one member, $\iota(S, a)$ is that sole member, so that $\iota(S, a) \in S$. If S has no members, or more than one member, then $\iota(S, a) = a$. In symbols*

$$\forall uv[(\exists x \in u)(\forall y \in u)[y = x] \text{ .implies. } \iota(u, v) \in u \; : and :$$
$$not(\exists x \in z)(\forall y \in z)[y = x] \text{ .implies. } \iota(u, v) = v]$$

$\lambda uv\iota(u, v)$ can be employed as a logical "description operator" as follows. The informal expression

$$\text{The } x \text{ in } a \text{ such that } \Phi(x)$$

can be rendered by

$$\iota(\{x \in a : \Phi(x)\}, \mathbf{t})$$

where **t** is any term. The original expression purports to name the unique member of a that satisfies Φ. If there is, indeed, such a unique member of a satisfying Φ then

$$\{x \in a : \Phi(x)\}$$

is the set whose sole member is that member of a, and therefore

$$\iota(\{x \in a : \Phi(x)\}, \mathbf{t})$$

actually names it. Otherwise ι takes the "don't care" default value $\mathbf{t}$. The function symbol "ι" is closely akin to the function symbol introduced by Frege for a similar purpose in §11 of the *Grundgesetze*, but I have followed the example of Bernays and allowed a "variable" default value as an extra argument. The only default values for $\iota(\mathbf{s}, v)$ that naturally suggest themselves are $v = \emptyset$ and $v = \mathbf{s}$. I shall sometimes suppress the second argument, and the reader may assume $\emptyset$ is meant as default value in such circumstances.

4.3 Pair Set, Replacement, Union, and Power Set

With the machinery of local quantification in place I can now present the axioms that, together with the Axiom of Comprehension, embody the most basic finiteness assumptions of set theory.

The first, and most obvious, of these is the Axiom of Pair Set. The finiteness principle on which it rests simply says that any two objects compose a finite plurality. Thus given any objects, a and b (which need not be distinct), the species of all objects x such that $x = a$ or $x = b$ is finite and therefore forms a set, which we denote by

$$\{a, b\} \text{ (read "the *pair set* of } a \text{ and } b\text{")}$$

The correspondence

$$a, b \mapsto \{a, b\}$$

thus defines a first level global function of two arguments.

The Axiom of Pairing can be formulated precisely as follows:

4.3.1 Axiom (*The Axiom of Pair Set*) *Let a and b be any objects. Then $\{a, b\}$ is the set whose members are precisely a and b. In symbols*

$$\forall xyz [z \in \{x, y\} \ .\textit{iff.}\ z = x \text{ or } z = y]$$

Perhaps it is worth pointing out that despite the utterly trivial nature of the finiteness principle embodied in this axiom, nevertheless the new global function constant introduced supplies us with the capacity to produce an unlimited number of names for distinct sets.

4.3.2 Definition *Let a be any object. Then by definition*

$$\{a\} = \{a, a\}$$

Thus for any object, a, $\{a\}$ (read " the *singleton* of a") is the set whose sole member is a.

The second of these axioms, the Axiom of Replacement, is rather more complicated to state, but rests on an underlying finiteness principle that is equally self-evident. Given a first level global function, σ, of one argument, and any set, S, then the species of all objects x such that $x = \sigma(y)$ for some $y \in S$ is finite and therefore forms a set, which we denote by

$$\{\sigma(x) : x \in a\} \text{ (read "the set of all } \sigma(x) \text{ such that } x \text{ lies in } a\text{")}$$

The idea here, as the name of the axiom suggests, is that if we "replace" the members x of a finite plurality, that is to say, of a set, by other objects, $\sigma(x)$, not necessarily mutually distinct, then the resulting plurality is also finite, and therefore forms a set. Surely this is utterly obvious.

The correspondence

$$\sigma, a \mapsto \{\sigma(x) : x \in a\}$$

defines a second level global function whose arguments are a first level global function, σ, of one argument, and an object a.

The axiom that embodies this finiteness principle can be stated as follows:

4.3.3 Axiom *(The Axiom of Replacement) Let σ be any first level global function of one argument, and let a be any set or individual. Then $\{\sigma(x) : x \in a\}$ is the set of all y for which $y = \sigma(x)$ for some $x \in a$. In symbols*

i.) $\forall y [\{\sigma(x) : x \in y\} \text{ is a set}]$
ii.) $\forall y \forall z [z \in \{\sigma(x) : x \in y\} \text{ .iff. } (\exists x \in y)[z = \sigma(x)]]$

As with the Axiom of Comprehension, I have opted to express the second level *replacement function* characterised by this axiom by using a variable binding operator. Thus we can form the *formal comprehension term*

$$\{s[x/\mathbf{a}] : x \in \mathbf{t}\}^5$$

obtained from $\{\sigma(x) : x \in a\}$ by replacing "a" by $\mathbf{t}$ and "σ" by $\lambda x s[x/\mathbf{a}]$.

---

[5] As with formal comprehension terms, "x" is a bound variable whose rightmost indicated occurrence is a *binding occurrence* whose *scope* is the indicated occurrence of $\mathbf{t}$ together with the indicated occurrence of the expression $s[x/\mathbf{a}]$. The indicated occurrences of x in that expression are *bound* by that binding occurrence of x. As usual we must stipulate that x is free for $\mathbf{a}$ in s.

4.3 Pair Set, Replacement, Union, and Power Set

The third of these axioms, the Axiom of Union, is the first of the basic axioms to embody a non-trivial claim about finiteness. The finiteness principle in question here asserts that, given any set or individual a, the species of all objects which are members of sets that are, in turn, members of a is finite, and therefore forms a set, which we denote by

$$\bigcup(a) \text{ (read "the union or sum set of } a\text{")}$$

Thus if we have a finite number of sets, that is to say, a *set* of them, then the totality composed of all the members of all those sets is finite, that is to say, forms a set. This is a very ancient idea – it underlies our conception of multiplication in simple arithmetic, for example – but the truth it embodies is perhaps not so obvious as the truths embodied by the previous axioms.

The correspondence

$$a \mapsto \bigcup(a)$$

defines a first level global function of one argument.

4.3.4 Axiom *(The Axiom of Union (or Sum Set))* Let a be any set or individual. Then $\bigcup(a)$ is the set of all objects that belong to sets that are members of a. In symbols

   i.) $\forall x[\bigcup(x) \text{ is a set}]$
   ii.) $\forall xy[y \in \bigcup(x) :\text{iff}: (\exists z \in x)[y \in z]]$

With the Axioms of Comprehension, Pair Set, and Union in place we can define the formal symbols that name the familiar Boolean operations.

4.3.5 Definition Let a and b be objects. Then by definition

  (i) $a \cup b = \bigcup(\{a,b\})$ *(read "a union b" or "the union of a and b")*
  (ii) $a \cap b = \{x \in a : x \in b\}$ *(read "a intersection b" or "the intersection of a and b")*
  (iii) $a - b = \{x \in a : x \notin b\}$ *(read "a minus b" or "the relative complement of b in a")*

The familiar properties of these operations can be easily established from their definitions.

The final axiom among these basic axioms, the Axiom of Power Set, embodies the most powerful, and least self-evident, of the finiteness principles that are common to both the Cantorian and the Euclidean standpoints. The principle in question asserts that, given any set or individual, a, then the species of all sets x that are included in a (i.e. for which $x \subseteq a$) is finite, and therefore forms a set, which we denote by

$$P(a) \text{ (read ``the \textit{power set} of } a\text{'')}$$

Thus a finite plurality (i.e. a set), however large it may be, has only finitely many subsets.

This is a strong assumption on whichever conception of "finiteness" we adopt: the number of such subsets, even for relatively small Euclidean finite sets, seems to lie far beyond our capacity to conceive in any concrete manner.

For example, let S be the set of inhabitants of Bloomington, Illinois on January 1, 1953, and suppose that the number of such inhabitants was $65,536$. Then the number of elements of $P(S)$ is $2^{65,536}$, a number which, if written down, would contain some $19,729$ digits. But $P(P(S))$ would contain $2^{2^{65,536}}$ elements, and that number cannot be said even to *have* a decimal representation in any concrete sense, that is to say, no decimal representation for it could conceivably be actually inscribed. Thus even in such a case we cannot conceive how $P(P(S))$ (or, indeed, $P(S)$) is partitionable, in any concrete way, into multiples of powers of 10 in the familiar manner described by decimal numerals, although it is a theorem that *there are* such partitions.

As in the previous cases, the correspondence

$$a \mapsto P(a)$$

defines a first level global function of one argument.

4.3.6 Axiom *(The Axiom of Power Set)* *Let a be any object. Then $P(a)$ is the set of all sets that are included ($\subseteq$) in a. In symbols*

$$\forall xy [y \in P(x) \text{ .iff. } y \text{ is a set and } y \subseteq x]$$

Notice that since the empty set is included in every object, we do not need to stipulate that $P(a)$ be a set, for it follows that for any object a

$$\emptyset \in P(a)$$

so that $P(a)$ must be a set in every case.

4.4 The status of the principal axioms of set theory

I have now presented the principal axioms of basic set theory, the ones that incorporate the finiteness principles of the theory. As I have

4.4 The status of the principal axioms of set theory

explained, however, these axioms do not explicitly assert the finiteness principles that they incorporate: on the contrary, they all seem merely to be fixing the membership conditions for certain sets; they all have a definitional air about them.

These observations raise an important question. It is not so much a question about mathematics or its foundations as about what we might call the "sociology" of mathematics. But it is a question that bears significantly on the foundations of mathematics nonetheless. The question is this: *why do mathematicians accept the principal axioms of set theory as true?* That they do accept them as true is attested beyond all possibility of doubt by their practice, if not always by their pronouncements. And actions speak louder than words here.

The question becomes more puzzling the more we reflect on the powerful finiteness assumptions that some of these axioms (Union and Power Set) embody; and if we add in the fundamental Cantorian assumption that the species of natural numbers (suitably defined) is finite and thus forms a set, we compound the puzzle manyfold. Nevertheless this question is, I believe, readily answered, and, moreover, the answer discloses a widespread and fundamental confusion about the nature of set theory and of its role in the foundations of mathematics. Let me explain.

Perhaps the best way to begin is by reflecting on the way in which set theory is taught. Apprentice mathematicians are rarely given a systematic and lengthy account of the theory in its foundational role. Rather they pick up their set theory on the job, so to speak, at the beginnings of courses of lectures, or in the introductory chapters of textbooks on algebra, analysis, and topology.

Typically the principles of set theory – the axioms – are presented in the guise of mere conventions of notation. A teacher announces "I shall use the letter 'N' to denote the set of natural numbers", and his audience duly take note, none of them stopping to ask whether there *is* a set of natural numbers, or what the significance of there being such a thing might be. In this way each of them simply accepts, without question, the most controversial of the finiteness principles underlying modern, Cantorian set theory, as a guard dog might accept a morsel of drugged meat from a housebreaker.

Think how the principles of Union and Replacement might be explained in the introductory chapter of a textbook on analysis or topology:

We may use the elements of a set to *index* a family of sets. Thus suppose that

I is such an *index set* (we need place no restriction on the members of such a set – they can be objects of whatever sort we please, even sets) and suppose we associate to each *index* $i \in I$, a set S_i. The set of all elements that belong to one or more of the sets S_i, for $i \in I$, is then called the *union*, over $i \in I$, of the sets S_i and is designated by

$$\bigcup \{S_i : i \in I\}$$

Thus for any x, $x \in \bigcup \{S_i : i \in I\}$ if, and only if, $x \in S_i$ for some $i \in I$.

In this way two existential assumptions, one of them quite powerful, the Axioms of Replacement and Union, are presented in the guise, not even of definitions, but of mere notational conventions. Indeed, *all* the principal axioms of set theory can be imposed on unsuspecting mathematical novices in exactly this way.

Of course, in almost all cases, the perpetrator of such a fraud is as innocent, or as deluded, as its victims. The ease with which the one proffers, and the others accept, such an account reflects a widespread misconception about the notion of set. The pre-theoretical notion of set that they all share is the *logical* notion of *class* or *species* rather than the *mathematical* notion of *number* (in the sense of *arithmos*). Any plurality that one can describe forms a class in this logical sense; but then such classes aren't really things – objects – themselves. The only pluralities that are legitimate things are *arithmoi*, and they are so precisely in virtue of being *finite*. So one really ought to worry whether the pluralities (e.g. the species of natural numbers) deemed to form sets really are finite, i.e. limited or definite in size.

The misconception here – and it is pretty nearly universal – thus stems from the failure to understand the true antecedents of our modern conception of set. Our notion of set arises, not out of the ancient *logical* notion of "class" (*genos*), but out of the ancient *mathematical* notion of "number" (*arithmos*). The latter has, as the former has not, the idea of finiteness built into it.

But then in modern mathematics the relevant notion of "finiteness" has been stretched to take in the natural numbers, the real numbers, higher function spaces, ... in fact, all the particular structures that mathematicians study. Indeed, one can always "insert" any particular mathematical structure, or set of structures, into a set sufficiently large (but, by definition of course, still finite in Cantor's sense) to serve as a generalised domain of discourse in which one can obtain free set-theoretical mobility, so to speak, carrying out such set-theoretical "constructions" as one requires, without ever going outside that domain.

4.4 The status of the principal axioms of set theory

This answers the "sociological" question that I posed above: *mathematicians accept the axioms of set theory, not as axioms, that is to say, fundamental propositions taken as true without proof, but as definitions or even as mere notational conventions.* The finiteness principles embodied by the axioms are not called into question simply because it is not, in general, even recognised that they are, indeed, embodied by the axioms.

Set theory, as a foundational theory, is seen, not a collection of propositions to be believed, but as a collection of definitions and conventions to be learned, and of techniques to be mastered.

The complacency about foundations that these facts both make possible and encourage is also bolstered by the resolute *localness of outlook* so carefully cultivated in modern mathematics. By this I mean the habit of mind that leads mathematicians invariably to concentrate their attention on particular structures and instinctively to shy away from global questions involving infinite species. One studies the real line, the complex plane, finite dimensional real manifolds, various function spaces associated with such structures, etc., focusing attention on these particular structures and structures easily "constructed" from them, as, for example, the L_p-spaces are constructed from the reals.

Although this preoccupation with the local pervades mathematics, it is not often commented upon, and the apprentice mathematician acquires it less by instruction from his elders than in imitation of them. Even where "localisation" is directly at issue it is rarely mentioned.

Consider the First Homomorphism Theorem of group theory, for example. One of its essential functions is to localise the global problem of finding epimorphic images of a group: it shows that every member of the infinite species of epimorphic images of a group G has an isomorphic copy in the *set* of factor groups of G. Yet ths localising aspect of the theorem is rarely mentioned.

But when global questions become unavoidable, as they notoriously do in category theory, for example, the various dodges by means of which mathematicians strive to avoid thinking about the set-theoretical presuppositions that underlie the concepts and techniques they habitually employ no longer suffice; the result, more often than not, is confusion.

The category theorist complains that "set theory" will not permit him to carry out natural constructions using "large" concrete categories, such as the category of *all* groups and *all* group homomorphisms, or the category of *all* topological spaces and *all* continuous maps. The techniques he wishes to employ in these constructions are, he claims, universally applicable to all structures. And so, indeed, they are. But the supposed "large" concrete categories are not structures! For a structure is a *set* equipped with a morphology.

None of this should be surprising. Mathematicians relish intellectual challenges, but quickly become impatient with what seems mathematically obvious or with what seems too amorphous to admit of a clear-cut mathematical analysis. Unfortunately, the problem of mathematical foundations manages to incorporate both of these mathematically undesirable features. Here is Halmos, for example, in the introduction to his *Naive Set Theory*:

> What is true is that the concepts [i.e. of set theory] are very general and very abstract, and that, therefore, they may take some getting used to. It is a mathematical truism, however, that the more generally a theorem applies, the less deep it is. The student's task in learning set theory is to steep himself in unfamiliar but essentially shallow generalities till they become so familiar that they can be used with almost no conscious effort. In other words, general set theory is pretty trivial stuff really, but if you want to be a mathematician you need some, and here it is; read it, absorb it, and forget it.

Halmos is, of course, absolutely right. The part of general set theory that must be mastered "if you want to be a mathematician" is, indeed, "pretty trivial stuff", though "trivial" is a bit hard perhaps, considering how long it took to distil what is really essential. But as an estimate of the amount of intellectual effort that must be expended in mastering such of the technicalities of the subject as are absolutely essential for the rest of mathematics it is probably accurate.

Surely, however, that is exactly as it should be. In mastering this part of set theory the apprentice mathematician is merely learning to handle the basic tools of his trade. But these simple tools suffice to construct the whole of mathematics – that is hardly a triviality! That all the truths of mathematics, that all the techniques of conceptual analysis which mathematics provides for the other sciences, that all these things arise out of the "shallow generalities" of general set theory, surely *that* is truly astounding.

To understand set-theory as a *foundational* theory, to see it from outside, as it were, to grasp its potentialities and its necessary limitations, to see its fundamental concepts for what they are, these things require more than a minimal mathematical aptitude. Indeed they require more than *mathematical* aptitude: what Pascal called "*l'esprit de la géométrie*" will not suffice here; "*l'esprit de finesse*" is required.

It may appear that in presenting these basic axioms of set theory so carefully and at such length I am belabouring the obvious. I am willing

4.4 The status of the principal axioms of set theory

to risk that appearance, however, because in these really fundamental matters the obvious is so frequently overlooked. How often has a presentation of "axiomatic set theory" been prefaced with a remark to the effect that the whole object of the enterprise is to lay down, with absolute formal precision, of course, axioms chosen on the basis that they will allow us to derive the theorems we wish to derive, but without allowing us to derive contradictions? And this is presented as a "foundation" for mathematics – what could be more absurd! Has it never occurred to these gentlemen that if your business is proof, and your aim is to convince others of the truth of your propositions, then you can scarcely afford to choose your first principles on such a basis? Truly, it is precisely the obvious that one must strive somehow to convey.

But are the basic axioms of set theory *obviously* true? Do they possess that *self-evidence* traditionally demanded of, and attributed to, axioms, that is to say, propositions chosen to serve as the starting points of all mathematical argument, propositions whose truth is accepted without proof? Here, of course, it is the finiteness claims that we must judge, not the propositions directly expressed by the axioms themselves.

Surely the axioms of Pair Set, Comprehension, and Replacement are as self-evident as self-evident can be. How could a plurality composed of two elements not be finite? How could a definite subplurality of a finite plurality not be finite? How could each element of a finite plurality be replaced by a single new element without the resulting plurality also being finite?

The remaining basic axioms, the Axioms of Union and of Power Set, are not self-evident in this immediate and overwhelmingly obvious sense. Indeed, I called attention to this fact when I introduced them. We can at least say that the finiteness claims laid down in these axioms are highly plausible – so it seems to me, in any event, but concerning such matters everyone must make up his own mind. In any case, the Axioms of Union and Power Set are self-evident in this regard, namely that there are no simpler or more plausible principles from which they can be derived: insofar as they are evident at all, they must be accounted *self*-evident. We must accept them as *axioms* or not at all.

It is well to bear in mind, when considering these matters, that none of the axioms actually *lays claim* to its own self-evidence: what each axiom expresses is a proposition about sets. Moreover, in laying down these axioms our primary claim, indeed, in the last analysis, our only claim is that they are true (or, more precisely, that their instances are true, since they are all Π_1 propositions). Self-evidence is a subjective matter – at least, there is a subjective element in judgements of self-evidence. But truth is altogether different. It is objective; and we cannot abandon our claims to truth without abandoning mathematics itself.

It might appear that the addition of Cantor's Axiom renders the truth claims embodied in the Axioms of Union and Power Set even more problematic. The other basic axioms retain their fully self-evident character even under Cantor's revolutionary assumption that the natural numbers, suitably defined, are finite in number and therefore form a set. But Power Set, in particular, seems to increase in strength, and thereby to decrease in plausibility, in the presence of Cantor's Axiom. I say "seems" here, because I am convinced that, on the contrary, the case for Power Set remains the same whether on Cantorian or Euclidean assumptions. The subsets of a plurality of determinate size are simply "there", in whatever multitude, definite or indefinite, they may compose. Surely it is implausible to suppose that multitude to be absolutely infinite. Surely we cannot conceive that the absolutely infinite could be rooted in, or could emerge from, a particular instance of the finite in such a manner. That is the force of the Power Set Axiom, or so it seems to me. And, of course, the case for the Axiom of Union is similar, but even stronger.

4.5 Ordered pairs and Cartesian products

With the basic axioms of set theory now in place, I can begin my general exposition of the theory. The first task that confronts me is to explicate certain fundamental notions, namely, the notions of *ordered pair, Cartesian product, local relation* and *local function*.

Now the definitions of these notions that I shall present are well known to all mathematicians. But those definitions, which can all be given explicitly in terms of the primitive notions already introduced, constitute solutions to problems that are of absolutely fundamental importance to the whole enterprise of set-theoretical foundations. And although the *solutions* are well known, the nature of the *problems* to which they are the solutions is not at all well understood.

The key to understanding the logical status of all these notions is to recognise that they are *symbol generated*: our conception of the "abstract objects" that fall under such concepts arises out of and reflects the operations we perform on those symbols, and the syntactic and algorithmic rules which govern those operations. In effect, we project our symbols and symbol combinations onto the world, positing "abstract objects" as the objective things named by them. The "real" properties of these projected

4.5 Ordered pairs and Cartesian products

entities are determined by the syntactic properties of the symbolism that gives rise to them.

The use of symbol generated concepts pervades the whole of mathematics. A straightforward and important technical example of this is the concept of a polynomial ring, $F[x]$, over a field F. The elements of such a ring are polynomials in the variable x over the field F, that is to say, intuitively, *expressions* of the form

$$a_n x^n + a_{n-1} x^{n-1} + \cdots + a_1 x + a_0$$

where n is a particular natural number (e.g. 17 or 297) and where "$a_0, \ldots, a_n$" are names for particular elements of the field F. The field operations here are "mathematical objects" that correspond to the syntactic rules that govern the addition and multiplication of polynomials.

Of course it cannot be literally true that we are dealing with expressions – syntactic assemblages – in such cases. If F is the field of real numbers, for example, then we should need uncountably many distinct symbols as names for elements in F in order to form the polynomials in $F[x]$, and that is clearly impossible. Still, there are various well known ways of making this sort of concept precise, standard set-theoretical "tricks", in fact, although when we resort to such devices in pursuit of logical rigour we inevitably leave the original intuition behind, and with it, perhaps, the feeling that what we are doing is straightforward and natural.

Other technical examples of such symbol generated concepts are easy to find: matrix rings and free groups come immediately to mind. In every case the concept in question can easily be given rigorous formulation using standard set-theoretical methods. But when we are dealing with really basic concepts of this symbol generated sort in set theory itself – when we are actually assembling our bag of tricks, so to speak – then these familiar methods are not yet available, and we must accordingly think things through more thoroughly.

Such is the problem that confronts us when we come to deal with the concept of ordered pair. For here we have a concept that occurs everywhere in mathematics and that plays a crucial, indispensable role in mathematical foundations; and yet it is quite clearly syntactic in origin.

This is a very important matter indeed: for *whoever does not grasp the fact that the concept of ordered pair is symbol generated, and does not understand the significance of that fact, cannot hope to understand how it is that set theory provides the foundations for contemporary mathematics.*

Consider how the notion of ordered pair arises naturally in co-ordinate geometry; that is certainly where most mathematicians first encounter it.

If we erect a pair of mutually perpendicular axes in a plane, then every point in that plane is uniquely determined by its (signed) distance from each of the two axes. *Syntactic* ordered pairs (if I may describe them in that way) are then naturally employed as *labels* for points in the plane. In the expressions which *are* those labels, the order of the components of the pair is manifest in the syntactic arrangement of the expression itself.

$$(3, -7)$$
$$(4, 2.6)$$
$$\vdots$$

where the first component of the pair gives the distance of the point labelled from the first axis (in some arbitrarily chosen order) and the second from the second. But in Arabic, perhaps, since it is read from right to left, the more natural[6] syntactic order would probably be

$$(-7, 3)$$
$$(2.6, 4)$$
$$\vdots$$

and in Chinese or Japanese it might be more natural to use a vertical arrangement:

$$\begin{pmatrix} 3 \\ -7 \end{pmatrix}$$

$$\begin{pmatrix} 4 \\ 2.6 \end{pmatrix}$$

$$\vdots$$

Which of these *notations* we employ is really just a matter of convenience: we are, after all, dealing with a mere convention for labelling, are we not? Well yes, of course, it is just a question of convention. And yet ... we are entering deep waters here; for we are looking for *objective correlatives* of these syntactic labels, things that remain invariant under all these changes of notation. These objective correlatives – ordered pairs rigorously defined set-theoretically – will provide the key to "arithmetising", and thereby rendering rigorous and objective *all* of

---

[6] Notice that Arabic numerals are written backwards by users of the Latin alphabet, although in Arabic, which is written right to left, the units come first, then the tens, then the hundreds, ...

4.5 Ordered pairs and Cartesian products

our symbol generated concepts: from ordered pairs we shall get Cartesian products, functions and relations; and from these we can construct the mathematical structures that form the subject matter of axiomatic theories.

Let us consider how this procedure of "objectifying" ordered pairs transforms geometry itself. Co-ordinate geometry leads, as is well known, first to the *analysis*, and then to the *replacement* of intuitive (*anschaulich*) geometrical notions by notions that are purely algebraic and arithmetic in character. The final stage in this process is reached when it is realised that the intuitive geometrical notion of the Euclidean plane can be replaced by the notion of the plane as the set, R^2, of all ordered pairs of real numbers: roughly speaking, each geometrical point can be replaced by its label.

But to speak this way is to speak very roughly indeed. For just as real numbers are not to be identified with the signs that designate them, so pairs of real numbers are not to be identified with syntactic pairs of real number names. Simple cardinality considerations rule out both of these identifications. But this presents us with a difficulty: even supposing that we know what real numbers are, do we then know what ordered pairs of real numbers are – I mean know *exactly*?

The operation of forming, say, the syntactic ordered pair "$(3, -7)$" out of the number names "3" and "-7" must somehow be made to correspond to the evaluation of a function at the arguments 3 and -7 (in that order) to obtain the ordered pair $(3, -7)$ which is the object designated by the syntactic pair "$(3, -7)$". Of course we shall need to be able to form ordered pairs of arbitrary objects, not just real numbers, so the pairing function must be a global function. But the question remains: what are to be the *values* of that global function?

The salient point here is this: since the intuition underlying the concept of ordered pair is syntactic, *there is no natural answer to this question*. Given objects a and b there is simply nothing in heaven or earth that naturally answers to the description "the ordered pair whose first member is a and whose second member is b". We are thus completely at liberty to *invent* an answer, for *it is not a question here of replacing a natural concept by an "artificially defined" set-theoretical concept: it is rather a question of laying down a suitable, rigorous definition in set theory in order to explicate a concept, itself artificial, that arises, not from the things themselves, but from our own, necessarily arbitrary, linguistic conventions for discourse about them.*

It would, of course, be possible to *axiomatise* the concept of ordered pair by laying down the conditions that a binary global function would have to meet in order to qualify as a pairing function. But then it would be necessary actually to exhibit a *particular* binary global function satisfying those conditions; for contrary to a surprisingly widespread misapprehension, one cannot conjure "mathematical entities" out of thin air simply by laying down axioms. Thus, in the light of what has been said, such a way of proceeding would be otiose: there would be no point in laying down axioms for a pairing function since there can be no principled objection to any particular definition that does the required job. Accordingly, I shall adopt the familiar Kuratowski definition.

The idea behind Kuratowski's definition of ordered pair is a very natural one, namely, that a linear ordering on a set can be identified with the set of all its non-empty initial segments. Thus if the set in question is $\{a,b,c,d\}$ and the ordering gives

$$b < c < a < d$$

then it can be represented by the set

$$\{\{b\}, \{b,c\}, \{b,c,a\}, \{b,c,a,d\}\}$$

The Kuratowski definition of ordered pair can be obtained by applying this definition to a two-element linear ordering.

4.5.1 Definition *Let a and b be any objects. Then by definition*

$$(a,b) = \{\{a\}, \{a,b\}\}$$

This is the principal definition, but we need two auxiliary definitions as well.

4.5.2 Definition *Let a be any set or individual. Then by definition*

$$a \text{ is an ordered pair .iff. } (\exists x \in \bigcup(a))(\exists y \in \bigcup(a))[a = (x,y)]$$

4.5.3 Definition *Let a be any set or individual. Then by definition*

(i) $1^{st}(a) = \iota(\{x \in \bigcup(a) : (\exists y \in \bigcup(a))[a = (x,y)]\}, \emptyset)$
(ii) $2^{nd}(a) = \iota(\{y \in \bigcup(a) : (\exists x \in \bigcup(a))[a = (x,y)]\}, \emptyset)$

Thus, informally, (a,b) is the ordered pair whose first co-ordinate is a and whose second co-ordinate is b. If a is an ordered pair then $1^{st}(a)$ is its first co-ordinate and $2^{nd}(a)$ is its second co-ordinate. If a is not an ordered pair then $1^{st}(a) = 2^{nd}(a) = \emptyset$.

4.5 Ordered pairs and Cartesian products

The essential property of ordered pairs is given by the following easily established theorem.

4.5.4 Theorem *Let a, b, c and d be any objects Then $(a, b) = (c, d)$ if, and only if, $a = c$ and $b = d$.*

Definitions 4.5.1, 4.5.2, and 4.5.3 specify the objective correlatives of the corresponding syntactic notions. But now the question naturally arises: do these definitions capture notions adequate to serve the purposes for which they were intended? This is a question that calls for a characteristic *kind* of mathematical judgement which is not a *mathematical* judgement in the strict sense.

A strictly mathematical question (Is 43 prime?) admits of a clear-cut, objectively determined answer. But these questions, though they are clearly mathematical questions in *some* sense (what else could they be?), do not have such clear-cut answers. They are more like questions of practical reason.

For example, since ordered pairs as defined in 4.5.1 are sets and not members of some separate category, we may get unexpected, and unwanted, identifications. Thus if we are dealing with a domain which contains an element, a, together with another element, c, which is the singleton of a, then we get the possibly embarrassing identification $(a, a) = \{c\}$. Whether these unintended identifications are sufficiently undesirable, either in general, or in specific cases, to cast discredit on 4.5.1, is not a question that has an objectively determined answer. The received wisdom is that they do not discredit 4.5.1, but the judgement on which that wisdom rests is practical, not apodeictic.

Now that the notion of ordered pair has been defined I can define the notion of the *Cartesian product* of two objects in the familiar way. Of course, the notion is interesting only when the objects in question are both sets.

4.5.5 Definition *Let a and b be any objects. Then by definition the Cartesian product, $a \times b$, of the objects a and b (in that order) is the set of all ordered pairs (u, v) where u lies in a and v lies in b. In symbols*

$$\forall xy[x \times y = \bigcup\{\{(u, v) : u \in x\} : v \in y\}]$$

Now we can turn our attention to the problem of finding objective correlatives for the symbol generated concepts of local function and local relation.

4.6 Local functions and relations

By a *local function* I mean, as a first approximation, a function whose domain of definition is a set (and whose range is therefore, by Replacement, also a set). It is thus to be contrasted with the Fregean conception of *global* function. This notion of local function plays so central a role in modern mathematics that it is scarcely possible to imagine what the subject might be without it.

One often hears it said that such a function is, or is given by, a *rule* that specifies how to associate to each possible argument of the function a value for that argument. Now it is perfectly true that when we wish to illustrate the notion we have in mind here we invariably begin with examples in which the functions are specified by rules. But that cannot be the essence of what it is to be a function in the sense intended here, as a little reflection will disclose. We speak, for example, of the real vector space $C[0, 1]$ whose "vectors" are continuous real-valued functions defined on the closed interval $[0, 1]$; but these functions are vastly too numerous – there are a continuum of them – for all of them to be given by rules, provided, that is, that the word "rule" is not to be stripped of the last vestige of its customary meaning.

The notion of a rule, although it has clear instances, is, as a general notion, too vague and too imprecise to be included among the *elements* of mathematics, where the requirement is for simplicity, clarity, and, above all, objectivity. In laying the foundations of mathematics we must not permit ourselves to become entangled in anthropological, cosmological, or theological fantasies: mathematics cannot rest upon speculations about what men could do if only they weren't subject to the limitations on their capacities that, in fact, they are subject to, or about what would be the case if the universe were different from what we believe it to be, on the best available evidence, or about how things might appear to an omniscient Deity. But any attempt simply to adopt the notion of function-as-rule as an *element* of mathematics (in the sense of Euclid or Aristotle) will, inevitably, involve us in one or more of these fantasies. No, to obtain a mathematically usable notion of local function, one that will serve the purposes of modern mathematics, we must somehow extract an objective essence from the naive conception of a function as a correspondence given by a rule.

Now it might be objected that in Section 3.4 I myself adopted a notion of function-as-rule in applying the Fregean notion of (global) function to set theory. But in fact I did *not* follow Frege in regarding those

4.6 Local functions and relations 131

global functions as *entities* of a special ("unsaturated") sort. On the contrary, I carefully explained that, although we cannot avoid talk about "them" without introducing an intolerable and tedious prolixity into our discourse, nevertheless, at bottom, and in the last analysis, "they" are merely *imaginary* correlatives of function "names", which, on the correct logical analysis, are not *names* at all, but rather *syncategorematic signs* of a special type that reflect and embody our linguistic conventions for naming genuine objects. Such "things" as global functions cannot be the material out of which mathematical structures, such as function spaces, are composed.

However, although we cannot simply *posit* functions-as-rules as elements – objects of a primitive, fundamental sort not requiring definition – nevertheless the idea of a function as a correspondence effected by means of a rule does suggest how a suitable objective correlative to that symbol generated concept can be defined. For we can imagine that in simple cases involving small sets a rule can always take the form of a *table* in which all the arguments of the function are listed and its values are placed opposite their corresponding arguments. Schematically, such a table might consist simply of a list of *syntactic* ordered pairs in which names of the arguments occupy the first place and names of the corresponding values the second.

Such a list for a function defined on a set S with values in a set T would contain, for every member x of S, exactly one pair whose first member was a name for x and whose second member was, of course, a name for the value, y, in T that the function takes at x. Conversely, any list of syntactic ordered pairs satisfying this condition would clearly define a unique function from S to T. Now we need only replace the syntactic ordered pairs in our list by their objective correlatives, the corresponding Kuratowski ordered pairs, to obtain the required objective correlative of the function itself. Thus local functions defined on S with values in T can be *identified with*, that is to say, *defined to be* subsets of the Cartesian product $S \times T$ of a certain special kind.

These considerations are very familiar to mathematicians, who use the notion of local function just described without giving the matter much thought. But it is worth pointing out that this account of local function effectively frees that notion of any but the most tenuous connection with the notion of a rule: it takes the *extensional* connection between the arguments and the corresponding values as the essence of the matter and regards the specification of rules describing that connection as irrelevant to the general concept of local function, though, of course, one can always *employ* such a rule if one wishes to single out a *particular* such function. One is perfectly able to maintain, therefore, that a certain set of local functions is non-empty without being able, even "in principle", to specify a single one of its members by means of a rule, however loose or attenuated a notion of "rule" one is prepared to accept.

The essential point here is that *there is nothing "artificial" about the conventional definition of a function as a set of ordered pairs*. In particular, there is no "natural" or "primitive" notion of function for which this definition of function goes proxy. This definition of local function really constitutes an *analysis* of what is *essential* in our naive notion of local function. The only problem raised by this definition is that of convincing ourselves that when we employ it in our definitions, theorems, and proofs, they actually mean what we think they ought to mean. It seems to me that the familiar story I have given above explains why the standard notion of (local) function has exactly the connection with the vague notion of "rule" that is required.

Everything I have said here about functions applies, *mutatis mutandis*, to relations, since relations, too, can, in simple cases, be specified by lists of syntactic ordered pairs. Here I am speaking about relations that are *local* and *extensional*: local in the sense that the field of objects to which such a relation applies is a set, and extensional in the sense that each such relation is determined solely and exclusively by which objects are related by it.

Although the notion of local function is perhaps the more important, the notion of local relation is logically prior in definition.

4.6.1 Definition *A (local) relation is a set of ordered pairs. In symbols*

$\forall x[x$ *is a local relation* $: iff: x$ *is a set and* $(\forall y \in x)[y$ *is an ordered pair*]]

I must now introduce some standard terminology and notation.

4.6.2 Definition *Let a be any set or individual. Then by definition*

(i) $Domain(a) = \{x \in \bigcup(\bigcup(a)) : (\exists y \in \bigcup(\bigcup(a)))[(x, y) \in a]\}$
(ii) $Range(a) = \{y \in \bigcup(\bigcup(a)) : (\exists x \in \bigcup(\bigcup(a)))[(x, y) \in a]\}$
(iii) $Field(a) = Domain(a) \cup Range(a)$

A local function can now be defined, in the usual way, to be a local relation that satisfies the functionality condition briefly described above.

4.6.3 Definition *A (local) function is a relation with the additional property*

4.6 Local functions and relations

that for each element of its domain there is exactly one element in its range to which that element stands in the given relation. In symbols

$$\forall x [x \text{ is a (local) function} : \textit{iff} : x \text{ is a local relation and}$$
$$(\forall u \in Domain(x))(\exists v \in Range(x))(\forall y \in Range(x))$$
$$[(u, y) \in x \textit{ iff } y = v]]$$

We need some standard notation which we can introduce by defining two global functions, the *evaluation* function

$$\lambda xy(x`y)$$

which assigns to a local function f and an argument a the *value*, $f`a$, of f at a, and the *image* function

$$\lambda xy(x``y)$$

which assigns to a local function f and a set S the *image*, $f``S$, *of S under f*.

4.6.4 Definition

(i) $\forall x \forall y [x`y = \iota(\{z \in Range(x) : (y, z) \in f\}, \emptyset)]$
(ii) $\forall x \forall y [x``y = \{z \in Range(x) : (\exists u \in y)[(u, z) \in x]\}]$

I shall use the standard notation

$$f : S \to T \text{ (read "} f \text{ maps } S \text{ to } T\text{")}$$

to express the proposition that f is a function, S is the domain of f and T is a superset of its range[7].

I shall use the usual notation for the *inverse* of a function or relation ("r^{-1}") and the composition of two functions ("$f \circ g$" – "first apply g, then f"), though I shall assume that these are defined universally

4.6.5 Definition *Let f and g be sets or individuals. Then by definition*

(i) $\forall x [x^{-1} = \{y \in Range(x) \times Domain(x) : (2^{nd}(y), 1^{st}(y)) \in x\}]$
(ii) $\forall xy [x \circ y = \{u \in Domain(y) \times Range(x) : (\exists vwz \in Field(x \cup y))[(v, w) \in y \text{ and } (w, z) \in x \text{ and } u = (v, z)]\}]$

Finally, we may define *set exponentiation*.

---

[7] $f : S \to T$ (1–1) if distinct arguments in S take distinct values in T; $f : S \to T$ (onto) if $Range(f) = T$. If f is (1–1) from S to T it is *injective (from S to T)*, if onto, *surjective*, and if both *bijective*.

4.6.6 Definition *Let a and b be any objects. Then by definition a^b is the set of all functions $f : a \to b$. In symbols*

$$\forall xy [x^y = \{z \in P(y \times x) : z : y \to x\}]$$

4.7 Cardinality

One of the most important, and most ancient, of mathematical ideas is that of the *size* or *cardinality* of a set. Indeed, the point of view we are taking here is that sets are precisely those pluralities that *have* sizes. The mathematical problem is to determine how to *compare* sizes[8].

The conventional view is that the size of a set (*arithmos*) is determined by counting out its elements (units) one by one in the familiar way. Modern mathematics began when Cantor noticed that equality of size could be *defined* independently of the notion of *counting out*.

4.7.1 Definition *Let S and T be any objects. Then by definition S is equinumerous with T if, and only if, there is a function $f : S \to T$ that is both one to one and onto. In symbols*

$$\forall XY [X \cong_c Y \text{ iff } (\exists z \in Y^X)[z : X \to Y (1\text{-}1, \text{ onto})]]$$

Thus S is equal to T in size or *cardinality* (in symbols, $S \cong_c T$) if there is a function from S to T which places S into *one-to-one correspondence* with T. It is straightforward to show that for sets S and T, if $S \cong_c T$ then $T \cong_c S$. Moreover, the identity function $id_S : S \to S$ bears witness to the fact that $S \cong_c S$, and if $S \cong_c T$ and $T \cong_c U$, then $S \cong_c U$; in short, $\lambda xy[x \cong_c y]$ is a *global equivalence relation* on the species of all sets.

Cantor was not the first to notice that *sameness of size* could be defined in terms of one-to-one correlation. But he was the first to grasp the fundamental significance of this fact and to exploit it in carrying out profound mathematical investigations. In this sense we may say that 4.7.1 was Cantor's discovery.

Why *discovery*? How can you *discover* a definition? In mathematics,

---

[8] The Greeks took the view that, in effect, *equality of size* is a primitive, given notion that does not require, or admit of, definition. That a particular *arithmos* is a ten or a three hundred and twenty-seven is just a brute fact about that *arithmos*, just as it is a brute fact about a particular animal that it is a horse or a man. *Arithmoi* are "the same" (i.e. in number) if they belong to the same numerical species, just as animals are "the same" (i.e. in kind) if they belong to the same animal species. Aristotle makes this point in *Physics IV* (224a2–16).

4.7 Cardinality

what you can discover is *the right way to analyse a fundamental concept*, a concept that may be only imperfectly grasped or whose real significance or utility may not be fully apparent before it is properly analysed. But how do we ever know that a proposed analysis is the correct one? There is no general answer to this question: each such analysis must be judged in terms of the particular aims it is intended to advance.

Cantor's *discovery* was that the relation, $\lambda xy[x \cong_c y]$, of "sameness of size", defined in terms of one-to-one correspondence, *is logically prior to the notion of counting or, indeed, to the notion of "number", either in the modern sense of "natural number" or in the Greek sense of "number species"*. This was the central insight that enabled him to invent modern mathematics by generalising the classical conception of finiteness.

Notice that operationalistic talk about *placing* the elements of one set into one-to-one correspondence with those of another, or of *counting out* sets to determine their sizes disappears when the rigorous definition of "sameness of size" is given. In that definition $S \cong_c T$ is stipulated to mean that a certain subset of the power set, $P(S \times T)$, of the Cartesian product, $S \times T$, of S and T is non-empty, once the meaning of the local existential quantifier is unpacked. Thus the definition in 4.7.1 is rigorous, static, set-theoretical – untainted by operationalism or anthropocentrism.

The question whether two sets are equal in size has a definite answer that is completely determined independently of anything that anyone has to do or to think or to calculate. This is as it should be: we cannot make the definition of cardinal equivalence depend on our capacities to correlate or to count, because those capacities are severely limited. Nor can we appeal to what we could correlate or count "in principle" or "in theory", because that only raises the questions "What principle?", "What theory?". Activities, such as correlating, counting, and calculating, may be important, even crucial, parts of the practice of mathematics, but they cannot form its actual subject matter: they may be what mathematicians *do*, but they cannot be what mathematics is *about*.

Of course, to grasp the significance of Definition 4.7.1 – to see that it represents the correct analysis of the concept *sameness of size* for sets – requires insight. One must understand the significance of the concept of local function, for example, and therefore of the concept of ordered pair, and so on.

Thus the chain of definitions leading up to 4.7.1 constitutes an indispensable part of the analysis. Cantor was not alone responsible for the end product: Zermelo and Kuratowski, in particular, had important contributions to make. Insight and judgement are required at every stage of the process leading up to the final analysis of cardinal equivalence, if the significance of Definition 4.7.1 is to be grasped. And, of course, these things cannot be supplied along with the axioms and definitions.

There is a natural definition of "smaller in size (cardinality) than" in the spirit of Definition 4.7.1.

4.7.2 Definition *Let S and T be any objects. Then by definition S is less than, or equal to T in cardinality if, and only if, S is equinumerous with some subset of T. In symbols*

$$\forall X Y [X \leq_c Y \ .\textit{iff.} \ (\exists Z \in P(Y))[Z \cong_c X]]$$

If $S \leq_c T$ but $S \not\cong_c T$, then we say that S is *stricly smaller in cardinality than* T $(S <_c T)$.

It is not possible, in general, to show that

$$S <_c T \ or \ S \cong_c T \ or \ T <_c S$$

without laying down further axioms. In the Cantorian version of the theory this follows from the Axiom of Choice; in the Euclidean version of the theory the Axiom of Choice itself can be established, but its proof rests on the Axiom of Euclidean Finiteness, which is the assumption that no set is of the same size as any of its proper subsets.

The next theorem is of fundamental significance, for it shows us that there is no largest size among sets[9]. Its proof requires only the axioms laid down so far, and thus it holds in both the Cantorian and the Euclidean conceptions of finiteness. In the Cantorian theory it gives us transfinite sets of differing sizes.

4.7.3 Theorem *Let S be any set. Then* $S <_c P(S)$.

Proof Let S be any set. Then $\{\{x\} : x \in S\} \subseteq P(S)$ is clearly equal to S in cardinality. But $S \not\cong_c P(S)$. For if $f : S \to P(S)$ is any function whatever from S to its power set, then the set

$$S_0 = \{x \in S : x \notin f`x\} \in P(S)$$

is not in the range of f, since if $x_0 \in S$ with $f`x_0 = S_0$, then $x_0 \in S_0$ if, and only if, $x_0 \notin S_0$. Thus no function $f : S \to P(S)$ is onto, *a fortiori*, no such function is one to one and onto. □

This simple and beautiful argument is pregnant with consequences.

---

[9] See Cantor's article "Über eine elementare Frage der Mannigfaltigkeitslehre" (p.278 in the *Gesammelte Abhandlungen*).

4.8 Partial orderings and equivalence relations

In the following chapters I shall make extensive use of the familiar notions of *equivalence relation* and *partial ordering*. I need to define them and to show that certain of their elementary properties can be established on the basis of the axioms laid down so far.

4.8.1 Definition *Let r be a local relation. Then by definition*

(i) *r is reflexive* : *iff* : $(\forall x \in Field(r))[(x,x) \in r]$
(ii) *r is symmetric* :*iff*: $(\forall xy \in Field(r))[(x,y) \in r \text{ iff } (y,x) \in r]$
(iii) *r is anti-symmetric* : *iff* :

$$(\forall xy \in Field(r))[(x,y) \in r \text{ and } (y,x) \in r \text{ .implies. } x = y]$$

(iv) *r is transitive* : *iff* :

$$(\forall xyz \in Field(r))[(x,y) \in r \text{ and } (y,z) \in r \text{ .implies. } (x,z) \in r]$$

(v) *r is an equivalence relation* : *iff* : *r is reflexive, symmetric, and transitive.*
(vi) *r is a partial ordering* : *iff* : *r is reflexive, anti-symmetric, and transitive.*

I shall adopt the convention that if **r** denotes an equivalence relation I shall usually write $s \equiv_r t$ rather than $(s,t) \in r$. Similarly, if **r** denotes a partial ordering I shall usually write $s \leq_r t$ or $s <_r t$ rather than $(s,t) \in \mathbf{r}$ or $[(s,t) \in \mathbf{r}$ *and* $s \neq t]$, respectively.

If *r* is an equivalence relation and $Field(r) = S$ then *r* is said to be an equivalence relation *on S*; if *r* is a partial ordering and $S = Field(r)$, then *r* is said to be a partial ordering *of S* or *on S*.

4.8.2 Definition *Let r be an equivalence relation and let S be a subset of Field(r). Then by definition*

(i) *S is an equivalence class of r* .*iff*.

$$(\exists x \in Field(r))[S = \{y \in Field(r) : y \equiv_r x\}]$$

(ii) $Field(r)/r = \{X \subseteq Field(r) : X \text{ is an equivalence class of } r\}$

$Field(r)/r$ can also be denoted by "$Field(r)/\equiv_r$". Clearly every element x of the field of an equivalence relation *r* belongs to exactly one equivalence class of *r*.

Now let us turn our attention to partial orderings. First I need to establish some notation.

4.8.3 Definition *Let r be a partial ordering and $a \in Field(r)$. Then by definition*

(i) $r_{\leq_a} = \{x \in r : 2^{nd}(r) \leq_r a\}$

(ii) $r_{<_a} = \{x \in r : 2^{nd}(r) <_r a\}$

It is obvious that if r is a partial ordering, then both $r_{\leq_a}$ and $r_{<_a}$ are partial orderings as well. In fact each is an *initial segment* of r in the sense of the following definition.

4.8.4 Definition *Let r and r_1 be partial orderings. Then by definition r_1 is an initial segment of r if, and only if,*

(i) $r_1 \subseteq r$

and

(ii) $(\forall x \in Field(r_1))(\forall y \in Field(r))[y \leq_r x \text{ implies } y \in Field(r_1)]$

Sometimes the term "initial segment" is used to refer to the *field* of an initial segment in the sense of Definition 4.8.4; in such cases I shall say that a set S is an initial segment of $Field(r)$[10].

The next theorem gives us an unlimited supply of partial orderings.

4.8.5 Theorem *Let S be any set all of whose members are sets. Then the relation*

$$\{x \in S \times S : 1^{st}(x) \subseteq 2^{nd}(x)\}$$

is a partial ordering.

Partial orderings of this sort arise with sufficient frequency to justify introducing a special notation.

4.8.6 Definition *Let S be any set or individual. Then by definition*

$$\subseteq_S = \{x \in S \times S : 1^{st}(x) \subseteq 2^{nd}(x)\}$$

Note that if S contains more than one member that itself has no members $\subseteq_S$ is not anti-symmetric; otherwise it is a partial ordering. In fact, these partial orderings $\subseteq_S$ where S is a set of sets are completely representative of all partial orderings, as I shall show below.

---

[10] So if S is an initial segment of $Field(r)$, then $r \cap S \times S$ is an initial segement of r (in the sense of Definition 4.8.4), and conversely.

4.8.7 Definition *Let r and q be partial orderings and*

$$m : Field(r) \to Field(q)$$

Then by definition m is an order morphism (or is order preserving) from r to q if, and only if,

$$(\forall xy \in Field(r))[x \leq_r y \text{ .implies. } m`x \leq_q m`y]$$

If, in addition, m is one to one and onto, and m^{-1} is also an order morphism, then m is an order isomorphism from r to q.

Clearly, if m is an order isomorphism from r to q, then m^{-1} is an order isomorphism from q to r. Thus it makes sense to speak of two partial orderings being (*order*) *isomorphic*. This notion is of sufficient importance to warrant the introduction of a special notation.

4.8.8 Definition *Let r and s be partial orderings. Then by definition*

$$r \cong_o s : iff : (\exists x \in Field(s)^{Field(r)})[x \text{ is an isomorphism from } r \text{ to } s]$$

The expression "$r \cong_o s$" is read "r is order isomorphic to s" or "r is order equivalent to s".

Isomorphic partial orderings are abstractly indistinguishable, so to speak, since their fields can be placed in one-to-one correspondence in a manner such that corresponding elements have exactly analogous order relations with the other members of the respective fields. From this standpoint the following theorem shows us that the "inclusion partial orderings" $\subseteq_S$, where S is a set of sets, are completely representative of partial orderings in general, in the way that transformation groups are completely representative of groups.

4.8.9 Theorem *Let r be a partial ordering and let $S = \{Field(r_{\leq_x}) : x \in Field(r)\}$. Then $r \cong_o \subseteq_S$.*

Of especial importance for set theory is the subspecies of the partial orderings consisting of the *well-orderings*. These form the subject matter of the next section.

4.9 Well-orderings and local recursion

A well-ordering is a partial ordering satisfying two additional conditions: linearity and well-foundedness.

4.9.1 Definition *Let r be a partial ordering relation. Then by definitio*

(i) *r is linear .iff.* $(\forall xy \in Field(r))[x \leq_r y \text{ or } y \leq_r x]$
(ii) *r is well-founded .iff.* $(\forall X \subseteq Field(r))$

$$[X \neq \emptyset : implies : (\exists y \in X)(\forall z \in X)[z \leq_r y .implies. z = y]]$$

(iii) *r is a well-ordering iff r is linear and well-founded.*

Much of the general theory of well-orderings is common to the Cantorian and Euclidean versions of set theory. Indeed, parts of that theory do not depend in any essential way on the linearity condition, and hence can be applied to arbitrary well-founded partial orderings. This is the case with the next theorem.

4.9.2 Theorem *(Proof by Induction along a Well-founded Partial Ordering) Let r be a well-founded partial ordering and let Φ be a first level global propositional function of one argument. Then*

$$(\forall x \in Field(r))[\;[(\forall y \in Field(r))[y <_r x \text{ implies } \Phi(y)] .implies. \Phi(x)]$$
$$: implies : (\forall x \in Field(r))\Phi(x)\;]$$

This theorem is proved by showing that the supposition that the set $\{x \in Field(r) : not\Phi(x)\}$ is not empty leads to a contradiction (consider a minimal element of that set).

Note that this theorem does *not* apply to arbitrary properties Φ, but only to propositional functions. In particular it does not hold if Φ is a Π_1 or Σ_1 property, for in those cases Brouwer's Principle dictates that we cannot, in general, infer the existence of $\{x \in Field(r) : not\Phi(x)\}$ as a *set*. This has important consequences for the foundations of set theory, as I shall show.

The first of these consequences is that we can establish the familiar technique of definition by recursion only for *local* functions, not for *global* ones. First we need a definition.

4.9.3 Definition *Let S be a set and f a local function. Then by definition f is a partial function on S if, and only if, Domain(f) $\subseteq$ S.*

I can now state the following fundamental theorem.

4.9.4 Theorem *(Definition by Recursion along a Well-founded Partial Ordering) Let r be a well-founded partial ordering, S be a set, T be the set of all partial functions on the field of r with values in S, and let $g : T \to S$.*

4.9 Well-orderings and local recursion

Then there is exactly one function $f : Field(r) \to S$ *which satisfies the recursion equation*

$$f`x = g`\{(y, f`y) : y <_r x\}$$

for all $x \in Field(r)$.

Proof The proof proceeds by defining a partial function $h \in T$ to be a *beginning segment* (i.e. of the function f sought for) if

(1) The domain of h is an r-initial segment of $Field(r)$
(2) $(\forall x \in Domain(h))[h`x = g`\{(y, h`y) : y <_r x\}]$

The desired conclusion can then be established by using induction (Theorem 4.9.2) to show that

(1) Any two beginning segments agree on the common part of their domains.
(2) Every element in $Field(r)$ lies in the domain of some beginning segment.

For then the required local function f can be taken to be the union of the set of all beginining segments. □

This argument depends essentially on the recursion being carried out with respect to a *local* function, g, whose values all lie in a set S given in advance, rather than on a *global* function, σ. If we were trying to prove the existence of an f satisying

$$f`x = \sigma(\{(y, f`y) : y <_r x\})$$

the standard proof I have just sketched would fail. For although the proof of (1) would go through essentially as before, the proof of (2) would founder on the fact that *lying in the domain of some beginning segment* is a Σ_1 property to which induction (Theorem 4.9.2) cannot be applied. Again, it is Brouwer's Principle that is operating here.

Note that the proofs of Theorems 4.9.2 and 4.9.4 are straightforward, set-theoretical proofs, and make no appeal to operationalistic fantasies of "running through" the elements of a well-ordering. The fact that these standard, set-theoretical arguments to establish transfinite induction and transfinite recursion work only for local functions has profound consequences for both the Cantorian and Euclidean versions of set theory. This is a theme to which I shall regularly return in later chapters.

Now we must turn to the general theory of well-orderings. If r is a well-ordering then the well-foundedness and linearity guarantee that each non-empty subset S of the field of r has a least element $\min(S)$. Moreover, if a lies in the field of r and a is not the greatest element of that field, then a has a unique *successor* element, $a +_r 1$, namely

$$a +_r 1 = \min\{y \in Field(r) : a <_r y\}$$

The following easily established lemma is fundamental.

4.9.5 Lemma *Let r be a well-ordering and let $f : Field(r) \to Field(r)$ be order preserving. Then*

$$(\forall x \in Field(r))[x \leq_r f`x]$$

This lemma follows from the simple observation that if

$$\{x \in Field(r) : f`x <_r x\} \neq \emptyset$$

and a is its minimal element, then $f`a <_r a$, and $f`f`a <_r f`a$ (since f is order preserving).

Now we need the notion of a *full* order preserving map, i.e. one whose image is an initial segment in its codomain.

4.9.6 Definition *Let r and s be partial orderings and let $f : Field(r) \to Field(s)$ be order preserving. Then by definition*

$$f \text{ is full .iff. } (\forall x \in Field(r))(\forall y \in Field(s))[y <_r f`x : implies : \\ (\exists z \in Field(r))[y = f`z]]$$

Clearly, $f : Field(r) \to Field(s)$ is full if, and only if, its range, $f``Field(r)$, is an initial segment of $Field(r)$.

4.9.7 Theorem *Let r and s be well-orderings, and let $f, g : Field(r) \to Field(s)$ both be order preserving and full. Then $f = g$.*

This follows from the simple observation that since f and g are order preserving, $f^{-1} \circ g$ and $g^{-1} \circ f$ are also order preserving. This allows us to apply Lemma 4.9.5 to obtain the desired result.

So given well-orderings r and s there is at most one full order morphism $f : Field(r) \to Field(s)$. In particular, since the identity function on $Field(r)$ is a full order morphism from r to r we have the following important corollary.

4.9.8 Corollary *Let r be a well-ordering and let $a \in Field(r)$. Then r is not order isomorphic to $r_{<_a}$.*

4.9 Well-orderings and local recursion

These results suggest the following definition.

4.9.9 Definition *Let r and s be well-orderings. Then by definition*

(i) $r \leq_o s \; : iff : \; (\exists x \in Field(s)^{Field(r)})$
    [x is full and order preserving from r to s]
(ii) $r <_o s \; : iff : \; r \leq_o s$ and $r \not\cong_o s$

The global relation $\lambda xy[x \leq_o y]$ induces a kind of partial ordering on the "equivalence classes" of the species of all well-orderings under the equivalence relation $\cong_o$, as the next theorem shows.

4.9.10 Theorem *Let r, s and t be well-orderings. Then*

(i) $r \leq_o r$
(ii) $r \leq_o s$ and $s \leq_o r$: implies : $r \cong_o s$
(iii) $r \leq_o s$ and $s \leq_o t$: implies : $r \leq_o t$
(iv) not $[r <_o r]$
(v) $r \cong_o s$ and $r \leq_o t$: implies : $s \leq_o t$
(vi) $r \cong_o s$ and $t \leq_o r$: implies : $t \leq_o s$
(vii) $r \leq_o s$ or $s \leq_o r$

Proof (vii) is easily established, for if we suppose that both r and s are non-empty and that $not[s \leq_o r]$, then we can prove by induction along r (Theorem 4.9.2) that $r_{\leq_x} \leq_o s$ for every $x \in Field(r)$. The union of the embeddings witnessing these inequalities then gives us an embedding witnessing the truth of $r \leq_o s$. The other parts of the theorem are even more straightforward. □

We can define the "arithmetical" operations of addition and multiplication on well-orderings.

4.9.11 Definition *Let r and s be well-orderings. Then by definition*

(i) $r +_o s$ is the relation on $(Field(r) \times \{0\}) \cup (Field(s) \times \{1\})$[11] *given by stipulating that for all $x, y \in Field(r)$ and $u, v \in Field(s)$:*

(a) $((x, 0), (y, 0)) \in r +_o s \;.iff.\; x \leq_r y$
(b) $((x, 0), (u, 1)) \in r +_o s$
(c) $((u, 1), (v, 1)) \in r +_o s \;.iff.\; u \leq_s v$
(d) $((u, 1), (x, 0)) \notin r +_o s$

---

[11] $0 = \emptyset$ and $1 = \{\emptyset\}$, here.

(ii) $r \times_o s$ is the relation on $Field(r) \times Field(s)$ given by stipulating that for all $x, y \in Field(r)$ and $u, v \in Field(s)$:

$$((x, u), (y, v)) \in r \times_o s \; :iff: \; [x <_r y \text{ or } [x = y \text{ and } u \leq_r v]]$$

It is not difficult to establish the basic properties of these operations.

Now we must turn to consider well-orderings of a special kind: the von Neumann well-orderings. These have an especially important role to play in both the Cantorian and the Euclidean versions of set theory. In the Cantorian theory we need to develop some of the theory of these well-orderings in order to state Cantor's Axiom (Axiom 5.2.2).

4.10 Von Neumann well-orderings and ordinals

If r is a well-ordering, then r is order isomorphic to the ordering $\subseteq_S$, where $S = \{Field(r_{<_x}) : x \in Field(r)\}$: the isomorphism associates each element of the field of r with the set of elements of the field of r that precede it. This suggests that we consider well-orderings r in which each element of the field of r simply *is* the set of elements of the field of r that precedes it. Such well-orderings are called *von Neumann well-orderings* and the elements of their fields are called *von Neumann ordinals*.

4.10.1 Definition *Let r be a well-ordering. Then by definition*

r is a von Neumann well-ordering
.iff. $(\forall x \in Field(r))[x = \{y \in Field(r) : y <_r x\}]$

Notice that if r is a von Neumann well-ordering, then $\leq_r \; = \; \subseteq_{Field(r)}$ and $<_r \; = \; \in_{Field(r)} \; = \; \{(x, y) \in Field(r) \times Field(r) : x \in y\}$. Conversely, if r is a well-ordering such that $<_r \; = \; \in_{Field(r)}$ and r is transitive (i.e. $(\forall x \in Field(r))[x \subseteq Field(r)]$) then r is a von Neumann well-ordering.

4.10.2 Definition *Let S be a set. Then by definition*

S is a von Neumann ordinal .iff. $\subseteq_S$ is a von Neumann well-ordering

The theory of von Neumann well-orderings belongs to that part of set theory that lies just beyond what is the common property of all mathematicians. Since there are many excellent treatments of that theory, I do not feel it necessary for me to carry out a detailed exposition of it here. I do, however, need to convince the reader that the elementary part of the theory can be carried out without Cantorian assumptions.

4.10 Von Neumann well-orderings and ordinals

Indeed, it is necessary to have the definition of the von Neumann natural numbers in place when Cantor's Axiom is laid down.

The following theorem is immediate from the definition.

4.10.3 Theorem *Let r be a von Neumann well-ordering.*
  (i) $\emptyset$ *is the least element of r.*
  (ii) *If x lies in the field of r and x is not the greatest element of r, then* $x +_r 1 = x \cup \{x\}$.
  (iii) *If S is a non-empty subset of the field of r and S is bounded above then* $\bigcup S$ *is the least upper bound of S in r.*
  (iv) *If S and T are both members of the field of r, then* $S \in T$ *or* $S = T$ *or* $T \in S$.

The definition of von Neumann ordinal I have given here is not the usual one, but it is equivalent to it, as the following theorem of R. M. Robinson shows.

4.10.4 Theorem *S is a von Neumann ordinal if, and only if, S is a set of sets and*
  (i) $(\forall X \in S)[X \subseteq S]$ *(i.e. S is transitive)*
  (ii) $(\forall XY \in S)[X \in Y \text{ or } X = Y \text{ or } Y \in X]$
  (iii) $(\forall U \subseteq S)[U \neq \emptyset : implies : (\exists X \in U)[X \cap U = \emptyset]]$

Proof If S is a von Neumann ordinal, then (i), (ii), and (iii) follow immediately. Conversely, suppose that S is a set of sets satisfying (i), (ii), and (iii).

Claim 1 If $Y \in X \in S$, then $Y \subseteq X$.

Pf. Suppose $Z \in Y \in X \in S$. Then $Y, Z \in S$, and, by (ii), $Z \in X$, $Z = X$, or $X \in Z$. But if $Z = X$ or $X \in Z$, $\{X, Y, Z\}$ violates condition (iii). Hence $Z \in X$, and the claim follows.

Claim 2 If $X \in S$, then $X \notin X$.

Pf. If $X \in S$ and $X \in X$, then $\{X\}$ violates condition (iii).

Claim 3 $(\forall U \subseteq S)[U \neq \emptyset : implies : (\exists X \in U)(\forall Y \in U)[Y \subseteq X \text{ implies } Y = X]]$

Pf. Let $\emptyset \neq U \subseteq S$. Using condition (iii), pick $X \in U$ such that $X \cap U = \emptyset$. Then X is $\subseteq_S$-minimal in U. For suppose $Y \in U$ and $Y \subseteq X$. Then by condition (ii), $Y \in X$, or $Y = X$, or $X \in Y$. But $Y \notin X$ as otherwise $Y \in X \cap U = \emptyset$. If $X \in Y \subseteq X$, then $X \in X$, and this has been ruled out by Claim 2. Hence, $Y = X$ as required. It follows that X is $\subseteq_S$-minimal, and the claim follows.

Claim 4 If $X, Y \in S$, then $X \subseteq Y$, or $X = Y$, or $Y \subseteq X$.

Pf. If $X \neq Y$ then by condition (ii), $X \in Y$ or $Y \in X$. If $X \in Y$ then $X \subseteq Y$ (by Claim 1), and if $Y \in X$, then $Y \subseteq X$.

Claim 5 If $X \in S$, then $X = \{Y \in S : Y \subseteq X \text{ and } Y \neq X\}$.

Pf. $X \subseteq \{Y \in S : Y \subseteq X \text{ and } Y \neq X\}$ by Claims 1 and 2. To prove the converse inclusion, suppose that $Y \in S$ with $Y \subseteq X$ and $Y \neq X$. Then by condition (ii), $Y \in X$ or $X \in Y$. But if $X \in Y$, then $X \in X$, and this contradicts Claim 2. Hence $Y \in X$, and the claim follows.

By Theorem 4.8.5, $\subseteq_S$ is a partial ordering; by Claims 3 and 4 it is a well-ordering; by Claim 5 it is a von Neumann well-ordering. Hence S is a von Neumann ordinal. □

The characteristic feature of von Neumann well-orderings is that they all start out the same: $\emptyset$ is always the first element in the field of a non-empty von Neumann well-ordering, $\{\emptyset\}$ the second element, $\{\emptyset, \{\emptyset\}\}$ the third, $\{\emptyset, \{\emptyset\}, \{\emptyset, \{\emptyset\}\}\}$ the fourth, and so on. If S is such an element, then $S \cup \{S\}$ is the next element in succession. Moreover, if S is an element of $Field(r)$, where r is a von Neumann well-ordering, then $\subseteq_S$ is a von Neumann well-ordering. It is as if there were a uniform "generating process" for all these well-orderings, as the following theorem shows.

4.10.5 Theorem

(i) $\emptyset$ *is a von Neumann ordinal.*

(ii) *If S is a von Neumann ordinal, then so is $S \cup \{S\}$.*

(iii) *If S is a non-empty set of von Neumann ordinals, then $\bigcap S$ is a von Neumann ordinal.*

(iv) *If S is a set of von Neumann ordinals, then $\bigcup S$ is a von Neumann ordinal.*

(v) *If S and T are von Neumann ordinals with $S \subseteq T$, then either $S = T$ or $S \in T$.*

(vi) *If S and T are von Neumann ordinals, then $S \in T$ or $S = T$ or $T \in S$.*

Note that in (iii) the ordinal $\bigcap S$ is the smallest element of S, and in (iv) the ordinal $\bigcup S$ is the *least upper bound* or *supremum* of S.

4.11 The Principle of Regularity

A set is an *extensional* plurality: there is nothing to the set whose members are $x, y, z, \ldots$ except that its members *are* $x, y, z, \ldots$, and, of course, that

4.11 The Principle of Regularity

they are finite in multitude, in the appropriate sense of "finite". Even though we may use a mathematical property to single out the members of a set, that property is extraneous to what the set is as an extensional plurality.

But the members of a set must be well-defined[12] objects, and this means that the sets among those members must, in their turn, be well-defined extensional pluralities composed of well-defined objects. If there are sets among the latter, they too must have well-defined objects for members, so the sets among *them*, in their turn, must be well-defined as extensional pluralities ...

These observations strongly suggest that there can be no *membership loops*

$$X \in Y \in Z \in \cdots \in W \in X$$

or *indefinitely descending membership chains*

$$\cdots \in W \in Z \in Y \in X$$

among sets. If either case arose we could form no coherent conception of what the set X was *as an extensional plurality*: it would violate the fundamental principle that there can be no vicious circles or indefinite regresses in essential definitions[13].

The question whether such loops or chains exist never arises in practice precisely because no one can form any conception of what the sets composing them could be. Besides, when we employ sets in the various branches of mathematics we never bother about what the members of our sets *are* unless they are special sets (e.g. local functions), in which case the case they will be related to "given" sets in some clearly specified way so that pathologies of the sort under discussion can be seen not to arise.

In fact, the quasi-constructive way in which sets are employed in mathematics means that questions of the sort discussed above will never arise naturally. Still, we cannot *prove*, on the basis of the axioms laid

---

[12] When I use the expression "well-defined" here I am alluding to *essential* definition, not *nominal* or *verbal* definition. Thus something may be *well-defined* in this sense, that is to say, objectively determined and delimited – separated off from everything else – without even *having* a verbal definition. (In Cantorian set theory cardinality considerations alone require that there must be sets well-defined in this sense which have no nominal definition.)

[13] What the pair $\{a,b\}$ is, as an *extensional* plurality, is completely determined by which object a is and which object b is. How, then, could there be such a pair $\{a,b\}$ in which, for example, $a = \{a,b\}$? What extensional plurality, what *arithmos*, could a be in those circumstances?

down so far, that these pathological cases cannot arise. Accordingly, we must lay down an additional axiom to that effect.

But here we encounter a technical difficulty. The characterisation of membership loops and chains given above is not, as it stands, mathematically precise, as the use there of the dots of ellipses discloses. The technical solution to this difficulty takes different forms depending on whether the Cantorian or the Euclidean version of finiteness is adopted. The following Principle of Regularity, however, holds true independently of which notion of finiteness we adopt, and is common to both versions of set theory.

4.11.1 Axiom *(The Principle of Regularity)*

$$\forall X [X \text{ is a set and } X \neq \emptyset \text{ .implies. } (\exists y \in X)[X \cap y = \emptyset]]$$

The condition laid down here seems quite different from the condition banning loops and chains discussed above; but if the latter is properly formulated they are equivalent, in the presence of suitable further assumptions.

The naive argument for this equivalence is easily put. Given a membership loop

$$x \in y \in z \in \cdots \in w \in x$$

the set

$$\{x, y, z, \ldots, w\}$$

violates Regularity. Similarly, given a descending membership chain

$$\cdots \in w \in z \in y \in x$$

the set

$$\{\ldots, w, z, y, x\}$$

violates Regularity.

Conversely, if X is a set which violates the Principle of Regularity, then for each $Y \in X$, $Y \cap X \neq \emptyset$. Hence there is $Y_1 \in X$, $Y_2 \in Y_1 \cap X$, $Y_3 \in Y_2 \cap X$, and so on. Thus

$$\cdots \in Y_3 \in Y_2 \in Y_1 \in X$$

is either an unlimited membership descent or gives rise to a membership loop (in the case $Y_i = Y_j$ for $i \neq j$).

Of course this is not a valid argument on the basis of the axioms laid

4.11 The Principle of Regularity

down so far[14], but it does give some idea of the reason for regarding the Principle of Regularity as true. I shall discuss this issue further when the relevant further axioms have been laid down.

Although I have formulated this principle for sets, and have put the case in its favour in set-theoretical terms, it is clear that these considerations apply, *mutatis mutandis*, to the more basic notion of *arithmos* in terms of which the notion of set is defined. Surely it is obvious that there can be no membership loops or unlimited descending membership chains among *arithmoi*[15].

---

[14] Not only am I employing natural numbers without having defined them, but I am using the Axiom of Choice as well.
[15] See my remarks at the end of Section 3.3.

Part Three
Cantorian Set Theory

No one shall be able to drive us from the paradise that Cantor has created for us.

Hilbert

5
Cantorian Finitism

5.1 Dedekind's axiomatic definition of the natural numbers

In this chapter I shall present the version of set theory that provides the foundations of modern mathematics as it is practised by the overwhelming majority of mathematicians. As I have explained, the characterising and fundamental assumption underlying this theory is that the species of natural numbers has a determinate size, that is to say, is *finite*, in the new Cantorian sense of "finite", and therefore forms a set.

But I haven't yet said what the natural numbers are. Indeed, in Chapter 2 I argued that natural numbers aren't anything at all, at least insofar as they are naively understood to be those abstract things, whatever they are, that are named by our number words and numerals. Nevertheless, even though we cannot take the natural numbers to be simply "given" as naturally occuring abstract objects, so to speak, we can *define* them, using the axiomatic method.

Thus I shall begin my exposition of this Cantorian version of set theory by presenting Dedekind's axiomatic definition of the system of natural numbers. First I need to specify the *structure type* to which Dedekind's axioms are to be applied and the notion of *morphism* for such structures.

5.1.1 Definition

(i) *A Dedekind structure is an ordered triple*

$$(N, s, a)$$

where N is a non-empty set, $s : N \to N$ is a function from N to N, and $a \in N$ is a distinguished element of N.

(ii) *Let (N, s, a) and (M, t, b) be Dedekind structures, and let $m : N \to M$ be a local function. Then, by definition*

m *is a morphism from* (N, s, a) *to* (M, t, b) : *iff* :
$m‘a = b$.*and*. $(\forall x \in N)[m‘s‘x = t‘m‘x]$

A bijective morphism is called an isomorphism.

If (N, s, a) is a Dedekind structure, the function s is called the *successor function* of the structure, and the element a its *zero element*.

A mathematical structure typically consists of an *underlying set* (or *sets*), often called the *universe* or *domain* (*of discourse*) of the structure, together with certain functions, relations, and distinguished elements defined on that universe which together constitute the *morphology*[1] of the structure. In the case of a Dedekind structure, the morphology consists of a single unary function (the *successor function* of the structure), a single distinguished element (the *zero element* of the structure), and no relations. In general, a morphism from one structure of a given *type* (or *category*) to another of that type is a local function whose domain is the universe of the first structure, whose range is a subset of the domain of the second, and which *respects* or *preserves* the morphologies of the structures.

It is easy to see that if m is an isomorphism of Dedekind structures, then so is m^{-1}. By virtue of this symmetry it makes sense to speak of two structures being *isomorphic*. When m is an isomorphism from one Dedekind structure to another it establishes a *one-to-one correspondence* or *correlation* between the universes of the two structures in which the zero elements are correlated, and if two elements of the repective structures are correlated, then so are their respective successors. Under this correlation, the "algebra" imposed on the universe of each of structures by its morphology is exactly mirrored by that imposed on the other by the other's morphology.

From an "abstract" point of view, that is to say, when we ignore the individuating peculiarities of the elements composing their universes, the two structures are indistinguishable. It is, in fact, the fundamental dogma of the modern axiomatic method that *mathematical structures that are*

---

[1] The terminology here is not standard. Etymologically, "morphology" ought to name the study of "form" (*morphe*) in the way that "topology" names the study of "place" (*topos*). Understood in that way and in that generality, however, it would be just another name for mathematics itself. I use the term "morphology" here in analogy with the use of "topology" to denote the family of open subsets of a topological space. In fact, topological spaces do not conform to the rough description of "mathematical structure" I have just given. For the morphology of a topological space is supplied by its topology, which is, in effect, a unary relation, not on the underlying set of the structure, but on the *power set* of that underlying set.

5.1 Dedekind's axiomatic definition of the natural numbers

isomorphic are essentially indistinguishable insofar as their mathematically significant properties are concerned.

From this dogma it follows that if we have specified the isomorphism class of a structure we have implicitly determined all of its mathematically relevant properties[2]. Accordingly mathematicians are satisfied if they succeed in specifying an isomorphism class by means of mathematically precise conditions, and in those circumstances they say that they have *defined a structure up to isomorphism*.

In fact, my immediate aim is to specify the isomorphism class to which the Dedekind structure determined by "the" natural numbers would belong, if only there were such a structure. The strategy is to lay down suitable "self-evident" truths about that supposed structure as *axioms* which suffice to characterise an isomorphism class of Dedekind structures. Of course, there cannot be truths of any kind, self-evident or otherwise, about a structure that does not exist; but the logic of axiomatic definition means that this doesn't matter. For the axioms have been converted from "self-evident truths" into defining conditions.

5.1.2 Definition *Let (N, s, a) be a Dedekind structure. Then (N, s, a) is a simply infinite system if it satisifies the following three axioms:*

I. $(\forall x \in N)[s`x \neq a]$

II. $(\forall xy \in N)[s`x = s`y \text{ implies } x = y]$

III. $(\forall S \subseteq N)[a \in S \text{ and } (\forall x \in S)[s`x \in S] \; : \text{implies} : \; S = N]$

These axioms are easily translated into formal mathematical logic.

In laying down the definition of "simply infinite system" I have, in fact, succeeded in my principal task of characterising an isomorphism class of Dedekind structures and thereby defining "the" natural numbers up to isomorphism. This, however, remains to be shown.

It is technically convenient to single out from Dedekind structures those that satisfy Axiom III of the definition just given. Let us call such Dedekind structures *induction systems*. Thus the principle of proof by mathematical induction is built into the definitions both of induction system and of simply infinite system. The crucial difference that is made in passing from induction systems to those induction systems that are

---

[2] In the case under consideration here, this rather vague formulation in terms of "mathematically significant properties" can be sharpened considerably, as I shall show in Section 6.6. Once this is done, what I have called the "fundamental dogma" can, in the case of natural number arithmetic, be stated and proved as a proper mathematical theorem.

also simply infinite is that in the latter, but not in all of the former, a principle of *definition* by induction, or, as I prefer to say, by *recursion* can be established. As I shall soon show, it is the principle of definition by recursion that is of fundamental importance in the axiomatic characterisation of natural number.

To establish definition by recursion in simply infinite systems I need first to prove two lemmas on induction systems.

5.1.3 Lemma *Let (M, t, a) be any Dedekind structure. Then there is exactly one induction system (N, s, a) such that $N \subseteq M$ and $s \subseteq t$.*

Proof Let $N = \bigcap \{S \subseteq M : a \in S \text{ and } t"S \subseteq S\}$ and let $s = t \cap N \times N$ (i.e. s is the restriction of t to N). The intersection is over a non-empty family of sets, since M itself belongs to that family. A straightforward argument[3] now shows that

$$\bigcap \{S \subseteq M : a \in S \text{ and } t"S \subseteq S\} = \bigcap \{S \subseteq N : a \in S \text{ and } t"S \subseteq S\}$$

so that $s : N \to N$ and (N, s, a) is an induction system. Given two induction systems of this sort, each can be shown to be contained in the other by induction, that is to say, by Axiom III of Definition 5.1.3. □

We shall also need the following lemma.

5.1.4 Lemma *Let (N, s, a) be an induction system. Then*

$$(\forall x \in N)[x = a \text{ or } x \in s"N]$$

Proof Suppose, by way of contradiction, that $b \in N$, but $b \neq a$ and $b \notin s"N$. Then $a \in N - \{b\}$ and $s"(N - \{b\}) \subseteq N - \{b\}$. Hence by induction along (N, s, a), $N - \{b\} = N$, i.e. $b \notin N$, and we have arrived at the required contradiction. □

I can now establish the theorem on definition by recursion along a simply infinite system. As Dedekind explains in his monograph (§§131–134) this is the key result in the axiomatic definition of the natural numbers.

5.1.5 Theorem *(Definition by Recursion along a Simply Infinite System) Let (N, s, a) be a simply infinite system and (M, t, b) an arbitrary Dedekind*

---

[3] The reader may find it instructive actually to carry out this argument.

5.1 Dedekind's axiomatic definition of the natural numbers 157

system. Then there is exactly one function $f : N \to M$ which is a morphism of Dedekind structures, that is to say, which satisfies the following equations:

(i) $f`a = b$
(ii) $(\forall x \in N)[f`s`x = t`f`x]$

Proof The equations giving the condition for f to be a morphism are called *recursion equations for f*. Consider the Dedekind structure

$$(N \times M, s \otimes t, (a, b))$$

where $s \otimes t : N \times M \to N \times M$ is given by

$$(\forall x \in N)(\forall y \in M)[(s \otimes t)`(x, y) = (s`x, t`y)]$$

and let $(f, p, (a, b))$ be the unique induction system guaranteed by Lemma 5.1.3 such that $f \subseteq N \times M$ and $p \subseteq s \otimes t$. Then the universe, f, of this structure is the unique function $N \to M$ required by the theorem.

For if we let

$$S = \{x \in N : (\exists y \in M)[(x, y) \in f]\}$$

and

$$T = \{x \in N : (\forall u, v \in M)[(x, u) \in f \text{ and } (x, v) \in f : implies : u = v]\}$$

then essentially straightforward (but surprisingly lengthy) induction arguments[4] along the simply infinite system (N, s, a) suffice to show that both $S = N$ and $T = N$, from which it follows that $f : N \to M$.

Moreover

$$f`a = b$$

(since $(a, b) \in f$) and

$$(\forall x \in N)[f`s`x = t`f`x]$$

(since $(\forall xy \in N)[(x, y) \in f \text{ implies } (s`x, t`y) \in f]$), and therefore f satisfies the recursion equations.

Finally, f is unique in so doing, as a simple induction argument along (N, s, a) shows. □

---

[4] Again, the reader may find it instructive to carry them out.

The proof here is surprisingly complicated when written out in full. For if we think of the simply infinite system (N, s, a) as "the" natural numbers, then 5.1.5 has all the appearance of a triviality. Naively, in the simply infinite system (N, s, a) the set N is obtained by starting with a and iterating the successor function s: whatever can be be obtained from a (Euclidean) finite number of such iterations belongs to N, and every member of N can be obtained by a (Euclidean) finite number of such iterations.

$$N = \{a, s`a, s`s`a, s`s`s`a, s`s`s`s`a, \ldots\} = \{0, 1, 2, 3, 4, \ldots\}$$

From this standpoint, the recursion equations

(1) $f`a = b$
(2) $f`s`x = t`f`x$ (for all $x \in N$)

are sometimes regarded as a kind of recipe, a set of instructions for calculating the function f. And for $n \in N$, $f`n$ is the *nth iterate* of the function t.

$$\begin{array}{cccccc} a & s`a & s`s`a & s`s`s`a & \cdots & s`s`s`s`a & \cdots \\ f\downarrow & f\downarrow & f\downarrow & f\downarrow & \cdots & f\downarrow & \cdots \\ b & t`b & t`t`b & t`t`t`b & \cdots & t`t`t`t`b & \cdots \end{array}$$

But this naive way of regarding the theorem on definition by recursion, however useful it may be as a heuristic aid to understanding, is completely inadequate as an interpretation of the real mathematical content of this important result. For the notion of Euclidean finite iteration on which this supposed explanation rests, and to which it appeals, is, in fact, the very notion that calls out for explanation here.

Theorem 5.1.5 is itself a mathematically precise formulation of the notion of "iteration", and its proof, whose surprising complexity is a measure of the logical and set-theoretical complexity of that seemingly straightforward notion, is what justifies our naive employment of it. To see the notion of Euclidean finite iteration as providing an adequate justification of the theorem is to confound *explicans* and *explicandum* here. For the naive idea of a simply infinite system given above, namely, that its universe is obtained by iterating the successor function, can itself be justified, indeed made intelligible, only by Theorem 5.1.5.

We can use Theorem 5.1.5 to establish the existence of the familiar arithmetical operations of addition, multiplication, and exponentiation.

5.1.6 Theorem *Let (N, s, a) be a simply infinite system. Then there are*

5.1 Dedekind's axiomatic definition of the natural numbers

unique binary functions $+ : N \times N \to N$, $\cdot : N \times N \to N$, and $exp : N \times N \to N$ such that, for all x, y in N,

$$x + a = x$$
$$x + s`y = s`(x + y)$$

$$x \cdot a = a$$
$$x \cdot (s`y) = (x \cdot y) + x$$

$$exp(x, a) = x$$
$$exp(x, s`y) = exp(x, y) \cdot x$$

Proof By 5.1.5 for each $x \in N$ there exists a unique morphism f_x from (N, s, a) to (N, s, x), so we may define $x + y = f_x`y$. Now for each $x \in N$ define $g_x : N \to N$ by setting $g_x`y = f_y`x$, for all $y \in N$. Then 5.1.5 guarantees the existence, for each $x \in N$, of a unique morphism $h_x$ from $(N, s, a)$ to $(N, g_x, a)$, so we may define $x \cdot y = h_x`y$. The function exp can be defined in a similar manner. □

This theorem leads us to the following definition.

5.1.7 Definition A *Peano system* is an ordered quintuple $(N, s, a, +, \cdot)$ such that

I. $(\forall x \in N)[s`x \neq a]$
II. $(\forall xy \in N)[s`x = s`y \ : implies : \ x = y]$
III. $(\forall S \subseteq N)[a \in S \ and \ (\forall x \in S)[s`x \in S] \ : implies : \ S = N]$
IV. $(\forall xy \in N)[x + a = x \ : and : \ x + s`y = s`(x + y)]$
V. $(\forall xy \in N)[x \cdot a = a \ : and : \ x \cdot (s`y) = (x \cdot y) + x]$

Axioms I, II, and III tell us that (N, s, a) is a simply infinite system. Axioms IV and V characterise the operations of addition and multiplication over that system. These axioms, the well-known *Peano axioms* for natural number arithmetic, are easily translated into formal mathematical logic.

From our present standpoint, the most important fact about Theorem 5.1.5 is that it leads directly to a proof of the following fundamental theorem.

5.1.8 Theorem (*The Categoricity Theorem for Simply Infinite Systems*) *Let (N, s, a) and (N_1, s_1, a_1) be simply infinite systems. Then they are isomorphic as Dedekind structures.*

Proof Since the identity morphism is the only automorphism of a simply infinite system, the unique morphisms $f : N \to N_1$, from (N,s,a) to (N_1, s_1, a_1) and $f_1 : N_1 \to N$, from (N_1, s_1, a_1) to (N,s,a) are easily shown to be inverses of one another and hence isomorphisms. □

As a corollary of 5.1.6 and 5.1.8 we have

5.1.9 Theorem *(The Categoricity Theorem for Peano Systems)* Let $(N, s, a, +, \times)$ and $(N_1, s_1, a_1, +_1, \times_1)$ be Peano systems. Then they are isomorphic as structures of their type.

By virtue of 5.1.8 and 5.1.9 we say that the axioms (I, II, and III) for a simply infinite system and the Peano axioms (I, II, III, IV and V) for a Peano system are *categorical*.

So any two simply infinite systems (or any two Peano systems) are isomorphic. But are there any simply infinite systems at all? Are the axioms for a simply infinite system consistent? We cannot answer these questions absolutely on the basis of the axioms that have been laid down so far. But the answer is *yes* if there are sets that are not Euclidean finite, in the sense of the following definition:

5.1.10 Definition *Let S be any set. Then by definition*

$$S \text{ is Euclidean finite .iff. } (\forall X \in P(S))[X \neq S \text{ .implies. } X \not\approx_c S]$$

A set that is not Euclidean finite is said to be *transfinite*.

This is a key definition. The idea behind it is to convert Euclid's *Axiom* or*Common Notion* 5 of *Elements*, Book I, namely

The whole is greater than the part

which for Euclid is a fundamental truth constitutive of the very notion of finiteness for quantities of all sorts, discrete or continuous, into a definition.

By assuming the existence of a transfinite set – a *Cantorian* finite plurality that is not finite in the original sense of Euclid – we can establish the existence of simply infinite systems.

5.1.11 Theorem *(Conditional Consistency Theorem for Simply Infinite Systems)* Let S be any set that is not Euclidean finite, $f : S \to S$ be 1-1 but not onto, and $a \in S - f``S$. Then there is a subset, N, of S and a function $s : N \to N$, such that (N, s, a) is a simply infinite system and $(\forall x \in N)[s`x = f`x]$.

5.2 Cantor's Axiom

Proof Let $N = \bigcap \{T \subseteq S : a \in T \text{ and } f\text{``}T \subseteq T\}$ and let $s = f \cap N \times N$. Then (N, s, a) is a simply infinite system. □

To convert this conditional consistency theorem into an absolute one we must somehow posit a transfinite set. This I shall do in the next section.

5.2 Cantor's Axiom

Cantor's Axiom is, roughly speaking, the assumption that "the" natural numbers form a set. It is, as I have explained, a finiteness assumption like the other basic axioms, and therefore it ought to be presented in conformity with the presentation given to those other axioms in Chapter 4. First we need a definition.

5.2.1 Definition *Let S be any set. Then, by definition, S is a von Neumann natural number if, and only if, S is a von Neumann ordinal and S is Euclidean finite.*

How can we justify calling these sets von Neumann *natural numbers*? A full justification will have to wait for later, but we can already determine that they have some of the expected properties. The idea behind von Neumann's definition is simply to identify each natural number, n, with the set, $\{0, 1, 2, \ldots, n-1\}$, of all its predecessors. In particular, the idea is that $0 = \emptyset$, and for each von Neumann natural number, n, $n+1 = n \cup \{n\}$. We already know that $\emptyset$ is a von Neumann ordinal; since it is clearly Euclidean finite it is a von Neumann natural number. Moreover, if S is a von Neumann natural number then so is $S \cup \{S\}$.[5]

Given distinct von Neumann natural numbers, one must be an element, and hence a proper subset, of the other.[6] Thus distinct von Neumann natural numbers have distinct cardinalities. Whether they represent all possible Euclidean finite cardinalities is something I am not yet in a position to assert. In any case let me press ahead with the statement of Cantor's Axiom.

Informally, Cantor's Axiom embodies the following finiteness assumption: *the species of von Neumann natural numbers is (Cantorian) finite and therefore forms a set, ω.* Formally, as with the other axioms, we must introduce a new constant symbol, "ω", and lay down a suitable

---

[5] That $S \cup \{S\}$ is a von Neumann ordinal has already been established. That it is Euclidean finite requires a separate argument, which I leave to the reader to supply.

[6] Theorem 4.10.5(v).

162   Cantorian Finitism

axiom giving necessary and sufficient conditions for membership in the set named by that new symbol:

5.2.2 Axiom *(Cantor's Axiom)*

$$\forall X[X \in \omega \; : \textit{iff} : \; X \text{ is a von Neumann natural number}]$$

ω is a transfinite set, for the function $f : \omega \to \omega$ defined by setting $f`x = x \cup \{x\}$, for all $x \in \omega$, is clearly injective but not surjective. From this observation we can infer the existence of a simply infinite system from Theorem 5.1.11. However, that tells us only that there is exactly one *subset* $N \subseteq \omega$, and a function $s : N \to N$ with $s`x = x \cup \{x\}$, for all $x \in N$, such that $(N, s, \emptyset)$ is a simply infinite system. But a straightforward induction along $(N, s, \emptyset)$ shows that N is a transitive set, so N itself is a von Neumann ordinal. It follows (by Theorem 4.10.5(v)) that $N \in \omega$, $\omega \in N$, or $N = \omega$. Since we can eliminate the first two possibilities, we have the following theorem.

5.2.3 Theorem *Let $s : \omega \to \omega$ be defined by $s`x = x \cup \{x\}$, for all $x \in \omega$. Then $(\omega, s, \emptyset)$ is a simply infinite system.*

Theorem 5.2.3 justifies our calling ω the set of von Neumann *natural numbers*; for they form a simply infinite system under their natural zero, $\emptyset$, and their natural successor function, $\lambda x(x \cup \{x\})$. To complete our analysis of natural number arithmetic and our justification of our "arithmetical" terminology, we still need to show that we can use "the" natural numbers (i.e. any simply infinite system) for the purpose of "counting out" Euclidean finite sets. But before we can do this we must first lay down another axiom, the famous Axiom of Choice.

5.3 The Axiom of Choice

The Axiom of Choice has had a curious history. First formulated and explicitly appealed to by Zermelo, in the course of proving the Well-ordering Theorem, the axiom has generated considerable controversy. I shall take the view, however – and I believe that it is the view of the overwhelming majority of mathematicians – that the axiom is an integral part of the Cantorian standpoint. Whether it has a reasonable claim to being self-evidently true is a more difficult question, but most mathematicians hold it to be so. To state the axiom we first need to define the notion of a *choice function*:

5.3 The Axiom of Choice

5.3.1 Definition *Let S be a set and f be a local function. Then by definition*

f is a choice function for S : iff :
$$[f : P(S) - \{\emptyset\} \to S \text{ and } (\forall x \in P(S) - \{\emptyset\})[f`x \in x]]$$

Thus a choice function f for a non-empty set S "chooses" an element $f`T$ from every non-empty subset $T \subseteq S$. Of course every such T *has* such an element. The problem is that, in the general case, we have no way of uniformly describing a particular element in each such set. This is why we must *posit* a choice function which makes the selection for us, so to speak. Of course the empty set is a choice function for the empty set. Now we can lay down the axiom.

5.3.2 Axiom *(The Axiom of Choice) Every set has a choice function.*

The Axiom of Choice does not have the form of a finiteness principle. In this respect it is anomalous, and for some time after Zermelo first stated it there remained the hope that it might be possible to prove the axiom from the other axioms of set theory. These hopes were finally dashed in 1963 by Paul Cohen[7], and we now know that we must assume the axiom itself or one of its equivalents.

The worry here must be, of course, that in laying down this axiom we are making an illegitimate appeal to operationalist intuitions[8]. For, on the face of it, we are basing the argument for the *existence* of choice functions for sets in general on our capacity actually to *define* them in the case of small Euclidean finite sets.

There are, indeed, various alternatives to the Axiom of Choice, that is to say, other propositions that can be proved equivalent to it. The first of these was Zermelo's Well-ordering Theorem.

5.3.3 Theorem *(The Well-ordering Theorem (Zermelo, 1904)) Every set can be well-ordered. More precisely*

$\forall X[X$ *is a set .implies.*
$(\exists Y \in P(X \times X))[Y$ *is a well-ordering and Field*$(Y) = X]]$

I shall not include a proof of the Well-ordering Theorem here as there are many excellent expositions of this result to which the interested reader can refer, including Zermelo's own[9].

---

[7] "The independence of the continuum hypothesis".
[8] See the discussion in Section 1.4.
[9] "Neuer Beweiss für die Möglichkeit einer Wohlordnung".

We can use the Well-ordering Theorem to show that every Euclidean finite cardinality is exemplified by a (in fact, exactly one) von Neumann natural number.

5.3.4 Theorem *Let S be a Euclidean finite set. Then there is a von Neumann natural number $X \in \omega$ such that $X \cong_c S$.*

Proof Let S be a Euclidean finite set. By the Well-ordering Theorem there is a well-ordering r whose field is S. Then for every $X \in \omega$

$$\subseteq_X <_o r \text{ or } \subseteq_X \cong_o r \text{ or } r <_o \subseteq_X$$

Suppose now, by way of contradiction, that

$$(\forall X \in \omega) not[\subseteq_X \cong_o r]$$

Then clearly

$$(\forall X \in \omega)[\subseteq_X <_o r]$$

From this we may conclude that there is, for each $X \in \omega$, a unique function

$$f_X : X \to Field(r)$$

which embeds $\subseteq_X$ in r as a proper initial segment. Hence

$$\bigcup \{f_X : X \in \omega\} : \omega \to Field(r) = S$$

is one to one, and this contradicts our assumption that S is Euclidean finite. The theorem now follows. □

This theorem completes our justification of Dedekind's axiomatic definition of the natural numbers. For now we have established that the von Neumann natural numbers, and therefore any other simply infinite system, can be used to "count out" any Euclidean finite set.

5.4 The extensional analysis of sets

In Section 4.11 I discussed the consequences of the extensional character of sets for the analysis of the relation of set membership. Now that we have developed the theory of natural number arithmetic (simply infinite systems) we can give the merely heuristic and informal arguments considered in that earlier section a rigorous mathematical treatment. In particular, we can now formulate the prohibition on "membership loops"

5.4 The extensional analysis of sets

and "indefinitely descending epsilon chains" precisely using the notion of an *analysing sequence* for a set S given in the following definition.

5.4.1 Definition *Let S be any set or individual. Then an analysing sequence for S is a local function f defined on a von Neumann natural number $n \in \omega$ such that*

(i) *If $0 \in Domain(f)$, then $f`0 \in S$*
(ii) *$(\forall x \in Domain(f))[(x+1) \in Domain(f)$ implies $f`(x+1) \in f`x]$*

If an analysing sequence for S has no proper extension that is an analysing sequence then it is said to be complete.

Naively, an analysing sequence for a set S begins by picking a member of S then a member of that member, then a member of the latter, in turn, and so on. When the empty set or an individual is reached then no further choices can be made and the analysing sequence is *complete*. There is nothing in the definition of an analysing sequence that excludes the possibility that there might be an ω-sequence of analysing sequences

$$f_0 \subseteq f_1 \subseteq \cdots \subseteq f_i \subseteq f_{i+1} \subseteq \cdots$$

where each f_i is defined on $i \in \omega$ and each f_{i+1} is a proper extension of f_i. But this is ruled out by the Principle of Regularity.[10] For otherwise we we may conclude that if $f = \bigcup\{f_i : i \in \omega\}$

$$(\forall x \in f``\omega)[x \cap f``\omega \neq \emptyset]$$

and this violates Regularity.

On the other hand, in Cantorian set theory this condition on ω-sequences of analysing functions logically entails the Principle of Regularity. For if Regularity fails for the non-empty set S, that is to say, if $(\forall x \in S)[x \cap S \neq \emptyset]$, and if $g : P(S) - \{\emptyset\} \to S$ is a choice function, then if we set $h`x = g`(x \cap S)$, for all $x \in S$, the recursion equations

$$f`0 = g`S$$
$$f`(n+1) = h`f`n$$

(with $(S, h, g`S)$ as the target Dedekind structure for the recursion) define an ω-sequence of the sort under consideration.

Thus we see that Cantor's Axiom together with the Axiom of Choice allow us to give an exact mathematical formulation of the informal

---

[10] See Section 4.11.

arguments for the Principle of Regularity employed in Section 4.11. For membership loops

$$x \in y \in z \in \cdots \in w \in x$$

and indefinitely descending membership chains

$$\cdots \in w \in z \in y \in x$$

both give rise to ω-sequences that violate Regularity.

In Cantorian set theory, therefore, we can formulate the prohibition on membership loops and indefinitely descending membership chains directly as an axiom from which the Principle of Regularity can be derived using the Axiom of Choice in the manner just indicated. Let us therefore adopt the following axiom.

5.4.2 Axiom *(The Axiom of Foundation – Cantorian Version)* *Let S be any set and let f be any ω-sequence, where for each $i \in \omega$, f_i is an analysing sequence for S and $f_i \subseteq f_{i+1}$. Then there exists an $n \in \omega$ such that $f_i \subseteq f_n$, for all $i \in \omega$.*

From this we obtain, as indicated above

5.4.3 Theorem *(The Principle of Regularity)*

$$\forall X [X \text{ is a set and } X \neq \emptyset \text{ .implies. } (\exists y \in X)[X \cap y = \emptyset]]$$

Note that, like the Axiom of Choice, but unlike the other principal axioms of set theory, the Axiom of Foundation is not a finiteness principle. Nevertheless, given the essential nature of sets as *extensional* pluralities, it does have that character of *self-evidence* that we require in an axiom, that is to say, a basic principle to which we may appeal without proof. But there is a further basic principle suggested by the extensional character of sets, as I shall now show.

If f is an analysing sequence for the set S and $x \in Range(f)$ then, as I argued in Section 4.11, the existence of x as a well-defined, determinate object is presupposed by the existence of S itself. I propose, therefore, to call those objects x that are terms of analysing sequences for S *constituents* of S. The constituents of the set S thus comprise all those objects (sets and individuals) that go, directly or indirectly, to make up the set S – all those things whose existence as well-defined objects is the necessary precondition for the existence of S itself as a well-defined object, viz. a set, that is to say, an extensional plurality definite in multitude.

5.4 The extensional analysis of sets

In the light of this observation it seems plausible to suppose that a set, as a (Cantorian) finite object, cannot require an absolutely infinite species of objects in its essential definition. Thus from the Cantorian point of view we ought to suppose that the constituents of S are (Cantorian) finite in multitude, and hence collectively form a set. This would imply, of course, that the species of analysing sequences for S also forms a set. I need, therefore, to lay down an additional finiteness assumption here in the form of an axiom. But since the species of all constituents of S has a Σ_1 defining property, it is technically necessary, if we wish our new axiom to conform to the pattern previously established, to posit, not the (Cantorian) finitude of the species of *constituents* of S, but rather that of the species of its *analysing sequences*, a species which is locally definable.

Accordingly we may say that, given any object a, the species of all analysing sequences for a is (Cantorian) finite and therefore forms a set which we denote by

$$\epsilon(a)$$

The correspondence

$$a \mapsto \epsilon(a)$$

thus defines a first level global function of one argument.

If we order the members of $\epsilon(S)$ in the natural way by extension[11], the resulting order structure is a tree. Accordingly, we may refer to the set $\epsilon(S)$ itself as the *epsilon tree* for S. We may think of the epsilon tree for a set S as completely determining the essential definition of S as a (Cantorian) finite, extensional plurality.

5.4.4 Axiom *(The Axiom of Extensional Analysis)* Let a be any set or individual. Then $\epsilon(a)$ is a set. Moreover

$$\forall xy [y \in \epsilon(x) \text{ .iff. } y \text{ is an analysing sequence for } x]$$

It follows that if a is an individual or the empty set, then $\epsilon(a) = \emptyset$. We can now define the set of constituents of a set S.

5.4.5 Definition *Let S be any set or individual. Then by definition*

$$Constituents(S) = \bigcup \{Range(f) : f \in \epsilon(S)\}$$

---

[11] So that if the finite sequences x and y are analysing sequences for S, then x precedes y in the natural ordering referred to precisely when x is an initial segment of y.

Recall that a set S is *transitive* if each of its set members is also a subset (i.e. $(\forall x \in S)[x \subseteq x]$). It is obvious that the intersection of any set of transitive sets is itself transitive, so if a given set S is included in *any* transitive set, it is included in a smallest such, which is called the *transitive closure* of S.

We have not yet established, however, that every set is included in a transitive set, so we cannot yet refer to the transitive closures of arbitrary sets. This, however, is easily remedied, for by the following theorem the set of constituents of a set S *is* its transitive closure.

5.4.6 Theorem *Let S be any set. Then*

(i) *Constituents(S) is a transitive set having S as a subset.*

(ii) *Constituents(S) is a subset of every transitive superset of S, that is to say*

$$\forall X [X \text{ is transitive and } S \subseteq X : \text{implies} : \text{Constituents}(S) \subseteq X]$$

The constituents of a set S that themselves have no constituents are what we might call the *ultimate constituents* of S. The only candidates for ultimate constituents are the empty set and individuals, since these are the only objects without elements. I shall call the set of ultimate constituents of S the *support* of S.

5.4.7 Definition *Let S be any set or individual. Then by definition*

$$\text{Support}(S) = \{x \in \text{Constituents}(S) : x = \emptyset \text{ or } x \text{ is not a set}\}$$

A set whose support does not contain any non-sets is called a *pure* set. Obviously any complete analysing sequence for a pure set must terminate in the empty set.

5.4.8 Definition *Let S be any set or individual. Then by definition*

$$S \text{ is a pure set } : iff : S = \emptyset \text{ or } \text{Support}(S) = \{\emptyset\}$$

Of course, the support of a set may, or may not, contain the empty set.

The existence, for each set S, of the set of constituents of S permits us to establish an important version of induction.

5.4 The extensional analysis of sets

5.4.9 Theorem *(The Principle of Epsilon Induction)* *Let Φ be any first level global propositional function of one argument. Then the hypothesis*

$$\forall X[(\forall y \in X)\Phi(y) \text{ implies } \Phi(X)]$$

entails the conclusion

$$\forall X \Phi(X)$$

The theorem follows from the observation that if S is any set and T is the set of constituents of S that do *not* satisfy Φ, then the hypothesis, together with Regularity, require that T be empty. For then every constituent, and therefore every element, of S satisfies Φ, so S itself must satisfy Φ, by the hypothesis.

Corresponding to the principle of proof by epsilon induction, there is, as in similar cases, a principle of definition by epsilon recursion.

5.4.10 Theorem *(The Principle of Definition by Epsilon Recursion – Local Version)* *Let S and T be sets with S transitive, let U be the set of all partial functions on S with values in T (functions $S_0 \to T$, where $S_0 \subseteq S$), and let $g : U \to T$. Then there is exactly one function $f : S \to T$ which satisfies the recursion equation*

$$f`x = g`\{(y, f`y) : y \in x\}$$

for all $x \in S$.

The argument required here is similar to that given in the proof of Theorem 4.9.4. If we define $U_0 \subseteq U$ to be the set of those partial functions in U whose domains are transitive and which satisfy the given recursion equation on their domains, then we can use epsilon induction to show that the functions in U_0 always agree on the common part of their domains, and that every member of S lies in the domain of some such function. The required f can then be defined to be $\bigcup U_0$.

Both of the axioms introduced in this section – the Axiom of Foundation (5.4.2) and the Axiom of Extensional Analysis (5.4.4) – lie outside that part of general set theory that is in common use in ordinary mathematics. Both, however, derive from the essentially *extensional* character of sets as *arithmoi*, and both have that character of self-evidence that we associate with axioms of the old-fashioned Euclidean sort, once it is realised that in set theory, as I explained in Section 4.4, we are using the arithmetical, rather than the logical notion of set.

We now have in place all of the axioms that underlie the ordinary practice of mathematics. Every axiom in the Zermelo–Fraenkel system, including the Axiom of Foundation, has now been put into place. But from the standpoint of the professional set theorist the theory is extremely weak. The central difficulty is that it does not contain a principle of global transfinite recursion. This means, for example, that usual proof that every well-ordering is order isomorphic to a von Neumann well-ordering will not go through. But that, in turn, means that we cannot develop Cantor's theory of the transfinite cardinal numbers – the ℵ numbers – nor can we show that the universe forms a cumulative hierarchy.

All of these results require global transfinite recursion. But as we have seen in Section 4.9 the standard proof of that principle requires Σ_1 instances of Comprehension which, by virtue of Brouwer's Principle, is not available to us.

Of course it would be possible simply to lay down an axiom that would assert that transfinite recursion is permitted. But such an axiom would not constitute a finiteness principle, and would have to appeal to supposed intuitions about carrying out "processes of definition" of transfinite length.

Accordingly, I shall devote the remainder of this chapter to presenting a series of stronger and stronger axioms, each of which has a claim to status as a finiteness principle – though the claims become weaker as we advance through the series – and the last of which enables us to establish definition by transfinite recursion.This procedure will seem otiose to anyone who thinks that definition by transfinite recursion is an "obvious" first principle. But my aim here is to see what that "obviousness" consists in. Besides, we are dealing with the basic principles that underlie all our proofs, and therefore we cannot afford to be cavalier about questions of self-evidence.

5.5 The cumulative hierarchy of sets

In axiomatic set theory it is possible to use the underlying classical logic of the theory to derive various *global* versions of definition by transfinite recursion[12]. Then, as is well-known, one can use transfinite recursion along the von Neumann ordinals of the theory[13] to prove that the universe of the theory can be arranged in a *cumulative hierarchy* whose *levels* are indexed by those ordinals.

---

[12] I mean first order Zermelo–Fraenkel set theory here: the theory ZF_1 described in Section 7.1.

[13] Whether these are genuine von Neumann ordinals depends on what model of the theory we have under consideration.

5.5 The cumulative hierarchy of sets

The strategy I shall follow in this section is to introduce the cumulative hierarchy directly by means of a suitable axiom[14] which, together with a further axiom to be introduced in the next section[15], will allow us to establish global transfinite recursion.

To lay down the required new axiom, we need to define the *rank* of a set, that is to say the level of the cumulative hierarchy at which it first appears as a subset. But since an axiom positing the hierarchy is not yet in place, we need to define the rank of a set independently of it. Fortunately, the means to do this are at hand.

The nature of sets as purely extensional pluralities encourages us to imagine each set as somehow "built up" from its constituents in uniform "stages": each constituent of S is "constructed" immediately after all of its constituents are "constructed". Of course, all talk of "construction" and "stages" here is purely metaphorical, but we can extract genuine mathematical content from it. For "stages" in this "construction" are well-ordered by set inclusion, and that well-ordering is a measure of the rank of the set S[16].

The definition of this well-ordering (whose field is a subset of the power set of the set of constituents of S) is essentially straightforward, but rather laborious, and therefore, in order to avoid interrupting the flow of my exposition, I shall relegate the details of this definition to an appendix[17].

For our present purposes it will suffice to say that we can define a binary global relation $\lambda xy[x \text{ is of rank } y]$ such that if r is a well-ordering and S is a set, then S is of rank r, if, and only if, r is order isomorphic to the well-ordering of the stages in the construction of *Constituents*(S) described informally above.

Accordingly, we can define relations $\lambda xy[x \cong_{rank} y]$, $\lambda xy[x \leq_{rank} y]$, and $\lambda xy[x <_{rank} y]$ with the obvious meanings. This enables us to formulate the first of our Hierarchy Principles.

The relation $\cong_{rank}$ is a global equivalence relation. But what is the nature of the $\cong_{rank}$ equivalence classes? In particular, given a set S do the sets which are of the same rank as S and whose support is a subset of the support of S compose an absolutely infinite species? It might seem,

---

[14] The *Strong Hierarchy Principle* (Axiom 5.5.5). In fact, I shall also consider another version of this axiom – the *Weak Hierarchy Principle* (Axiom 5.5.1) – to provide a point of comparison.
[15] The *Strong Closure Principle* (Axiom 5.6.8).
[16] In fact, it is order isomorphic to the natural ordering $\subseteq_a$ of the von Neumann ordinal, a, which *is* the rank of S (Definition 5.5.2).
[17] Appendix 2.

at least on the face of it, natural to suppose that they do not and that, on the contrary, the species of those sets is Cantorian finite and therefore forms a set. We thus have a candidate for a new Finiteness Principle, which we may formulate in the standard way for such principles, as follows:

Given any object S, the species of all sets T such that $Support(\{T\})$ is a subset of $Support(\{S\})$ and $T <_{rank} S$ is (Cantorian) finite and therefore forms a set which we denote by

$$R(S)$$

The correspondence

$$S \mapsto R(S)$$

thus defines a first level global function of one argument which can be characterised as follows:

5.5.1 Axiom *(The Weak Hierarchy Principle)*

$$\forall X Y [Y \in R(X) : iff :$$
$$Support(\{Y\}) \subseteq Support(\{X\}) \text{ and } Y <_{rank} X]$$

It will soon become clear why I have called this the *Weak* Hierarchy Principle. Notice that every *pure* set of smaller rank than S will lie in the set $R(S)$ whereas the set of *constituents* of S need not contain any pure sets at all.

First I must lay down an important definition.

5.5.2 Definition *Let S be any object. Then by definition*

$$rank(S) = \{x \in R(S) : x \text{ is a von Neumann ordinal}\}$$

It is not difficult to show that $rank(S)$ is always a von Neumann ordinal.
A straightforward transfinite induction proves the following theorem.

5.5.3 Theorem *Let a be a von Neumann ordinal. Then*

(i) *a is of rank $\subseteq_a$*

(ii) *$rank(a) = a$*

In fact, it can be shown that both S and $R(S)$ are of rank $\subseteq_{rank(S)}$. Note that each von Neumann ordinal a is, in an obvious sense, the simplest

5.5 The cumulative hierarchy of sets

possible set of rank $\subseteq_a$, since each "stage" in its "construction" is a von Neumann ordinal $< a$.

We can summarise the essential properties of the global function R in the following theorem.

5.5.4 Theorem *Let U be a (possibly empty) set of individuals, and a be a von Neumann ordinal. Then*

(i) $R((a+1) \cup U) = R(a \cup U) \cup P(R(a \cup U))$.
(ii) *If a is a limit number, then* $R(a \cup U) = \bigcup \{R(x \cup U) : x < a\}$.
(iii) *Let S and T be sets with*

$$rank(T) \leq rank(S) \leq a \text{ and } Support(T) \subseteq Support(S) \subseteq U$$

Then $R(T) \subseteq R(S) \subseteq R(a \cup U)$
(iv) *For any set S, $S \subseteq R(rank(S) \cup Support(S))$.*
(v) *For any set S, $a < rank(S)$ implies $S - R(a \cup Support(S)) \neq \emptyset$.*
(vi) *For any set S, $S \subseteq R(S)$.*
(vii) *For any set S, $S \in R((rank(S) + 1) \cup Support(S))$.*

In formulating the Weak Hierarchy Principle we start from a given *set* and posit that the sets which are of smaller rank than the given set and whose support is a subset of its support are Cantorian finite in multitude and therefore compose a set.

But what if we start with a given *well-ordering*, r, and posit that, for each set S, the sets whose supports are subsets of $Support(S)$ and each of which is ranked by some initial segment of r together compose a set, $V_r(S)$? Of course if S is a set of rank r, then $V_r(S) = R(S)$. But there is no way to establish on the basis of the axioms that have been laid down so far – even if we include the Weak Hierarchy Principle among them – that every well-ordering ranks some set (or what amounts to the same thing, that every well-ordering is order isomorphic to a von Neumann ordinal).

Thus if we posit the global function $\lambda rxV_r(x)$ just described we obtain a stronger assumption than the Weak Hierarchy Principle. We can do this by laying down a new finiteness principle in the manner we have used hitherto as follows.

Given any object S and any well-ordering, r, the species of all sets T such that $Support(\{T\}) \subseteq Support(\{S\})$ and T is of rank $<_r$ is (Cantorian) finite and therefore forms a set which we denote by

$$V_r(S)$$

The correspondence

$$r, S \mapsto V_r(S)$$

thus defines a first level global function of two arguments which can be characterised as follows:

5.5.5 Axiom *(The Strong Hierarchy Principle) Let r be a well-ordering, and let S and T be sets or individuals. Then* $T \in V_r(S)$ *if, and only if*

(i) $Support(\{T\}) \subseteq Support(\{S\})$
(ii) $(\exists u \in Field(r))[T \text{ is of rank } r_{<u}]]$

The connection between the two Hierarchy Principles is is expressed by the following theorem.

5.5.6 Theorem *Let S be a set and r a well-ordering. Then*

$$S \text{ is of rank } r \; : \text{implies} : V_r(S) = R(rank(S) \cup Support(S))$$

An important consequence of the Strong Hierarchy Principle is the following theorem, which can be established by transfinite induction.

5.5.7 Theorem *Every well-ordering is order isomorphic to a von Neumann well-ordering.*

This follows since S and $rank(S)$ are both of rank r. In fact, this theorem, if taken as an axiom together with the Weak Hierarchy Principle, suffices to establish the Strong Hierarchy Principle.

Theorem 5.5.6 justifies the following definition.

5.5.8 Definition *Let r be a well-ordering. Then by definition*

$$Ordertype(r) = rank(V_r(\emptyset))$$

Thus the von Neumann ordinals can serve as *order types* for well-ordered sets. Note that $Ordertype(\subseteq_a) = a$ for von Neumann ordinals a.

The arithmetic of ordinal numbers can be transferred from that of well-orderings[18]. Indeed, the Strong Hierarchy Principle allows us to develop a large part of Cantor's theory of the transfinite ordinal and cardinal numbers.

---

[18] See Definition 4.9.11.

5.5 The cumulative hierarchy of sets

It is easy, for example, to define the *cardinal number*, $||S||$, of a set, S: just define it to be the smallest ordinal number in the set

$$\{Ordertype(r) : r \text{ well-orders } S\}$$

What is conspicuously missing is the possibility of defining global functions by transfinite recursion. This means, for example, that although we can define cardinal numbers (initial ordinal numbers) *individually*, so to speak, Cantor's function $\lambda x(\aleph_x)$, which *enumerates* the cardinal numbers, cannot be defined in the standard way using transfinite recursion along the von Neumann ordinals. We can, however, establish the possibility of *bounded* recursion, and this will suffice for the definition of $\lambda x(\aleph_x)$.

By a *global ordinal function* I mean a global function that always takes ordinal numbers as values for ordinal numbers as arguments[19].

5.5.9 Theorem *(Definition by Limited Transfinite Recursion)* Let ϕ, σ, and τ, be global ordinal functions and c be a von Neumann ordinal such that

(i) ϕ satisfies the global recursion equations

$$\phi(0) = c$$
$$\phi(x+1) = \sigma(\phi(x)), \text{ for all von Neumann ordinals } x$$
$$\phi(l) = \bigcup\{\phi(x) : x < l\}, \text{ for all limit ordinals } l$$

(ii) For all von Neumann ordinals x and y

$$x < y \text{ implies } \sigma(x) < \sigma(y)$$

(iii) For all von Neumann ordinals x

$$\tau(x) \leq \sigma(x)$$

Then we can define a global ordinal function ψ which satisfies the global recursion equations

$$\psi(0) = c$$
$$\psi(x+1) = \tau(\psi(x)), \text{ for all von Neumann ordinals } x$$
$$\psi(l) = \bigcup\{\psi(x) : x < l\}, \text{ for all limit ordinals } l$$

---

[19] This is, *prima facie*, a Π_1 definition. But any global function, σ, can be "converted" into an ordinal function by replacing it by $\lambda x \, rank(\sigma(x))$, which doesen't change the values of genuine ordinal functions.

and for which

$$\psi(x) \leq \phi(x), \text{ for all von Neumann ordinals } x$$

Notice that each of the three hypotheses of this theorem is Π_1, as is the conclusion.

Now the enumerating ordinal function $\lambda x(\beth_x)$ is easily defined outright:

5.5.10 Definition *Let a be any von Neumann ordinal. Then by definition*

$$\beth_a = \|V_{\omega+a}(\emptyset)\|$$

The global function $\lambda x(\beth_x)$ can easily be shown to satisfy the standard recursion equations

$$\beth_0 = \omega$$
$$\beth_{a+1} = \|P(\beth_a)\| \text{ for all ordinals } a$$
$$\beth_l = \bigcup\{\beth_a : a < l\} \text{ for all limit ordinals } l$$

We may therefore apply Theorem 5.5.9 with $\beth$ in the role of ϕ, $\lambda x(\|2^x\|)$ in the role of σ, $\lambda x(x^+)^{20}$ in the role of τ, and $\aleph$ in the role of ψ, to obtain a definition of Cantor's function $\aleph$ which enumerates the cardinal numbers.

But even though Theorem 5.5.9 allows us, as we have just seen, to carry out certain important transfinite recursions, most set theorists would find it unnatural not to have unrestricted transfinite recursion available as a technique of definition. In the next section I shall introduce a new finiteness principle that will allow us to derive it as a theorem.

5.6 Cantor's Absolute

Let us briefly take stock. When we come to consider the axioms and principles introduced in Sections 5.4 and 5.5, we leave the world of ordinary mathematics behind. All of them treat of the structural properties of the universe of sets, and therefore do not have the air of "invisibility" that the earlier axioms possess. Of these later axioms and principles, only the Axiom of Foundation (5.4.3) and the Axiom of Extensional Analysis (5.4.5) have that character of self-evidence shared, in varying degrees, by the axioms of Chapter 4.

---

[20] a^+ is the first cardinal number beyond the ordinal number a.

5.6 Cantor's Absolute

Once we arrive at the Hierarchy Principles of Section 5.5, however, we have left the domain of immediate self-evidence behind – or so it seems to me. These principles deal directly with the global structure of the universe of sets as a cumulative hierarchy, and they encourage us to regard that universe as somehow "generated" by arbitrary iterations of the operation "set of" (to use Gödel's turn of phrase). Such talk about transfinite "processes" has been quite common among set theorists ever since Cantor, and indeed it has a certain natural appeal: in Cantor's own case, it was, I believe, grounded in his theological views.

What distinguishes the Hierarchy Principles from the earlier axioms is that the naive justification for both of them seems to rest on this concept of transfinite "generation" but taken literally rather than as a mere metaphor. Thus we "construct" $R(S)$ by carrying out all possible iterations of Gödel's "set of" operation along any well-ordering which ranks S, and we "construct" $Ordertype(r)$ by iterating the operation $\lambda x(x \cup \{x\})$ along the well-ordering r.

Of course anyone who studies general set theory soon develops "intuitions" about "transfinite processes". But what is the epistemological status of these intuitions, and how seriously are we to take them? The analogy with natural number arithmetic may be instructive here.

We all have intuitions about "infinite processes" in arithmetic as well. But Dedekind and Frege, when they came to investigate the logical foundations of arithmetic, took it as the very starting point of their investigations, that *appeal to such intuitions should not play any role in their analysis.*[21] Indeed, Dedekind took it to be his central task to give an account of the concept of the *iteration* of a function without appealing to the notion of the *repeated application* of an operation[22].

The problem with appealing to such notions of the "generation" of sets by transfinite "processes" is that such "processes" are so far beyond our *actual* experience, that we can form only the vaguest idea of what they might *be*, and therefore no clear idea of what is *meant* by such an appeal.

Is it not something of an intellectual scandal that our predecessors strove so hard to eliminate such notions from the foundations of arithmetic, analysis, and geometry, only for us place them right at the heart of our deepest foundational theory?

---

[21] I shall discuss this point at greater length in Section 8.2.
[22] See the discussion in following the proof of Theorem 5.1.5 and the passage from Dedekind's letter to Kefferstein quoted in Section 8.2.

But matters are not quite so simple, for there is an important general principle at stake in all this. When we are considering the fundamental principles upon which all mathematics rests – the axioms of set theory – we have to accommodate two fundamentally opposing interests. On the one hand, since these principles ground our proofs, it is in our interest that they be self-evident. On the other hand, since we are searching for the truth, we must take care not to be so severe in the standards we apply to proposed axioms that we fail to acknowledge genuine principles that express true insights into the nature of the universe of sets.

That these interests are in conflict is obvious. Our problem is, perhaps, to achieve the right balance between them in any given case. When the question of self-evidence is paramount we are concerned with the finitude of the finite: are we justified in concluding that the species (implicitly) declared to be sets by a proposed axiom can legitimately be regarded as finite, in the extended Cantorian sense of "finite".

But when the question under consideration is the extent of the set-theoretical universe, then it is the infinitude of the infinite that is at issue: what axioms must we lay down in order adequately to capture the *absolute* character of the infinity of the universe of sets? There is, of course, no sharp dividing line between these two attitudes, the one shading almost imperceptibly into the other, but there are fairly clear instances of each.

By virtue of the Hierarchy Principle (in both forms), the "length" of the von Neumann ordinal sequence is a measure of the extent of the universe of sets. Moreover, the Well-ordering Theorem guarantees, in the presence of the Strong Hierarchy Principle, that for each set there is a von Neumann ordinal having exactly the same size. The consequence of these facts is that, from Cantor on, set theorists have been obsessed with the problem of the length of the sequence of ordinal numbers, and have proposed various *axioms of (strong) infinity* to characterise it.

The simplest of these take the form of assertions to the effect that the absolutely infinite sequence of all von Neumann ordinals cannot be captured "from below" using the particular methods for defining "new" ordinals that are given by applying particular global ordinal functions to ordinals already "given". I shall now show how we can derive global versions of the principle of definition by recursion from simple axioms of this sort.

5.6.1 Definition *Let σ be a global ordinal function of one argument and a*

5.6 Cantor's Absolute

be any von Neumann ordinal. Then by definition

$$a \text{ is simple closure point for } \sigma \;.: \text{iff} :. \; (\forall x < a)[\sigma(x) < a]$$

Given a global ordinal function σ, a von Neumann ordinal, a, that is not closed under σ is, in some sense, "definable from below" by σ. Let me make this rather vague observation more precise.

5.6.2 Theorem *Let σ be an increasing ordinal function and let $a > b > \omega$ be von Neumann ordinals such that a is a limit ordinal and*

$$(\forall x \leq a)[b < x \text{ .implies. } x \text{ is not a simple closure point for } \sigma]$$

Then there exists a von Neumann natural number, n, and a function $f : n + 1 \to a$ such that

(i) $f\text{'}0 = b + 1$
(ii) $f\text{'}(m+1) = \sigma(f\text{'}m)$ *for all* $m < n$
(iii) $\sigma(f\text{'}n) \geq a$

Proof Let S be the set of all functions f with

(a) $Domain(f) \in \omega$
(b) $f\text{'}0 = b + 1$
(c) $f\text{'}(m+1) = \sigma(f\text{'}m)$ for all $m < Domain(f)$
(d) $Range(f) \subseteq a$

Clearly, if g and h both lie in S with $Domain(g) \subseteq Domain(h)$, then $g \subseteq h$. There are thus two possibilities:

$$(\forall x \in \omega)(\exists g \in S)[Domain(g) = x + 1] \text{ (Case 1)}$$

and

$$(\exists f \in S)(\forall g \in S)[g \subseteq f] \text{ (Case 2)}$$

I claim that Case 1 is impossible. For suppose otherwise, and let $f = \bigcup S$. Then $f : \omega \to a$ and $b < \bigcup\{f\text{'}n : n < \omega\} \leq a$, so $\bigcup\{f\text{'}n : n < \omega\}$ gives us a closure point for σ, which contradicts the hypothesis of the theorem.

Hence Case 2 obtains. Let $f = \bigcup S$ be the maximal member of S and let $n + 1 = Domain(f)$. Then $\sigma(f\text{'}n) \geq a$, as otherwise $f \cup \{(n+1, \sigma(f\text{'}n))\}$ is a proper extension of f in S, which is impossible. The theorem now follows. $\square$

Could there be an increasing ordinal function σ for which *no* von Neumann ordinal was a simple closure point for σ beyond some ordinal b? Theorem 5.6.2 tells us that if that were the case then *every* von Neumann ordinal $a > b$ could be reached by a Euclidean finite number of iterations of σ starting from $b + 1$. This seems to violate our intuition of the absolute infinity of the sequence of all von Neumann ordinals.

But we cannot eliminate this possibility on the grounds of the axioms so far considered, for if ϕ is defined at ordinals a by setting

$$\phi(a) = \sup\{Ordertype(x) : x \in Wo(a)\}$$

where

$$Wo(a) = \{x \in V_a(\emptyset) : x \text{ is a well-ordering}\}$$

and if $b > \omega$ is the first simple closure point for ϕ beyond ω then $V_b(\emptyset)$ satisfies all those axioms, but there is no simple closure point for ϕ in $V_b(\emptyset)$ beyond ω.

The alternative is to suppose that the ordinal a and the global ordinal function σ together implicitly define the first simple closure point, $CL_0(\sigma, a)$, for σ beyond a.

5.6.3 Axiom *(The Principle of Simple Closure)* Let σ be any increasing global ordinal function, and a any ordinal. Then

(i) $CL_0(\sigma, a)$ is a simple closure point for σ.
(ii) $a < Cl_0(\sigma, a)$.
(iii) $(\forall x < Cl_0(\sigma, a))[a < x \text{ .implies. } x \text{ is not a simple closure point for } \sigma]$.

As an immediate consequence of this assumption we obtain the following important theorem.

5.6.4 Theorem *(Definition by Global Recursion along ω)* Let σ be a global ordinal function and a be a von Neumann ordinal. Then there is a unique local function f with $Domain(f) = \omega$ that satisfies the following recursion equations:

$$f`0 = a$$
$$f`(n+1) = \sigma(f`n) \text{ for all } n \in \omega$$

Proof Since $CL_0(\sigma, a)$ is a simple closure point for σ, the restriction of σ to $CL_0(\sigma, a)$ is a local function $g_{\sigma,a} : CL_0(\sigma, a) \to CL_0(\sigma, a)$, so we can define f by a local recursion on $g_{\sigma,a}$. □

5.6 Cantor's Absolute

But the construction in the proof of Theorem 5.6.2 suggests a further consideration. The functions considered there clearly represent Euclidean finite *iterations* of the function σ. Of course we cannot know, on the basis of the axioms adopted so far, whether the "iteration" of σ can be extended into the transfinite. But we can lay down the following definition.

5.6.5 Definition *Let σ be any global ordinal function of one argument, let a and b be von Neumann ordinals with $a > 0$, and let f be a local function with $\text{Domain}(f) = a$. Then by definition*

f *is an iteration of σ along a (starting) from b .iff.*

(i) $f\text{`}0 = b$.
(ii) $f\text{`}(x+1) = \sigma(f\text{`}x)$ *for all x for which $x + 1 < a$.*
(iii) $f\text{`}l = \bigcup \{f\text{`}x : x < a\}$ *for all limit ordinals l for which $l < a$.*

We can now lay down the definition of a *strong closure point* for a global ordinal function σ.

5.6.6 Definition *Let σ be any global ordinal function of one argument and let $a > \omega$ be a limit ordinal. Then by definition*

a is a strong closure point for σ . : iff : . $(\forall x, y < a)(\forall f : x \to a)[f$ is an iteration of σ along x from y .implies. $\bigcup f\text{``}x < a]$

Note that a strong closure point for σ is also a simple closure point.

It seems reasonable to suppose that this gives us one possible formulation of the notion of an ordinal's *not* being "definable from below" by the global ordinal function σ. The following theorem lends plausibility to this account of "definability from below".

5.6.7 Theorem *Let σ be an increasing global ordinal function, let $a > \omega$ be a limit ordinal, and let $\omega < b < a$ such that*

$(\forall x \leq a)[b < x \text{ .implies. } x \text{ is not a strong closure point for } \sigma]$

Then there exists a von Neumann natural number, $c < a$, and a function $f : c \to a$ such that

(i) f *is an iteration of σ along a starting from b.*
(ii) *If c is a successor ordinal $c = d + 1$ then $\sigma(f\text{`}d) \geq a$.*
(iii) *If c is a limit ordinal then $\bigcup f\text{``}c = a$.*

Proof Let S be the set of all functions f such that

(1) f is an iteration of σ along $Domain(f)$ starting from $b+1$.
(2) $Domain(f) < a$.
(3) $Range(f) \subseteq a$.

Clearly, if g and h both lie in S with $Domain(g) \subseteq Domain(h)$, then $g \subseteq h$. There are thus two possibilities:

$$(\forall x < a)[0 < x \text{ implies } (\exists g \in S)[Domain(g) = x]] \text{ (Case 1)}$$

and

$$(\exists f \in S)(\forall g \in S)[g \subseteq f] \text{ (Case 2)}$$

I claim that Case 1 is impossible. For suppose otherwise, and let $f = \bigcup S$. Then $f : a \to a$ and, since σ is increasing and a is a limit ordinal, $\bigcup \{f`x : x < a\} = a$. Hence a is a strong closure point for σ, which contradicts the hypothesis of the theorem.

Hence Case 2 obtains. Let $f = \bigcup S$ be the maximal member of S, and let $Domain(f) = c < a$. Suppose first that c is a limit ordinal. Then $\bigcup f``c \geq a$, as otherwise $f \cup \{(c, \bigcup f``c)\}$ is a proper extension of f in S, which is impossible. But then $\bigcup f``c = a$ as required, since $f``b \subseteq a$. On the other hand, if $c = d+1$ is a successor ordinal, then $\sigma(f`d) \geq a$ as required, since otherwise $f \cup \{(c, \sigma(f`d))\}$ is a proper extension of f in S, which is impossible. The theorem now follows. □

Note that if a is given as in the theorem, then the local function f whose existence is guaranteed by its proof can be defined by a *local* transfinite recursion along the ordinal c. Such functions are simply *there*, as it were, and bear witness to the existence of local "processes" of iteration which define ordinals beyond the ordinal a.

As before we may ask: could there be an ordinal function σ and an ordinal $b \geq \omega$ for which *no* limit ordinal $a > b$ is a strong closure point for σ? Let us imagine a process of "generating" von Neumann ordinals starting with $b+1$ and "iterating" the function σ. As larger and larger ordinals are "generated" in this fashion, they can serve, in their turn, as the domains for further and further "iterations". Theorem 5.6.7 tells us that in the case that *every* von Neumann ordinal beyond the particular von Neumann ordinal b is *not* a strong closure point for σ, every such ordinal would be reachable from below by σ; in this sense we could say that the whole of the absolutely infinite sequence of von

5.6 Cantor's Absolute

Neumann ordinal numbers could be generated by a single (Cantorian) finite procedure, namely that given by the definition of σ[23].

These considerations seem to suggest that a fixed global ordinal function σ and a von Neumann ordinal a together implicitly define a first ordinal $CL_1(\sigma, a) > a$ that is a strong closure point for σ. This is the content of the Principle of Strong Closure.

5.6.8 Axiom *(The Principle of Strong Closure)* *Let σ be a global ordinal function of one argument, and let a be a von Neumann ordinal. Then $CL_1(\sigma, a)$ is a von Neumann ordinal such that*

(i) $a < CL_1(\sigma, a)$.
(ii) $CL_1(\sigma, a)$ *is a strong closure point for σ.*
(iii) *For all ordinals x, if $a < x < CL_1(\sigma, a)$, then x is not a strong closure point for σ.*

The Principle of Strong Closure allows us to establish a theorem on definition by global transfinite recursion along the von Neumann ordinals.

5.6.9 Theorem *(Definition of a Global Ordinal Function by Transfinite Recursion along the von Neumann Ordinals)* *Let σ be a global ordinal function of one argument and let c be a von Neumann ordinal. Then we can define a global ordinal function, ϕ, of one argument satisfying the following recursion equations:*

$$\phi(0) = c$$
$$\phi(a+1) = \sigma(\phi(a)) \text{ for all von Neumann ordinals } a$$
$$\phi(l) = \bigcup \{\phi(x) : x < l\} \text{ for all limit ordinals } l$$

Proof Since every strong closure point of σ is a simple closure point, the restriction of σ to $CL_1(\sigma, a)$ gives a local function $g_{\sigma, a} : CL_1(\sigma, a) \to CL_1(\sigma, a)$. Local recursion with respect to $g_{\sigma, a}$ gives us a function f_a defined on a and satisfying the recursion equations. (Here we must use the fact that a is a *strong* closure point for σ.) We can then define $\phi(a)$ to be $f_{a+1}`a$. □

Thus we see the global form of transfinite induction along the von

---

[23] σ may be obtained from a global function τ by assigning fixed sets as values to all but one of its arguments. But those fixed arguments contribute only a *Cantorian finite* amount of information to the definition of σ, so to speak.

Neumann ordinals can be justified by arguments that rest on considerations of the absolute infinity of the universe of sets. In fact, the proof ultimately rests on both the Strong Hierarchy Principle and the Strong Closure Principle. On the other hand, both the latter principles can be derived from the Principle of Global Transfinite Induction.

All three of these principles rest on intuitions about the possibilities of *iterating* certain kinds of "constructions". Indeed, in the case of Global Transfinite Induction we seem to have a direct expression of these intuitions. My aim in presenting these principles in the way, and in the order, that I have has been to draw distinctions among these various intuitions of iteration, and to show how considerations of the *absolute* infinity of the ordinal number sequence can be employed in justifying the introduction of these principles.

In particular, it seems to me significant that in formulating the Strong Closure Principle in terms of iterations (Definition 5.6.5) we don't have to *assume outright* that transfinite iterations of global functions are possible: we have only to assume that *insofar as they are possible*, we cannot use them to generate the absolute from below, so to speak. This surely makes it preferable to take *it* as an axiom and to derive the Principle of Definition by Transfinite Recursion as a theorem, rather than the other way around.

Clearly this question of what role intuitions about iteration should play in the foundations of Cantorian set theory is of central importance[24]. Moreover it seems to me to be beyond question that these intuitions arise out of our experience of dealing with symbolic notation.

This is perhaps most easily seen in connection with the Principle of Global Recursion along ω (Theorem 5.6.4). We can clearly *syntactically* iterate a global function σ "without limit", so to speak:

$$S, \sigma(S), \sigma(\sigma(S)), \sigma(\sigma(\sigma(S))), \ldots$$

and this tempts us into supposing that a *proof* is unnecessary here. Indeed, if we are Cantorians can we not form the union

$$\bigcup \{S, \sigma(S), \sigma(\sigma(S)), \sigma(\sigma(\sigma(S))), \ldots\}$$

and then start the process all over again? What is to prevent our iterating process from passing beyond *any* particular, fixed ordinal? But how literally can we take this sort of conception? Surely it is clear that here,

---

[24] The question is even more important in Euclidean set theory, except there the answer is clear: they should play no role whatsoever. In the Cantorian case matters are not so clear-cut.

5.7 Axioms of strong infinity

for better or for worse, we are extrapolating our own, rather feeble *actual* capacities beyond all conceivable limits. Is this legitimate? *That* is the question. And it arises even with global recursion along ω.

But let us leave this difficult question and ask: can intuitions of "transfinite processes" be pressed further to yield yet stronger axioms of infinity? Or is there a clear resting place here?

5.7 Axioms of strong infinity

There is a widespread view (Zermelo and Gödel both held it) that considerations of the absolute infinity of the universe of sets give rise to natural axioms of strong infinity that go far beyond the axioms considered so far. Here is Gödel on the subject:

[T]he axioms of set theory by no means form a system closed in itself, but, quite on the contrary, the very concept of set on which they are based suggests their extension by new axioms which assert the existence of still further iterations of the operation "set of". These axioms can be formulated also as propositions asserting the existence of very great cardinal numbers (i.e. of sets having these cardinal numbers). The simplest of these strong "axioms of infinity" asserts the existence of inaccessible numbers (in the weaker or stronger sense) $> \aleph_0$. The latter axiom, roughly speaking, means nothing else but that the totality of sets obtainable by use of the procedures of formation of sets expressed in the other axioms forms again a set (and, therefore, a new basis for further applications of these procedures)[25]. Other axioms have first been formulated by P. Mahlo[26]. These axioms show clearly not only that the axiomatic system of set theory as used today is incomplete, but also that it can be supplemented without arbitrariness by new axioms which only unfold the concept of set explained above [i.e. the concept of set obtained by the iteration of the operation "set of"].[27]

Let us attempt to mount, in some detail, the kind of argument that Gödel has in mind here.

Recall that a von Neumann ordinal, a, is *singular* if there is an ordinal $b < a$ and a function $f : b \to a$ such that $\bigcup f``b = a$, and *regular* otherwise. Now it seems that, *in some sense*, a singular ordinal is reachable "from below". This suggests that we might augment our existing theory by introducing a stronger analogue of Strong Closure.

5.7.1 Definition *Let σ be a global ordinal function of one argument and $a > \omega$ be a von Neumann ordinal. Then by definition*

---

[25] Zermelo, "Über Grenzzahlen und Mengenbereiche".
[26] "Über lineare transfinite Mengen".
[27] "What is Cantor's continuum hypothesis?", Section 3.

> a is an inaccessible point for σ : iff : a is regular and a is a simple closure point for σ

Note that the condition that a be regular guarantees that an inaccessible point for σ is also a strong closure point for σ. Using the notion of an inaccessible point we can formulate a version of Mahlo's Principle as an obvious analogue of the Principle of Strong Closure.

5.7.2 Axiom *(Mahlo's Principle) Let σ be a global ordinal function of one argument and let a be a von Neumann ordinal. Then*

(i) $a < M(\sigma, a)$.

(ii) $M(\sigma, a)$ is an inaccessible point for σ.

(iii) $(\forall x < M(\sigma, a))[x \leq a \text{ .or. } x \text{ is not an inaccessible point for } \sigma]$.

Note, for example, that a regular ordinal is just an inaccessible point for the global ordinal function $\lambda x(x+1)$; in particular, $\omega = M(\lambda x(x+1), n)$, for any von Neumann natural number n. An inaccessible point for the global ordinal function $\aleph$ is called a *weakly inaccessible number* and an inaccessible point for the global ordinal function $\beth$ is called a *strongly inaccessible* number[28]. Thus what Gödel refers to above as "the simplest" of these strong infinity axioms are, in effect, particular cases of this axiom.

Mahlo's Principle is much stronger than the Strong Closure Point Principle – the pure sets of rank less than the first strongly inaccessible ordinal are easily seen to model all the axioms and principles we have considered up to the statement of Mahlo's Principle. In fact, we can use Mahlo's Principle in conjunction with global transfinite induction to define global ordinal functions that grow at a stupefying rate.

But can we mount the same sort of argument for Mahlo's Principle that we gave for the two Closure Principles? This depends upon the extent to which we can legitimately claim that an ordinal a which is *not* an inaccessible point of a global ordinal function σ is "reachable from below" by σ. Obviously this is the case if $\sigma(b) \geq a$ for some ordinal $b < a$. The critical case is when a is not an inaccessible point for σ by virtue of being singular[29]. Does a's being singular mean that it can be "reached from below"?

---

[28] Strongly inaccessible numbers are the order types of the ordinals in models of second order Zermelo–Fraenkel set theory. See Section 7.1.

[29] An ordinal a is *singular* if a is a limit ordinal and there is an ordinal number $b < a$ and a function $f : b \mapsto a$ such that $\bigcup f``b = a$.

5.7 Axioms of strong infinity

Let us try to "generate" the ordinal a: as we carry out our principal process of generating the ordinals $< a$

$$0, 1, \ldots; \omega, \omega + 1, \ldots, b, b + 1, \ldots$$

we set up auxiliary processes of generation

$$f\text{'}0, f\text{'}1, \ldots; f\text{'}\omega, f\text{'}(\omega + 1), \ldots, f\text{'}c, \ldots, \text{ for } c < b$$

for each ordinal $b < a$ and each local function $f : a \to a$, adding the ordinals thus generated to the principal list.

But it is precisely here that we see the difficulty. For what we need to generate is not so much the ordinals $b < a$ as the set, S, of all pairs (b, f) for which $b < a$ and $f : b \to a$ – if, that is, we are to see *a in advance* of its complete generation; but until we have generated the *whole* of a we shall not have succeeded in generating a *single* element of the set S which witnesses the singularity of a. We do not have access to *all* the potential witness in S to the singularity of a until *after* we have generated the *whole* of a: we are not given all those functions *in advance* in the way that we are given σ in advance. Thus we discover that a is singular only after the fact, so to speak: we cannot see a in advance of its generation.

Of course, if a is described in such a way as to make it obvious that it is singular (e.g. if $a = \aleph_\omega$), then we *can* see in advance that a is singular, for the function witnessing its singularity is, in this example, the restriction of the global function $\aleph$ to ω. But such cases are atypical, for, in general, the local functions witnessing the singularity of an ordinal are not given by anything remotely resembling a "rule"[30].

Note the contrast here with the analogous argument for the Strong Closure Principle. In that case the auxiliary processes of generation for ordinals $b < a$

$$\sigma(0), \sigma(1), \ldots; \sigma(\omega), \sigma(\omega + 1), \ldots, \sigma(c), \ldots, \text{ for } c < b$$

are all with respect to a *fixed σ given in advance of the whole process of generation*.

Of course such arguments can only be suggestive. But what the argument I have just given suggests is that the jump from the informal justification for the Strong Closure Principle to an analogous justification

---

[30] Consider the ordinal number $\aleph_1^L$ which is the first *constructible cardinal* beyond $\aleph_0$ in Gödel's constructible universe L. Although this is uncountable in L many set theorists believe it to be countable absolutely. In that case $\aleph_1^L$ provides an example of a singular ordinal a – in fact, an ω-cofinal one – whose singularity cannot be witnessed by any local function of which we can form a clear conception, let alone one specifiable in advance of the generation of the whole of a.

for Mahlo's Principle is a very great one indeed. Indeed, I submit that the Strong Closure Principle itself constitutes a precise formulation of Gödel's claim, quoted above, that

the totality of sets obtainable by use of the procedures of formation of sets expressed in the other axioms forms again a set ...

and therefore the latter does *not*, as Gödel claims, entail the existence of strongly inaccessable numbers. So we cannot, *pace* Gödel, make a case for Mahlo's Principle analogous to that made for Strong Closure. But, in fact, among the strong infinity axioms that have actually been considered Mahlo's Principle is very modest.

None of these strong infinity axioms, not even the weakest of them, the Strong Closure Principle, belong to that "invisible" part of set theory that underlies the everyday practice of mathematics. But that doesn't mean that they are irrelevant to that practice. On the contrary, there is a vast gap between the set-theoretical machinery required to *define* the various structures studied by mathematicians and that required *prove* facts about them.

It is a consequence of the famous Incompleteness Theorems of Gödel that postulating various axioms of strong infinity allows us to prove facts about diophantine equations that would otherwise not be provable. The most familiar examples of this are propositions that represent complicated encodings in natural number arithmetic of propositions of general syntax – propositions asserting the consistency of certain formal theories, for example. Perhaps it might be thought that these examples are too artificial to belong to "ordinary" mathematics. But there are examples that are much more natural.

The circumstances surrounding the discovery of Borel Determinacy are especially instructive here. The theorem that every Borel game has a winning strategy was proved by Martin[31]. It is a theorem of real analysis and can be easily *stated* using only such set-theoretical machinery as is the common property of all mathematicians.

But some years before Martin discovered his proof, Friedman[32] had proved that the theorem cannot be proved without appealing to the existence of sets of power $\beth_\omega$, so that, for example, it is unprovable in formal Zermelo set theory[33], which is much stronger proof-theoretically

---

[31] "Borel Determinacy".
[32] "Higher set theory and mathematical practice".
[33] This is first order Zermelo–Fraenkel set theory without the Axiom of Replacement. See Section 7.1.

5.7 Axioms of strong infinity

than the formal theory of real analysis in which the theorem is most naturally formulated[34].

Thus we need to know the truth about these axioms of strong infinity if we are to know the truth about superficially simpler propositions in combinatorics and number theory. But how are we to gain this knowledge?

It is conceivable that, in Gödel's words, we may discover

> other (hitherto unknown) axioms of set theory which a more profound understanding of the concepts underlying logic and mathematics would enable us to recognize as implied by those concepts.

But then an equally interesting possibility is that we might discover that the simplest of these axioms – the existence of an inaccessible number – is inconsistent[35]. Either of these eventualities would certainly simplify the foundations of Cantorian set theory.

Perhaps we should not regard axioms of strong infinity beyond Strong Closure in the same way that we regard genuine axioms like Comprehension or Replacement. For we are not really dealing with *axioms* here, that is to say, *self-evident* principles with which to augment our basic theory.[36]

Surely it is better that we openly acknowledge the *hypothetical* and *speculative* character of these "axioms". We use them to investigate *possibilities* for the size of the *absolutely* infinite, and since they are (as far as we know) *consistent* with the axioms of set theory, we have good reason to regard them, at least provisionally, as *true*, for they tell us that the absolute is larger than we might otherwise have supposed. But we have no reason to describe these *hypotheses* as *axioms*.

---

[34] Friedman has also established a series of results showing the equivalence of natural problems in the combinatorics of the natural numbers to the consistency of certain axioms of strong infinity in the theory of sets.

[35] Indeed, even the inconsistency of formal first order Zermelo–Fraenkel set theory is not, after all, inconceivable, however unlikely. The latter would be perfectly compatible with the consistency of our theory with all the axioms up to and including the Strong Closure Principle.

[36] I believe that our technical terminology is playing us false here. When we deal with axiomatic theories as understood in the modern sense – e.g. group theory – the "axioms are not to be regarded as fundamental *truths* but rather as components in a *definition*. So if set theory is taken to be a conventional axiomatic theory, then when it is augmented to include, say, a version of Mahlo's Principle, it is natural to call the latter an "axiom of set theory", just as it is natural to call the Associative Law an "axiom of group theory". But set theory, in its capacity as a foundation for mathematics, is not, and cannot logically be, an axiomatic theory in this modern sense. The use of the term "axiom" in these contexts is most unfortunate and productive of serious confusion. I shall take up these issues in the next chapter.

In any case, the most powerful of these hypotheses are best formulated as propositions to the effect that there exist models of certain extensions of *axiomatic* Zermelo–Fraenkel set theory, formalised, following Zermelo's example, as a *second order* theory. I shall return to this point in Section 7.2.

6
The Axiomatic Method

6.1 Mathematics before the advent of the axiomatic method

Up to now I have been treating set theory as the general theory of finite pluralities and my central preoccupation has been with the concept of finiteness, for that is what lies behind the basic principles. Now that the theory, in its Cantorian version, is in place, I can turn to consider its central role in mathematics as a foundation for the modern axiomatic method. But let me begin with some observations on the older approach to mathematics that the modern axiomatic method has supplanted.

Before the advent of modern axiomatics mathematicians were happy to call upon what they fancied to be their capacity for creative abstraction. Newton, as we have already seen[1], defined numbers as "abstracted ratios of concrete quantities". This is somewhat mysterious, since ratios are not things, but *relations* between things, a difficulty already pointed out by Newton's predecessor, Isaac Barrow.

In Lecture XX of his Lucasian lectures, which bears the title *That Reasons* (i.e. ratios) *are not Quantities*, Barrow, after first apologising for entering "upon a Thing so subtle and intricate, which, either by the Nature of the Thing, or thro' the Fault of those that treat of it, is involved in a most thick Mist", goes on to explain why he rejects views similiar to those that Newton was later to adopt.

Since Reason [i.e. *ratio*] is, and is acknowledged to be a pure perfect Relation, how can it pass into another Category, and constitute a Genus of Quantity? Since it is nothing but a respect of two *Quantums* founded in Quantity, how can it be conceived a *Quantum* itself, or a Subject of Quantity? Since it is abstractly a Relation how can it be concretely a Thing related? Is this not to confound Things absolute with Things respective, and concrete Names with absolute? Logicians

---
[1] See Section 2.1.

have hitherto taught that *Relations are inherent to, are attributed to, and depend upon absolute Things*; but absolute Things, I believe, are never said or heard by any to inhere in Relations. So neither does it seem plausible to be said, nor possible to be conceived, that Relations respect themselves, that Habitudes have themselves this or that Way, that Distances are distant, Similitudes similar, or Comparisons compared.[2]

I have quoted Barrow at length in order to convey some of the flavour of his objections to the innovations that his successor was prepared to embrace.

The scholastic terminology, ultimately Aristotelian in origin, in which Barrow's objections were couched was already falling into disfavour, especially with mathematicians and natural philosophers. All these careful distinctions do, indeed, constitute a barrier to exploiting the possibilities latent in the then new techniques of symbolization; they also constitute a barrier to metaphysical confusion, exactly as they were designed to do. We have inherited both the powerful symbolic techniques and the metaphysical confusion.

In any case, encouraged by the success of their science in describing nature, and emboldened by the prestige such success conferred, Newton's successors laid claim to a licence for coining all sorts of abstract mathematical objects: real numbers, imaginary numbers, geometrical points at infinity, ... The mathematician, it would seem, has a creative facility limited only by what is logically absurd.

In his essay on the foundations of the theory of natural numbers[3], Dedekind insisted on maintaining that the natural numbers are "free creations of the human mind". But when he comes to explain, in §73, the exact nature of that "creation" he gives us something all too familiar:

If in the consideration of a simply infinite system, N, set in order by a transformation, ϕ[4], we entirely neglect the special character of the elements: simply retaining their distinguishability and taking into account only the relations to one another in which they are placed by the order-setting transformation ϕ, then are these elements called *natural numbers* or *ordinal numbers* or simply *numbers*, and the base-element 1 is called the *base-number* of the *number series N*. With reference to this freeing of the elements from every other content (abstraction) we are justified in calling numbers a free creation of the human mind.

---

[2] Isaac Barrow, *The Usefulness of Mathematical Learning explained and demonstrated: being Mathematical Lectures read in the Publick Schools at the University of Cambridge*, Lecture XX *That Reasons are not Quantities*, pp.368–369.

[3] *Was sind und was sollen die Zahlen?*

[4] In the terminology of Section 5.1, $(N, \phi, 1)$ is a simply infinite system (starting with "1" instead of "0").

6.1 Mathematics before the advent of the axiomatic method

The simply infinite system N and the transformation ϕ are clearly just *given* here, so the "creation" of the numbers is effected simply by "neglecting", that is to say, "failing or refusing to attend to" the special characters and individuating peculiarities of the elements of N. Now refusing to attend to inessential features of a structure is, without doubt, a sensible thing to do: but it hardly counts as an act of *creation*.

In fact, all this talk of "creating" mathematical objects – by abstraction or by simple fiat – is unalloyed nonsense. And it is the great virtue of the modern axiomatic method that it renders all such talk unnecessary.

The irony here is that Dedekind saw what was required with greater clarity than any of his contemporaries, and, indeed, the monograph from which the above remarks on "creative" abstraction are taken is perhaps the first mathematical work in which the modern axiomatic method is properly and self-consciously employed. It is as if Dedekind, boldly striding out into the unknown, felt he had to perform one last act of obeisance to the received ideas of the philosophy of his day.

It would be difficult to overestimate the importance of the invention of the modern axiomatic method, for philosophy as well as for mathematics itself. For mathematics, it has been the central and indispensable technique underlying the tremendous expansion in the scope of the subject that has taken place in the twentieth century. By cutting our mathematics free of its former *logical* dependence on geometrical and kinematical intuition, it has made it possible, for example, for us to invent a "geometry" of infinite dimensional "spaces". And this geometry is not a mere flight of mathematical fancy, for it plays a central role in modern physics.

Where does this method get its power? It is a commonplace that mathematics is now much more "abstract" than it was in the past. But this stands the facts squarely on their heads, for the axiomatic method is precisely the method that permits mathematicians to *avoid* trafficking in abstractions.

"Abstraction" is a red herring: the natural numbers are not what they are because of some ineffable quality each natural number has in its role as a particular abstract mathematical object; it would be closer to the truth to say that the *system* of natural numbers is what it is because its morphology is governed by the conditions laid down in the axioms for a simply infinite system.

The system of natural numbers, considered in the traditional way as a system of particular "abstract objects", is like the imaginary economy in which everyone earns his living by taking in other people's washing:

each natural number takes its being from its relations, algebraic and order-theoretic, with all other natural numbers.

But, as I explained in Section 5.1, in the axiomatic approach there is no unique, particular system of natural numbers: there is instead an absolutely infinite species of mutually isomorphic *simply infinite systems*, each of which has as much (and therefore as little) claim to unique status as *the* system of natural numbers as any other. Moreover, in any particular simply infinite system each element has its own individual personality, so to speak: it can stand on its own as a *bona fide* object, and its existence does not depend upon its relations with the other elements of that system.

The advent of the axiomatic method, in its modern form, encouraged, indeed required, the development of set theory as a foundation for mathematics. But, as I explained at length in Sections 2.5 and 2.6, sets are not "abstract objects" like natural numbers or real numbers (as conventionally conceived).

The ontological extravagance in modern mathematics does not consist in its taking sets *per se* to be genuine objects – the set of books in my office is not an ineffable abstraction, or, indeed, an abstraction of any sort. The extravagance, if extravagance it be, consists, rather, in our following the example of Cantor in extending the boundaries of the finite so as to include among finite pluralities, that is to say, among sets, pluralities hitherto universally regarded as genuinely infinite.

It is, if you like, an extravagance of *extrapolation* rather than an extravagance of *postulation*. It is not the existence of *sets*, but the existence of sets *which are not Euclidean finite*, that should arouse our suspicions.

The notion that mathematics deals with a special class of "mathematical objects" is an *idée reçue* which precludes an understanding of the place of mathematics in the intellectual economy, and those who are caught up in that idea will, no doubt, continue to exercise their wits in explaining how we stand with respect to those objects, how they manage to affect the circuitry of our brains, whether they are invented or discovered, whether they are inside or outside our minds, etc., etc. But the axiomatic method has rendered all such speculations superfluous, and surely that represents a significant advance.

Mathematicians under the *ancien régime* set a very poor example to philosophers: they were in complete agreement that their propositions were true, indeed necessarily so, but quite incapable of saying what their propositions were true *of*. The prestige which accrued to mathematics in consequence of its unprecedented success in applications, both practical and theoretical, swept all before it.

Thus Berkeley's acute criticisms of the logical shortcomings of the Calculus were noted and then ignored: of what consequence could such quibbles be when weighed against the stupendous advances in the understanding of nature which the Calculus made possible? (It wasn't until the twentieth century that physics advanced beyond the point where traditional Newtonian methods in the Calculus were adequate.)

But the introduction of the axiomatic method, in its modern guise, has redeemed these shortcomings, making it possible to settle philosophical disputes of great antiquity concerning the nature of mathematics, of its subject matter, and of its method. It has also set a standard of rigour in argument and in definition of quite unprecedented severity, thus providing philosophers and mathematicians alike with an illuminating and edifying example of the intellectual precision that is within the grasp of human reason.

6.2 Axiomatic definition

Let us examine how it is that the axiomatic method allows us to dispense with the "mathematical objects" of tradition. Every mathematician understands this, at least up to a point, but it is hardly ever spelled out in any but the sketchiest detail. In fact, the axiomatic method shows us how old-fashioned talk about "mathematical objects" can be rendered harmless, at least in most *mathematical* contexts, despite being without content when taken literally. The mathematician thus has no incentive to purge his discourse of reference to these "things", or to worry about how he might set about doing so. Moreover, in many mathematical contexts talk about certain kinds of "mathematical objects" – real numbers, for example – greatly simplifies that discourse.

But when foundational or philosophical issues are under discussion, as they are here, these loose, if convenient, ways of talking can lead to serious confusion. It is essential, therefore, that we draw careful distinctions – distinctions that in an ordinary mathematical context might seem unnecessarily pedantic – in order to attain a clear conception of what mathematicians are actually and literally talking about when they employ the axiomatic method.

Traditional mathematical objects are of two basic kinds: the *ideal objects* of traditional geometry and rational mechanics (points, lines, planes, triangles, circles, pyramids, spheres, etc., all ideally exact, and continuous motions of these things, again all ideally exact), and the *abstract objects* of traditional algebra and arithmetic (numbers – natural, rational, real, and complex).

The axiomatic method, however, requires us to regard, not mathematical *objects*, but rather mathematical *structures*, as the principal bearers of mathematically significant properties, and moreover, to regard as mathematically significant only those properties a structure shares with all structures isomorphic to it. This means that when we attempt to characterise a *particular* traditional domain (the system of real numbers, say, or three-dimensional Euclidean geometry), we must do so by laying down axioms for the appropriate "intended" mathematical structure; and the most we can hope for is to capture that structure *up to isomorphism*, so that our axioms are satisfied by all and only those structures that are isomorphic to it[5].

Thus if we are to understand *how* the axiomatic method frees us of the need to employ the "mathematical objects" of tradition we have to look carefully at the notions of *mathematical structure* and *isomorphism*.

As I observed in Section 5.1, a mathematical structure is typically composed of an *underlying set* or *universe of discourse* equipped with a *morphology*, which, in its turn, typically consists of one or more operations and/or relations defined on the underlying set of the structure[6]. A group, a ring, a Dedekind structure, a partial ordering – all these are mathematical structures in this straightforward sense.

I have already discussed the notion of isomorphism for Dedekind structures in Section 5.1 in connection with the axiomatic definition of the natural numbers. But natural numbers, as *abstract* objects, present themselves to us through their mutual relations and the algebraic operations to which they are subject. With "mathematical objects" of this abstract, symbol generated sort, which do not correspond to particular representations in pure intuition (what Kant called "*reine Anschauung*"), it requires, perhaps, no great alteration in our conception of them to see them as "given" or "determined" in some way by an axiomatic theory.

The *ideal* objects of geometry are quite another matter, however; for,

---

[5] We have no reason to believe, and strong reason to doubt, as we shall see, that there really is an intended structure in any of these critical cases, so the assumption that our axioms are true of "it" can only be provisional.

[6] The morphology can also be given by higher order relations on the underlying set, as in topological spaces, where the morphology consists in a unary relation on its power set. Sometimes the morphology of a structure involves other structures, as in the case of a differentiable manifold or a real vector space. These, however, are complications that we need not consider here.

unlike the *abstract* objects of traditional algebra and arithmetic, they can be visualised or imagined: represented as particulars in pure (sensual) intuition (Kant's *reine Anschauung*). This has the important consequence that these ideal mathematical objects are directly related to perception and therefore seem to have a more robust claim on existence than the abstract ones do[7].

Nevertheless, the logic of the axiomatic method is exactly the same in both these cases. It will prove useful, however, to have an example of both sorts of axiomatic "exorcism" in hand. Let us turn, therefore, to three-dimensional Euclidean geometry for a close analysis of the notion of isomorphism.

Suppose that the mathematical structure ***E*** constitutes the *intended* (or *standard*) *model* for a suitable system of axioms for three-dimensional Euclidean geometry, so that ***E*** lays out for inspection, as it were, the traditional elements of that geometry – points, lines, and planes, say, regarded in the traditional way – as members of its underlying set, and singles out the key traditional concepts of that geometry – the concepts *point, line, plane, lies on*, etc. – by means of its morphology. Thus the structure ***E*** represents a mathematical embodiment, so to speak, of our intention to interpret the Euclidean axioms in the traditional way[8].

Now subject the elements of the underlying set, E, of ***E*** to an arbitrary permutation $f : E \to E$ ($1-1$, onto). In general such a permutation will send elements of E to elements of E in so random a fashion that the points, lines, and planes of E are jumbled all higgledy-piggledy after transformation by f. However, the permutation f naturally induces a permutation on the n-ary functions over E and the n-ary relations on E for every n[9].

Now let ***E_f*** be the structure obtained from ***E*** by replacing each item of morphology by its image under the relevant permutation induced by f. Then f is an *isomorphism* from ***E*** to ***E_f***, and in these circumstances mathematicians say that ***E*** and ***E_f*** are "abstractly" identical, that they have the same "form", and that the mathematically significant properties of the one structure are, *mutatis mutandis*, those of the other.

---

[7] Indeed, as we have seen, Newton regarded real numbers as being obtained by "abstracting" from relationships of size between ideal lines, ideal surfaces, etc.

[8] Surely that is one way (among many) to interpret them, is it not? That remains to be seen, but let us suppose so for the time being.

[9] For example, the function f induces the permutation $h \mapsto h_f$ on the set $E^{E \times E}$ of binary operations on E given by $h_f = f \circ h \circ (f^{-1} \otimes f^{-1})$, where $(f^{-1} \otimes f^{-1})\text{'}(u,v) = ((f^{-1})\text{'}u, (f^{-1})\text{'}v)$.

Nothing turns on the fact that f is a permutation of the underlying set of E, for all these observations would continue to apply if we were to replace the permutation f here by *any* bijective function. Thus given any set T with $T \cong_c E$ and any bijection $f : E \to T$ we could obviously define a structure T_f, with underlying set T, which was "abstractly" identical to the structure E, which "shared" its "form", in essentially the same way that E_f does.

Now to maintain that E and E_f (or E and T_f) are "abstractly identical", or share the same "form", or have all their "mathematically significant" properties in common, is to maintain that the individual natures of the elements that make up the underlying set of a structure are irrelevant to the essential mathematical form manifested by that structure.

Moreover, to say that the function f is an *isomorphism*, that, as the etymology of the word "isomorphism" suggests, it actually bears witness to the fact that the structures E and E_f have the *same form*, means, among other things, that for each element x of the set E, $f'x$ plays the very same role in the structure E_f that x plays in E, and the image, h_f, of the binary operation, h, on E plays the same role in E_f that h plays in E, and similarly for other functions and relations of varying degrees over E.

The genuine Euclidean point x may be mapped by the permutation f to a genuine Euclidean plane $f'x$ (genuine Euclidean points, lines, and planes being, by assumption, the elements out of which the set E is composed); but the morphology of E_f forces us to regard $f'x$ as a "point" from the standpoint of E_f, despite its real individual character as a plane. And for every mathematically significant proposition, Ψ, about the point x (interpreted in the "world" determined by E) there is a cognate, and logically equivalent, proposition Ψ_f, about the "point" $f'x$ (interpreted in the "world" determined by E_f). The actors are interchangeable as individuals: it is the *roles* in the drama that are fixed.

Notice the sense in which this constitutes an *arithmetisation* of traditional geometry: the *elements* of Euclidean space – the points, lines, and planes – have been separated out and stripped of their "cohesion" so that they are treated simply as units in an *arithmos* (in modern terminology, as elements of a set), and the "internal" relations among them, which arise out of (or perhaps constitute) their cohesion, have been *extensionalised*.

These, then, are the ideas that underlie what I called (in Section 5.1) the "fundamental dogma of the axiomatic method". As we shall see in Section 6.6, we can reformulate this "dogma" as a proper mathematical theorem using mathematical logic.

6.2 Axiomatic definition

Like natural number arithmetic, three-dimensional Euclidean geometry satisfies a simple set of axioms that are *categorical*, in the sense that any two models of them are isomorphic. If we suppose that the morphology of E, which fixes its structure type, suffices to define all geometrically significant notions[10], and if we accept the fundamental dogma, then in laying down those axioms we fix the mathematically significant properties of three-dimensional Euclidean space once and for all, and this in a manner *logically* independent of the points, lines, and planes of the traditional theory. In this sense perhaps it is legitimate to say that it is the system of geometrical axioms alone that determines the meanings of the primitive notions of the theory as that theory is now understood, quite independently of the "intended" structure E[11].

It is especially significant that we do not need to posit the traditional objects of geometry in order to establish the consistency of our geometrical axioms. For we can follow the example of Hilbert in his *Foundations of Geometry* and construct a *Cartesian* model whose "points" are ordered triples of real numbers, and whose lines and planes are defined using the traditional methods of Descartes' analytical geometry.

But doesn't that simply replace the *ideal* objects of traditional Euclidean geometry with the *abstract* objects of traditional algebra, the real numbers? Yes, that is perfectly true; but we have not yet finished, because "the real numbers" are not essential here either. For the theory of real numbers is itself characterised by a categorical system of axioms for *complete ordered fields*.

The consistency of *those* axioms, that is to say, the existence of complete ordered fields, follows, as is well known, from the existence of simply infinite systems. But the existence of simply infinite systems follows, as I showed in Section 5.2, from the mere *existence* of transfinite sets, which, in turn, follows from Cantor's Axiom.

These existence proofs are effected by constructing models for the relevant axiom systems. But nowhere in this chain of constructions do we require recourse to traditional mathematical objects or to structures built up out of them. The only models required for our existence proofs can be defined using straightforward, exclusively set-theoretical methods,

---

[10] By the very nature of the case, this is not something that can be established by a straightforward mathematical proof. After all, much of Hilbert's *Foundations of Geometry* is devoted to showing that the primitive notions employed in his axiomatic geometry are adequate in this respect.

[11] We can sharpen this observation using formal mathematical logic, as we shall see in Section 6.5 below.

as in the proof of the existence of a simply infinite system from the existence of a transfinite set (Theorem 5.1.11).

This is, of course, essential to the success of the whole enterprise of axiomatic definition. For once we have acquired the standards of precision and clarity fostered, and indeed required, by the modern axiomatic method, we come to doubt that sufficient sense can be made of the notion of structures that are composed of traditional mathematical objects (like the supposed Euclidean structure E discussed above) to meet the exacting requirements of mathematical rigour.

What *are* Euclidean points, lines, and planes? What is the space that these "mathematical objects" compose? If we are to believe current physics it is not the physical space we actually inhabit. Our best physical theories suggest that physical space may not be infinite either in extent or in divisibility; and, in any case, it does not conform to the laws laid down in Euclid's axioms.

But if Euclidean three-dimensional space isn't actual physical space it must be some sort of ideal simulacrum, not indeed of actual space, but of space as we mistakenly supposed it to be before Planck and Einstein. How can an exact mathematical science like Euclidean geometry be based – *logically* based – on something which seems so vague and so insubstantial?

I am making a simple and straightforward *logical* point here. In mathematics we must take our truth claims, and therefore our existence claims, literally. And we can't do that if we are claiming to deal with structures composed of traditional mathematical objects, whose essential natures escape our understanding, and whose claims to objective existence are feeble in the extreme.

On the other hand, to deny that there is *some* sense in which the traditional elements of Euclidean geometry – points, lines, planes, etc. – strike us as well-defined, and the axioms strike us as *a priori* true of space as we experience it, is to deny an obvious fact of personal and collective experience. Without this experience we should have no reason to consider Euclidean geometry as a formal axiomatic theory.

There is a serious epistemological problem here[12], and the modern, axiomatic approach to geometry does not provide us with a solution.

---

[12] This puzzle about the status of the "intended model" of Euclidean geometry calls into question the central role played by the notion of real number in our mathematics. For the traditional notion of real number, which our modern, axiomatic theory of the reals makes precise, is closely tied to traditional geometry. Is it, after all, *obvious* that we should always use the traditional reals in formulating the basic theories of physics?

What it does provide us with is a means of carrying on with rigorous definition and proof in geometry in the absence of such a solution. And that is no mean thing.

Notice that in none of these cases – Euclidean structures (if I may call them that), complete ordered fields, and simply infinite systems – does the construction of a particular model of the relevant axioms in the course of the consistency proof constitute a *definition* of the traditional "mathematical objects" (points, lines, planes; real numbers; natural numbers) with which the axioms in question ostensibly have to do.

When the consistency of the axioms for a complete ordered field is established, for example, there is no claim that the real numbers really just *are* Dedekind sections of rationals (or whatever the model we construct says they are). Any stucture whatever, however "artificial" it may seem, will suffice for the purposes of that proof, so long as it satisfies the axioms for a complete ordered field.

What is essential here – what the axioms are really capturing – is the *isomorphism type* of the reals, or more precisely, *an* isomorphism type, to which the ordered field composed of "the" real numbers themselves would belong, if only there were such an ordered field. It is this feature of the axiomatic theories we are considering here – their *generality* – that corresponds to the *abstractness* or the *ideality* of the putative "objects" treated in the corresponding traditional theories. Moreover, once the Categoricity and Consistency Theorems are established in each of these cases we immediately see that the existence of a "standard" or "intended" model constructed out of traditional "mathematical objects" is *logically* superfluous, even though "it" plays an essential role *psychologically,* so to speak, in guiding us as we search for the right system of axioms for the theory under consideration.

There are thus two senses in which these three basic axiomatic theories must be justified. On the one hand we must convince ourselves that the axioms do express, or correspond to, "truths" about the traditional "objects"[13] that constitute the subject matter of these theories as traditionally conceived; this we may call the *psychological* or (perhaps better) *phenomenological* justification. But the *logical* justification moves in the opposite direction, for it is the Consistency and Categoricity Theorems

---

[13] I enclose these key words in quotation marks since, after all, it is not at all clear that there *are* such objects, and thus not clear that any proposition about "them" can, in the last analysis, even be meaningful, let alone true.

that guarantee that all our chatter about the "mathematical objects" in question is mathematically (if not philosophically) harmless.

The axioms are used, we might be inclined to say, to characterise a particular abstract mathematical *form* rather than a particular *instance* of that form. But what do we mean when we say such things, that is to say, what do we mean *literally*? Are there really such *things* as "abstract mathematical forms"?

To ask these questions is to confront a traditional philosophical problem – the problem of *universals* – in an unfamiliar setting. And surely it is of philosophical significance that mathematicians have come up with a precise, mathematically exact solution to their version of that problem, a solution, moreover, which is essentially *nominalistic* in spirit. All this confusing and vague talk about "abstraction" and "form" is replaced by mathematically precise discourse involving the key concept of *isomorphism*. Instead of saying that E and E_f are "abstractly" identical, or that they have, or "share" the same "form", the mathematician says that there is an isomorphism from the structure E to the structure E_f, effected, in this case, by the function f. There simply is no need for mathematicians to invoke "abstract forms" in laying down their definitions and constructing their proofs.

Now of course it does not follow from the fact that we need not *invoke* abstract mathematical forms explicitly in our mathematical discourse that there *are* no such forms. We may still maintain that there *are* such forms, and that they are what determine the mathematical properties of the structures that mathematicians study. Indeed, this is a point of view with which many mathematicians – and I include myself in their number – may find themselves in sympathy: there is, as Plato himself knew well, a natural affinity between mathematics and the Platonic philosophy. But surely intellectual economy dictates that if we aren't *forced* to invoke such forms then we are forced *not* to invoke them, at least in mathematics, where, as a matter of principle, we must keep our assumptions and our posits to a minimum.

In any case, anything that we can say truly – indeed, intelligibly – in mathematics using notions like "abstract form", we can express in the language of mathematical structures, their morphologies, and the morphisms between them. It is only when we try to explain the significance of mathematical discourse, to connect our exact talk about isomorphisms with the traditional idiom which employs notions like "abstraction" and "form", that we encounter difficulties.

Having banished mathematical *objects* (and abstract "mathematical

6.2 Axiomatic definition

forms", into the bargain) we are left with mathematical *structures*. Are we any the better off? From a technical standpoint within mathematics we are very much better off indeed. Mathematical structuralism and the modern axiomatic method are so much clearer, so much more precise, so much more flexible, so much more *intelligible*, than the old-fashioned, traditional methods they have replaced that it is inconceivable we shall ever go back to the traditional outlook. If we had literally to employ abstraction, as Newton fancied he did in his definition of real number, we should be utterly stymied, as, indeed, our predecessors were stymied until they developed the methods we now use.

Abstraction belongs to the psychological aspect of our theories. It is not, as Newton's definition of number suggests, a tool with which to fashion "mathematical objects". Its proper function is to allow us to *clarify* our naive intuitions to the point at which we can determine which of our naive concepts ought to be reflected in the morphologies of the structures to which the corresponding axiomatic theories apply, and which axioms ought to be laid down to capture the appropriate isomorphism type.

As its etymology suggests (*abs+trahere* to draw or pull out or away from), abstraction in modern mathematics is a selecting out, from complexes in imagination and experience, of those features that can be made to correspond to items of morphology in a precisely defined type of mathematical structure. But how this works, in detail, though familiar to mathematicians, is not easy to describe exactly. Perhaps a familiar example will illustrate this.

Consider the axiomatic theory of three-dimensional real vector spaces. This theory is categorical so all of its models are isomorphic to one another. Typically we represent the vectors of such a space in our imaginations as "arrows" emanating from a point (the *origin* or the *null vector*). But this familiar way of representing such vectors is misleading. For the "concrete" vectors of the representation have properties that have no correlates in any model of the axioms. Thus in the concrete representation distinct non-zero vectors are separated by a definite angle, although nothing corresponds to angle in a real vector space. Similarly, in the representation, non-collinear vectors can be compared as to size, but nothing corresponds to relative size in a real vector space unless the vectors to be compared are collinear.

Thus even in our concrete, "intuitive" representations of three-dimensional real vectors we are forced to abstract from manifest properties of their representatives in order to arrive at a suitable conception of them as representations in imagination. *Logically speaking*, however, the axioms do the "abstracting" for us, so that what is significant is what is "common" to all models of the axioms: it is the *generality* of the interpretability of the axioms that replaces the abstraction performed on the elements of our particular, concrete representations.

Abstraction provides us with a guide to specifying our structure types and laying down our axioms, but the structures which belong to those types and satisfy those axioms are neither abstract themselves, nor composed of abstractions. Abstraction, however important its psychological or heuristic role in the production of our drama, does not itself actually appear on the stage during performance.

The real gain here is ontological. For when we employ the axiomatic method, the only special subject matter that we need acknowledge as belonging especially to mathematics is that of sets – *arithmoi* – finite pluralities composed of objects. All that was required, historically, to replace the traditional methods completely was to lay down three axiomatic theories: the theory of complete ordered fields (real numbers), the theory of simply infinite systems (natural numbers), and the theory of Euclidean structures (three-dimensional Euclidean space), and then to show how traditional geometry, kinematics, and, more generally, mathematical analysis (the Calculus) could be reconstructed by straightforward algebraic and set-theoretical methods applied to models of those theories. And, of course, we need Cantor's radical new analysis of what it is to be finite – what it is to be an *arithmos* – in order to carry out this *arithmetisation* of these traditional theories.

But we can go no further along these lines. In particular, it is obvious that we cannot use the axiomatic method to eliminate the notion of *set* in the way we have used it to eliminate the notions of *natural number* and *real number*: the concepts of *structure* and *isomorphism* that play so central a role in the axiomatic method as I have described it here are essentially and uneliminably set-theoretical in character. Any attempt to give an axiomatic account of *set* along these lines would thus involve us in an obvious vicious circle.

But in any case, there is no *need* to eliminate the notion of set in such a way. For sets are neither abstract, like natural numbers and real numbers, nor ideal, like geometrical figures. By comparison with those traditional "mathematical objects", sets have an absolutely transparent character, and there is no difficulty about the genuine existence of sets, *in general*, although, as I have never attempted to disguise, there *is* a difficulty about the existence of *transfinite* sets. It is a difficulty in determining the true meaning of *finiteness*, and *that* is the deepest problem in the foundations of mathematics.

Up to now I have been considering only axiomatic theories of a special kind, so-called *categorical* theories in which the axioms single out a *particular* isomorphism class of structures of the appropriate type. This is because my principal concern here has been with the significance of the modern axiomatic method for the problems of "mathematical objects" and "mathematical forms". But most of the axiomatic theories studied by mathematicians are not of this sort. They single out a species of structures which have merely analogous morphologies, so to speak, structures which, though not "mathematically indistinguishable", nevertheless have important morphological properties in common: groups, rings, fields, topological spaces, categories, vector spaces, manifolds, ... all these things are defined axiomatically, and in these cases the logical function of the axioms is to pick out common and salient features of what are otherwise quite different structures with different mathematically significant properties. In considering these non-categorical axiomatic theories, however, the notion of isomorphism continues to play an essential role and what I have called the "fundamental dogma" – that isomorphic structures are indistinguishable in their essential mathematical features – continues to obtain[14].

All this naturally gives rise to the question: how can the notions of *mathematical structure* and *morphism* be characterised in full generality with mathematical precision? Bourbaki attempted a general definition of mathematical structure[15]; but it has proved less useful to mathematicians than the axiomatically defined notion of *category* introduced by Eilenberg and Mac Lane[16]. Indeed, the notion of category was designed to capture, among other things, the idea of a system of structures together with the morphology-preserving maps between them.

Of particular significance for axiomatics has been the development of *topos theory*. A topos is a category with additional morphology which allows us to regard it as a system of higher order logic/set theory; but the logic embodied in a topos is, in general, constructive rather than classical, and this means that the set theory "internal" to a given topos is quite different in character from ordinary set theory. This theory thus allows us to extend the conventional axiomatic method by carrying out

---

[14] Mathematicians speak of *the* cyclic group of order 4, for example, when, of course, they are actually dealing with an isomorphism class.
[15] *Elements* I, Chapter IV.
[16] "General theory of natural equivalences".

our axiomatic definitions *inside* a given topos, thereby employing an alternative to conventional set theory in formulating the underlying logic of our axiomatic theories. This represents a considerable enrichment of the conventional axiomatic method as I have described it here.

A particularly striking use of topos theory has been to provide the logical foundation for F. W. Lawvere's smooth infinitesimal analysis, a theory that constitutes a serious alternative to conventional real analysis (the theory of complete ordered fields discussed above). Inside a suitable topos the analogue of the field of real numbers loses its "arithmetical" (or "set-theoretical") character because of the undecidability of the internal analogue of the identity relation: the units of the "*arithmos*" of real numbers are no longer sharply distinguished as they are in a conventional structure.

This theory reminds us that the conventional approach to geometry and the Calculus, based as it is upon Cantorian finitist assumptions, is not free of difficulties, even though it is accepted by the overwhelming majority of contemporary mathematicians[17].

A topos is itself a mathematical structure with an underlying domain and a morphology characterised by axioms in the same way that a group, ring, or field is characterised by axioms. But, in strong contrast to conventional algebraic theories, topos theory, and, indeed, category theory in general, has made deliberate use of "structures" whose underlying domains are absolutely infinite species, rather than sets. The logic of such "large" structures is not well understood, and, in particular, it is not at all clear how we should define them or what mathematical operations and constructions we can legitimately apply to them[18]. What is clear, however, is that we cannot treat them as ordinary structures.

In any case, we cannot use topos theory, *considered as an axiomatic theory*, to provide a foundation for the the axiomatic method itself: if we regard toposes, or, more generally, categories, as models of the appropriate system of axioms, then to investigate structures and morphisms using topos theory or category theory is to *employ* the axiomatic method rather than to *explain* it.

---

[17] See F. W. Lawvere, "Categorical dynamics" and also Moerdijk and Reyes, *Models for Smooth Infinitesimal Analysis*. There is an excellent elementary account of Lawvere's theory in J. L. Bell's *A Primer of Infinitesimal Analysis* which discusses these issues in the foundations of real analysis.

[18] I shall return to this topic in Section 7.2.

6.3 Mathematical logic: formal syntax

Up to now I have been considering *non-formal* axiomatic theories, not the formal axiomatic theories of mathematical logic. In fact, everything I have said about the non-formal theories applies, *mutatis mutandis*, to formal theories as well. But there is a surprisingly widespread view that a profound gulf separates formal from non-formal axiomatics, and that, indeed, formal logic is an autonomous discipline, prior to, and independent of, mathematics in general, and set theory in particular.

Such a view might suggest the possiblity of eliminating, not just the traditional notion of mathematical *object*, but even the notion of mathematical *structure* from the foundations of mathematics, and of achieving a purely "linguistic" or "logical" foundation for the subject using an axiomatic method based on formal mathematical logic. That, of course, would require us to develop formal mathematical logic independently of the notion of mathematical structure, indeed, independently of the notion of set itself.

This way of regarding of mathematical logic was once widespread, and still has adherents. It is, however, untenable. For it is not possible to give a coherent account of formal mathematical logic, an account that that can be defended as an account *of logic*, without making use of set theory.

Mathematical logic is a branch of mathematics, like group theory or real analysis, and, as such, calls upon the resources of the theory of sets for its development. It follows that set theory must already be established before a rigorous development of mathematical logic can be carried out in a properly motivated, coherent manner. Indeed, not only formal semantics, but even formal syntax, requires set-theoretical methods.

The point I am making here is neither subtle nor obscure, but simple and obvious, and it is scandalous that it is so often ignored. Mathematical logic does not provide mathematicians with a set of ready-made tools which they can use to construct the foundations of their science. On the contrary, the tools it provides must first themselves be forged with painstaking care using the concepts and techniques of set theory, antecedently understood.

Many a would-be account of the foundations upon which mathematics rests has come to grief because its author has granted himself permission to appeal to the fundamental results of mathematical logic without stopping to ask what are the even more fundamental principles upon which those results themselves depend.

This issue is so important, and the confusion surrounding it is so gross and so widespread, that I need to address it directly. Besides, formal mathematical logic does have a crucial role to play in any exposition of the foundations of mathematics, even though it is perhaps a more modest role than that often assigned to it: we need it in order to complete our account of the axiomatic method. I must, therefore, discuss its foundations in some detail.

What I mean by "mathematical logic" here is the study of standard formal languages of first or second order, based on conventional ("classical") logic[19]. I do not consider it part of my brief to give a detailed account of the logic of these languages, so I shall confine myself to matters that are of direct relevance to the foundations of mathematics. In particular, I intend to show that mathematical logic presupposes the theory of sets in the same way that real analysis, group theory, and general topology do. Indeed, this is as true of formal syntax[20] as of formal semantics.

Let me turn first to a discussion of the role of set theory in formal syntax; I shall consider the role of set theory in semantics in the next section.

When logicians treat of formal syntax they often assume, implicitly or explicitly, that their formal expressions and formal proofs are, or could be, actual concrete inscriptions. They appeal not only to our intuitions of what inscriptions are like, but also to our intuitions of what sorts of instructions for the writing down of such inscriptions we could carry out. This gives the theory of syntax a reassuring (to some) air of concreteness and even practicality, which, however, is entirely spurious.

In fact, the notion of "concreteness" cannot be taken literally in this context without risking serious confusion. For example Schoenfield in Section 1.2 of his textbook *Mathematical Logic*, emphasises the "concrete" nature of his symbols as inscriptions, and, accordingly, in Section 1.3 describes the formation of new expressions by adding a symbol "to the right of" an expression already given. But since he does not tell us what to do when the expression to be so augmented already extends to the right-hand edge of our paper, and since, in any case, it is clear that among his "concrete" expressions he intends to include "expressions" (e.g. strings composed of 10^{729} juxtaposed symbols) which, to the best

---

[19] The logic of languages of higher order can be reduced to second order logic.
[20] As I am using the term here, formal syntax deals not only with formal expressions (terms, formulas, etc.) but also with formal proofs.

6.3 Mathematical logic: formal syntax

of our knowledge, are physical impossibilities in the space we actually inhabit, we may reasonably infer, I think, that these things are "concrete" only in a Pickwickian sense.

Now all this may seem quite harmless, but when we invoke this peculiar *attenuated concrete* (if I may call it that) we run the risk of deluding ourselves. For example, Schoenfield tells us[21] that

> One point is apparent: there is no value in studying concrete (rather than abstract) objects unless we approach them in a concrete or constructive manner. For example, when we wish to prove that a concrete object with a certain property exists, we should actually construct such an object, not merely show that the non-existence of such an object would lead to a contradiction.

Is it really true that we must *actually construct* such objects? I know, for example, that in a given formal language there are formulas that contain more than 10^{729} symbol occurrences, because the non-existence of such formulas would obviously lead to a contradiction with the definition of formula. Clearly I could not *actually* construct such a formula. The best I could do would be to provide schematic instructions for writing it down[22]. But we wouldn't credit an architect with having *actually constructed* a mile-high skyscraper if all he had done was to provide us with schematic instructions, in the form of a blueprint, for building it.

From a pedagogical standpoint, it would be hard to improve on Schoenfield's treatment of syntax. Obviously the best way to introduce logical syntax to the beginner is to treat formulas, terms, and proofs in the way that Schoenfield does, namely, *as if* they were actual inscriptions, and thus to take advantage of his natural geometrico-combinatorial intuitions. But if we are looking for a theoretical analysis of the assumptions that underlie formal syntax this naive approach simply will not do.

What *kinds* of things are the expressions and proofs of our formal languages? Do we ever actually write them down? Or do the things we actually write down stand to them in the same sort of relation that the triangles we actually draw with our rulers and pencils stand to the Euclidean triangles of traditional geometry?

Surely *that* can't be right: it can't be a matter of our written expressions approximating ideally exact ones, as in geometry. On the other hand, there are clearly borderline cases in written inscriptions where it is not obvious which symbol has been written down.

---

[21] *Mathematical Logic*, p.2.
[22] My instuctions would be schematic in this sense, for example: I should not explain how to assemble the necessary quantities of paper and ink.

Perhaps we are dealing with a type–token distinction here? But if so, what is the status of a type for which there couldn't possibly be a token, even "in principle", in the physical universe we actually inhabit?

We mustn't let the fact that we customarily and naively think of our formal languages as *written* languages mislead us, for that is not, in fact, their salient characteristic. What is essential to formal syntax is that it deals with the combinatorics of discrete, Euclidean finite configurations. The fact that these configurations are usually conceived "geometrically" as assemblages of signs is quite beside the point. Indeed, that way of regarding them distracts our attention from what is essentially problematic, namely, their *Euclidean finiteness*. It is precisely this that requires us to employ set theory in our theoretical syntax.

It is a serious mistake to suppose that spatial or temporal intuition (*Anschauung*) can provide us with the conception of Euclidean finiteness in syntax that we require: it does not help us to imagine our syntactic objects as "completed" geometrical configurations confined to bounded regions in space, or as "constructions", of a geometrical or more general sort, but each effected by a discrete succession of actions in time bounded by a first and last act. What such imaginings leave out is precisely the chief and essential matter, namely, Euclidean finiteness, which is an arithmetical (i.e. *set-theoretical*) notion, not a geometrical or phoronomical one.

To ground the notion of Euclidean finiteness in spatio-temporal intuition in such a manner would require us, in effect, to *construct* the concept of a *Euclidean finite syntactic configuration* or of a *Euclidean finite temporal sequence of acts* in the precise sense of "concept construction" given by Kant in the *Critique of Pure Reason* (A712 ff).

Kant developed this notion in order to answer a central question about mathematical method, namely, how can mathematical proof and definition rest upon *intuition* when intuition is always of *singulars*? How can we, for example, deduce a *general* proposition about all triangles by contemplating (or intuiting) a *single* geometrical diagram?

Kant's answer was: because rightly viewed, the diagram represents the construction of the appropriate concept in intuition. To construct a concept in intuition is to exhibit, in intuition, a *single* instance of that concept which nevertheless captures the concept in its entirety.

Consider the concept *straight line segment* in intuitive (*anschaulich*) geometry, for example. Each particular straight line segment captures, or illustrates, in a way that is obvious but cannot easily be put into words, the entire content of the concept.

But this is not quite the whole story (and the example I have chosen may be misleading in its simplicity): it is not the single instance itself that alone captures the entire concept, but the instance *together with the universal conditions that govern its construction.*

Thus I construct a triangle by exhibiting an object corresponding to this concept, either through mere imagination, in pure intuition, or on paper, in empirical intuition, but in both cases completely *a priori*, without having had to borrow the pattern for it from any experience. The individual drawn figure is empirical, and nevertheless it serves to express the concept without damage to its universality, for in the case of this empirical intuition we have taken account only of the action of constructing the concept, to which many determinations, e.g., those of the magnitude of the sides and the angles, are indifferent, and thus we have abstracted from these differences, which do not alter the concept of the triangle.(*Critique of Pure Reason* A713–14.)

Kant's own example (*Critique of Pure Reason* A714) is the construction of the familiar diagram to prove that the angles of a triangle sum to two right angles. Here the concept to be constructed is that of *a triangle, one of whose sides is extended, and with a line parallel to the given side passing through the vertex opposite that side.*

The terminology "concept *construction*" is apt, for we are dealing with an actual geometrical construction here. Moreover, the act of constructing the figure *shows*, in a way that is readily experienced, but not so readily put into words, that the particularising conditions of that figure – the lengths of its sides and magnitudes of its angles – are irrelevant to its capacity to demonstrate the truth that it does, in fact, demonstrate.

Kant's theory of the nature of geometrical concepts provides his explanation for the efficacy of *particular* diagrams in establishing *general* geometrical truths. But it is notorious that diagrams don't work in arithmetic, and Kant's desultory attempt to extend his theory to arithmetic (*Critique of Pure Reason* A717) founders on confusion over what we should now call the use of free variables in logical argument[23].

Surely it is obvious that the *general* concept of Euclidean finiteness cannot be captured by a single one of its instances; you cannot capture the concept of *Euclidean finite set*, in its full entirety, by exhibiting a particular set with, say 29 elements; *a fortiori*, you cannot capture the general concept of *Euclidean finite string of symbols* by exhibiting a particular string composed of 29 juxtaposed symbols either.

---

[23] In fact, diagrammatic proofs of general propositions in arithmetic can be given, e.g. of the proposition "$7+5 = 12$"; but such propositions are not usually acknowledged to be the general propositions that they unquestionably are. Indeed, Kant himself explicitly characterises them as singular propositions and therefore denies them status as axioms, in his special sense (*Critique of Pure Reason* A164). See the discussion in Section 2.8.

The contrast between this arithmetical example and the earlier geometrical example of a straight line segment is highly instructive. Straight line segments are so alike that we can scarcely say what it is that distinguishes them; with Euclidean finite sets – numbers (*arithmoi*) in the traditional sense – we (quite literally) cannot *imagine* (represent in pure intuition) what it is that they have in common. Of course we can *say* what they have in common: Euclidean finiteness. But that is merely a statement of the problem, not an explanation. For Euclidean finiteness is what Kant calls a "discursive" concept, and cannot be constructed in intuition in his precise sense.

When we employ old-fashioned geometrical combinatorics in our theoretical syntax we smuggle the notion of Euclidean finiteness into our theory without noticing that we have done so. For practical purposes this is perfectly all right, provided we are prepared to accept the Cantorian assumptions that underlie Dedekind's axiomatic treatment of natural number, which I presented in Section 5.1 above. For that is where the Cantorian account of Euclidean finiteness is to be found. If, however, we do not accept those assumptions, then the general theory of syntax becomes problematic. I shall return to this topic in Part Four when I take up the exposition of the Euclidean version of set theory.

What is not legitimate is to treat logical syntax in a naive, *geometrico-combinatorial* way – as if one were dealing with actual inscriptions – and then to pretend that the formal theories described in this loose manner can be used in a *foundational* capacity. This is what old-fashioned formalists do, and it is without any justification whatsoever.

You are not allowed to draw cheques on a bank in which you have no funds deposited. And you are not in a position to boast of financial independence if someone else is paying all your bills.

But how should formal syntax be presented? One possibility would be to replace the *geometrical* notion of inscription (i.e. a linear juxtaposition of individually inscribed tokens of the primitive signs) by the *set-theoretical* notion of a finite sequence of signs (i.e. a function defined on an initial segment of some specified simply infinite system, with values in the set of primitive signs).

Alternatively, we could develop the theory of syntax axiomatically, treating it as a *generalised arithmetic* in the sense of Kleene[24]. Such an axiomatic treatment of syntax would be along the lines of Dedekind's theory of simply infinite systems: particular models of the axioms would

---

[24] *Introduction to Metamathematics*, Chapter X, §50.

then correspond to particular simply infinite systems in Dedekind's theory.

The axiomatic approach is, in principle, superior, for it makes plain, from the outset, that the expressions of a formal language, like the natural numbers or the real numbers, are not particular "abstract", or perhaps in this case "ideal", mathematical objects, but that, as with "the" natural numbers and "the" real numbers, our notion of the abstract or ideal character of expressions as particular "things" arises out of the fact that in studying the syntax of a particular formal language we are really studying an absolutely infinite species of structures that constitute an isomorphism class, so that the "abstractness" or "ideality" are really disguised *generality*.

It is clear, in any case, that set theory has an uneliminable role to play in formal syntax. Even if we take the axiomatic approach, it is still necessary to construct a *model* of the axioms of formal syntax. For, contrary to a widespread misapprehension, not even logicians can conjure mathematical structures out of thin air by the mysterious act of "postulation". You can call spirits from the vasty deep; but will they come when you do call for them?

6.4 Global semantics and localisation

Every formal language $\mathscr{L}$ is naturally associated with a certain species or *type* consisting of all those mathematical structures that are suitable to provide an interpretation for $\mathscr{L}$, the $\mathscr{L}$-*structures*. *Local* semantics is concerned with the notion of *formal truth* or *satisfaction* in individual $\mathscr{L}$-structures given by Tarski's famous truth definition. I shall use the customary notation

$$M \vDash A(f)$$

to express the proposition that the $\mathscr{L}$-structure M satisfies the $\mathscr{L}$-formula A at the assignment f of values to the free variables of $\mathscr{L}$. If $\mathscr{L}$ is a first order language, its free variables range over the elements of the universe of M; if $\mathscr{L}$ is a second order language, it has, in addition, free variables ranging over relations and functions of all degrees defined over that universe. An assignment f of values in M to the free variables of $\mathscr{L}$ is then a function defined on those free variables (considered as symbols in $\mathscr{L}$) giving each a value of the appropriate sort. I shall follow the common custom of using the same symbol (e.g. "M") for both the $\mathscr{L}$-structure and its underlying universe of discourse[25].

In studying the *global* semantics of a language $\mathscr{L}$ we consider the semantic properties of $\mathscr{L}$ relative to the absolutely infinite species of all $\mathscr{L}$-structures. The central logical notions of global semantics are those of *universal* (or *logical*) *validity, logical consistency,* and *logical consequence.*

An $\mathscr{L}$-sentence[26] A is *universally* (or *logically*) *valid* if A holds (i.e. is satisfied, in Tarski's sense) in *every* $\mathscr{L}$-structure; a set, Γ, of $\mathscr{L}$-sentences is *logically consistent* if some $\mathscr{L}$-structure (simultaneously) satisfies every $\mathscr{L}$-sentence in Γ; an $\mathscr{L}$-sentence A is a *logical consequence* of a set, Γ, of $\mathscr{L}$-sentences if A holds true in *every* $\mathscr{L}$-structure which satisfies all the sentences in Γ. More formally,

6.4.1 Definition *Let $\mathscr{L}$ be a formal language of first or second order.*

(i) *Let A be any $\mathscr{L}$-sentence. Then, by definition,*

$$A \text{ is logically valid}$$

means

$$\forall x[x \text{ is an } \mathscr{L}\text{-structure .implies. } x \vDash A]$$

(ii) *Let Γ be any set of $\mathscr{L}$-sentences. Then, by definition,*

$$\Gamma \text{ is logically consistent}$$

means

$$\exists x[x \text{ is an } \mathscr{L}\text{-structure .and. } (\forall A \in \Gamma)[x \vDash A]]$$

(iii) *Let A be any $\mathscr{L}$-sentence, and Γ be any set of $\mathscr{L}$-sentences. Then, by definition,*

$$A \text{ is a logical consequence of } \Gamma$$

means

$$\forall x[x \text{ is an } \mathscr{L}\text{-structure } : \text{implies} :$$
$$(\forall y \in \Gamma)[x \vDash y] \text{ .implies. } x \vDash A]$$

These notions, together with Tarski's definition of truth, are the fundamental notions of mathematical logic. Their definitions disclose that these basic notions involve global quantification in an essential way: logical validity and logical consequence are both Π_1, logical consistency Σ_1.

---

[25] Note that a given set of individual, function and predicate constants suffices to determine both a first and a second order formal language based on those constants, and those two languages give rise to exactly the same species of suitable interpretations.

[26] i.e. an $\mathscr{L}$-formula without free variables (of any sort).

6.4 Global semantics and localisation

In each of these basic definitions the presence of a global quantifier in the *definiens* means that, in accordance with Brouwer's Principle, the *definiendum* is not of a form that admits of receiving a determinate truth value, *true* or *false*. This, in turn, means, for example, that, *prima facie*, the species of universally valid $\mathscr{L}$-sentences is a sub-*species* of the set of all $\mathscr{L}$-formulae which does not correspond to any of its sub-*sets*: it is thus a natural candidate for what Vopenka calls a *semi-set*.

This applies, in principle, to both the first and second order languages, but in the first order case these definitions can be *localised*, in a sense I shall soon make precise; and this means that in first order logic these three basic logical notions can be given "orthodox" (i.e. local) characterisations.

In second order logic, however, there are good reasons to believe that localisation is not possible, so that, for example, the species of logically consistent sets of $\mathscr{L}$-sentences is a proper semi-set of the power set of the set of all $\mathscr{L}$-sentences, and therefore provides us with an example of a species that is *indefinite* without being *infinite*[27].

Let us turn our attention first to the problem of localising the notion of logical consistency in first order logic. The key result here is the Löwenheim–Skolem Theorem. Given a first order language $\mathscr{L}$ it is possible to define a *set*, $Can_{\mathscr{L}}$, of *canonical $\mathscr{L}$-structures*[28] which contains, for every $\mathscr{L}$-structure M, a structure which satisfies exactly the same set of $\mathscr{L}$-sentences.

6.4.2 Theorem *(Löwenheim–Skolem Theorem for First Order Logic)* Let M be any $\mathscr{L}$-structure. Then there is a canonical $\mathscr{L}$-structure M_0 such that for all $\mathscr{L}$-sentences A, $M \vDash A$ if, and only if, $M_0 \vDash A$.

This leads immediately to the localisation of the notion of logical consistency.

6.4.3 Theorem *(Localisation Theorem for Logical Consistency in First Order Logic)* Let Γ be any set of $\mathscr{L}$-sentences. Then each of the following propositions entails, and is entailed by, the other:

(i) Γ *is logically consistent*.
(ii) $(\exists x \in Can_{\mathscr{L}})(\forall A \in \Gamma)[x \vDash A]$.

---

[27] See Section 7.3.
[28] The structures can be defined as those $\mathscr{L}$-structures whose universes of discourse are composed of equivalence classes of $\mathscr{L}$-terms under suitable equivalence relations.

Proof That (i) entails (ii) follows from the Löwenheim–Skolem Theorem. That (ii) implies (i) is immediate. □

By virtue of this mutual entailment, we are justified in taking the local proposition

$$(\exists x \in Can_{\mathscr{L}})(\forall A \in \Gamma)[x \vDash A]$$

as an alternative *characterisation* of logical consistency in first order logic. This gives a precise meaning to propositions which *deny* the consistency of a set of sentences, and allows us to combine statements of logical consistency with other propositions to form complex propositions having determinate truth values.

Now let us turn to the problem of localising the other two basic notions of logical validity and logical consequence. Notice that there is no straightforward way of obtaining these results from Theorem 6.4.2. For since, *prima facie*, neither the proposition that A is logically valid, nor the proposition that A is a logical consequence of Γ has a conventional truth value, neither of them has a conventional negation. But even if we could provide them with negations in some way, it is by no means obvious that we could reason from *not* $\forall x \Phi(x)$ to $\exists x$ *not* $\Phi(x)$, which would be required to establish the localisations in question in the way that immediately suggests itself.

To establish localisation theorems for first order logical validity and first order logical consequence we have to lay down a system of formal proof and establish Soundness and Completeness Theorems for that system. Such theorems can be proved for all the familiar systems of formal proof for first order logic. Let us suppose that $\mathscr{P}$ is one of these systems.

6.4.4 Theorem (*Soundness Theorem for the System $\mathscr{P}$ of Formal First Order Proof*) *Let $\mathscr{L}$ be a first order language, let M be an $\mathscr{L}$-structure, let Γ be any set of $\mathscr{L}$-sentences, and let A be any $\mathscr{L}$-sentence. If there is a formal proof of A in $\mathscr{P}$ from the hypotheses Γ, and if $M \vDash \Gamma$, then $M \vDash A$.*

It is worth pointing out here that the Soundness Theorem is a perfectly straightforward mathematical theorem. Its proof is by induction on the structure of proofs in the system $\mathscr{P}$ and employs Tarski's definition of satisfaction in an $\mathscr{L}$-structure. In particular it does not make any appeal to the ordinary, non-formal notions of meaning or truth.

The first proof of a Completeness Theorem was given by Gödel[29].

---

[29] "Die Vollständigkeit der Axiome des logischen Functionenkalküls".

6.4 Global semantics and localisation

6.4.5 Theorem *(Completeness Theorem for the System $\mathscr{P}$ of Formal First Order Proof)* *Let $\mathscr{L}$ be a first order language, let Γ be any set of $\mathscr{L}$-sentences, and let A be any $\mathscr{L}$-sentence. Then either there is a formal proof of A in $\mathscr{P}$ from the hypotheses Γ or there is a canonical structure $M \in Can_\mathscr{L}$ such that $M \vDash \Gamma \cup \{\neg A\}$.*

These two fundamental results allow us to localise the notions of logical validity and logical consequence for first order logic.

6.4.6 Theorem *(Localisation Theorem for Logical Consequence in First Order Logic)* *Let Γ be any set of $\mathscr{L}$-sentences and A be any $\mathscr{L}$-sentence. Then each of the following three propositions entails, and is entailed by, each of the others:*

(i) *A is a logical consequence of Γ.*
(ii) *$(\forall x \in Can_\mathscr{L})[(\forall y \in \Gamma)[x \vDash y] \text{ implies } x \vDash A]$.*
(iii) *A is formally provable in $\mathscr{P}$ from the hypotheses Γ.*

Proof That (i) entails (ii) is immediate. That (ii) entails (iii) follows from the Completeness Theorem for the system $\mathscr{P}$. That (iii) entails (i) follows from the Soundness Theorem for the system $\mathscr{P}$. □

Of course localisation for first order logical validity is a special case of Theorem 6.4.6, the case in which Γ is empty.

What are the consequences of Theorem 6.4.6 for the foundations of mathematics? As with the Localisation Theorem for first order logical consistency (Theorem 6.4.3) the entailments established in 6.4.6 justify our taking either of the local propositions

(1) $(\forall x \in Can_\mathscr{L})[x \vDash A]$
(2) A is formally provable in $\mathscr{P}$

as a characterisation of A's being logically valid. But this raises a question of deep foundational significance.

Surely it would not occur to anyone to take (1) as a *definition* of logical validity for a sentence A, nor, for that matter, to take

$$(\exists x \in Can_\mathscr{L})[x \vDash A]$$

as a *definition* of the logical consistency of A. Any attempt to take these *as definitions* would immediately elicit the response: why are we allowed to confine our attention to *canonical* models here? Of course the answer to this is that two deep and important theorems – the

Completeness Theorem and the Löwenheim–Skolem Theorem – connect these characterisations to the natural, *semantic* definitions of logical validity and logical consistency.

But with condition (2) the case seems, at first glance, to be entirely different. There is, after all, a very old tradition more or less identifying formal logic with demonstration. And indeed, it wasn't until Tarski gave his definition of truth in an $\mathscr{L}$-structure that mathematical logicians became convinced that the natural definition of logical validity (truth under all interpretations) could even be *formulated* with mathematical precision, so that before Tarski, (2) was the only definition of logical validity available. Even now there are logicians who regard (2) as a perfectly adequate definition of logical validity. Nevertheless such a *formalist* account of logical validity is untenable.

Formalism in mathematical logic is the doctrine that the semantical definitions of the basic concepts of *universal validity, logical consistency,* and *logical consequence* given in Definition 6.4.1 can be dispensed with, or, at least, degraded to the status of "motivational" material. The doctrine is based on solid fact; for, as we have seen in the case of first order logic, all of these notions can, indeed, be given purely syntactic characterisations. Of course this is notoriously not the case with second order logic, but for that very reason second order logic has been relegated to a subordinate status by the formalists.

Now very few mathematical logicians are formalists in their mathematical practice. They are, for the most part, quite happy to employ the semantical notions, and indeed all the technical apparatus of modern set theory, when laying down their definitions and proving their theorems. This means nothing more than that, as mathematicians, they belong to the orthodox mainstream.

But logicians do find the formalist doctrine that first order logic is the essential core of mathematical logic congenial. For the study of that logic constitutes by far the largest part of their subject. From a technical, professional standpoint, therefore, this bias is perfectly understandable. After all, the very limitations of first order logic give rise to the important general theorems – the Completeness, Compactness, and Löwenheim–Skolem theorems – that constitute the principal logical tools logicians employ[30].

---

[30] A logic sufficiently powerful to axiomatise natural number arithmetic (the theory of simply infinite systems) cannot be complete or compact; a logic sufficiently powerful to axiomatise real analysis (the theory of complete ordered fields) cannot satisfy the Löwenheim–Skolem property.

6.4 Global semantics and localisation

Formalism, however, is foundationally incoherent. For the *global* semantic definitions of universal validity, logical consistency, and logical consequence are *logically prior* to their syntactic counterparts. It therefore follows that even though we can *characterise* those notions syntactactically, we cannot *define* them that way. Nor can we relegate second order logic to a secondary status, for it has a role to play in elucidating the foundations of the axiomatic method for which first order logic is completely inadequate.

These points are not unconnected. For once we see why the semantic definitions of universal validity, logical consistency, and logical consequence are primary, even in first order logic, we can see why second order logic has every bit as strong a claim to the status of formal logic.

Why, then, are these semantic definitions primary? Why can they not be replaced, as the formalists claim, by definitions given in terms of the notion of formal proof? Why, in short, is formalism untenable?

I have already pointed out that there are very serious difficulties even in describing the syntax of formal languages without having recourse to the technical apparatus of Cantorian set theory. Once you see this, once you realise that a logician has no more right to lay the foundations of his theory of syntax in Kantian *reine Anschauung* than does a geometer or an analyst, then you see that the whole formalist project is simply based on a misconception. Its pretensions to rest on the familiar and the concrete are thoroughly bogus.

But let us leave aside such considerations, with which the formalists might take issue, in any case, and grant, for the purposes of argument, that the old-fashioned geometrical treatment of formal syntax is legitimate. Let us assume, moreover, that we have under consideration any one of several equivalent systems of formal proof, presented in the "constructive" geometrical style, and further grant that formal proofs are the sorts of symbol configurations we could actually write down ... *in principle*, that is to say, supposing the laws of psychology, physiology, and physics to be other than we now think them to be. Let us, I say, grant all of this, however implausible it may be.

The question we must now ask ourselves is this: how are we to justify our system of formal proof as an account of *logical* proof? What justifies our claim that the set of sentences formally provable in our system is precisely the set of *logically valid* ones? This question splits into two parts:

(I) What justifies our claim that a sentence is logically valid *if* it is derivable in our system of formal proof?

(II) What justifies our claim that a sentence is logically valid *only if* it is derivable in our system of formal proof?

The first of these is the question of the intuitive *soundness* of our system of formal proof, the second that of its intuitive *completeness*. I say "intuitive" soundness and completeness here, for we cannot, of course, appeal to the soundness and completeness *theorems* if we wish to remain formalists.

Now question (I) is easily dealt with. We can convince ourselves of the soundness of our system of proofs by considering each of its axioms and rules of inference individually in order to convince ourselves that it is logically valid (if it is an axiom) or preserves truth (if it is a rule of inference).

Of course these arguments based on our naive intuitions and conforming to formalist restraints on the employment of set-theoretical concepts are not mathematically precise. But then mathematical precision is not called for in such arguments, where we are attempting to justify a mathematical account of an ordinary notion: there is a parallel to be drawn here with the problem of justifying the familiar ϵ–δ definition of continuity in analysis. In fact, with all of the commonly used systems of formal proof for first order logic it is easy to carry out a case by case justification of its soundness in this way.

But when we come to question (II), the question of the *completeness* of our system of formal proof, things are altogether different. For in this case we cannot confront our system piecemeal, considering the axioms and rules individually and severally: we must confront it as a whole. For we must convince ourselves that we have not left out some vital principle whose absence from our system will make it impossible to give formal proofs of certain sentences that we should want to count as logically valid.

This is a thoroughly practical problem that will confront anyone who ventures to invent, or to assess, a system of formal proof. However, for a formalist it is unsatisfactorily vague, for it cannot be formulated mathematically in terms that he would be prepared to accept.

Nevertheless, the formalist must somehow contrive to confront it: how otherwise can he hope to convince mathematicians that the whole of their science should be recast in conformity with his Procrustean policy. Bourbaki tries to duck the issue by insinuating that he has arrived at *his*

6.5 Categoricity and the completeness of theories

formal system "by analysis of the mechanism of proofs in suitably chosen mathematical texts"[31]. However, even if we grant, as we must, Bourbaki's unrivalled familiarity with the relevant literature, we are entitled, I believe, to be sceptical about this. Who has ever gone about the task of laying down axioms in such a fashion? And, in any case, could we trust even Bourbaki not to have overlooked something essential?

There is simply no way to make sense of first order proof theory *as an account of logical consequence* unless we know that our system is complete. But a formalist, if he is consistent, cannot even say, in a mathematically satisfactory way, what completeness consists in. If we do not know that our system is complete we cannot attach theoretical significance to a proof that a certain proposition is *not* formally provable. But many of the deepest theorems in mathematical logic – Gödel's Incompleteness Theorems, for example – are theorems to the effect that certain entailments are *not* formally provable.

This is why old-fashioned formalism in the style of Bourbaki is intellectually incoherent, and, indeed, why that particular way of attempting to avoid difficult issues in the foundations of mathematics is utterly sterile. What is more, once we accept the logical prority of the semantic definitions of logical validity, logical consistency, and logical consequence even in first order logic, we have no reason to degrade second order logic from its status as *logic*, properly so called.

Of course it is notorious that second order logic does not admit of a complete system of formal proof procedures, nor is it compact, nor does it satisfy a Löwenheim–Skolem property. Perhaps more significantly, there is no straightforward way to localise the basic logical notions of logical validity, logical consistency, and logical consequence in that logic.

This is a matter of considerable importance for the foundations of mathematics, and I shall discuss it at greater length in Chapter 7. But second order logic is logic nonetheless, and, as I shall show in the next section, has an important role to play in explaining, and justifying, the axiomatic method.

6.5 Categoricity and the completeness of theories

Let us now discuss categorical axiom systems[32] in the context of formal logic. This is essentially a second order phenomenon, since the only

---

[31] *Elements* I, Introduction.
[32] Recall that these are axiom systems all of whose models are isomorphic. Note that the proposition "Γ is categorical" is Π_1.

categorical first order theories are theories all of whose models are of a fixed, Euclidean finite cardinality.

Given a formal, second order language $\mathscr{L}$, the notions of $\mathscr{L}$-structure and of isomorphism for $\mathscr{L}$-structures are completely determined by the primitive vocabulary of $\mathscr{L}$. Of course these notions do not reflect the fact that $\mathscr{L}$ is a second order language: they would be exactly the same for the first order language based on the same primitive vocabulary. Nevertheless, an isomorphism m from an $\mathscr{L}$-structure S to an $\mathscr{L}$-structure T, which effects a one-to-one correspondence between their universes of discourse and sends the morphology of S to that of T, can be extended in an entirely natural way, to a one-to-one correspondence between the n-ary functions on both structures and the n-ary relations on both structures, for every n, as I have already remarked in Section 6.1. Let us call the function so extended "m" as well.

Given the isomorphism m from S to T, and an assignment, f, of values with respect to S to the free variables (of all sorts) of $\mathscr{L}$, $m \circ f$ is an assignment of values with respect to T to the free variables of $\mathscr{L}$. Indeed, the correspondence $f \mapsto m \circ f$ establishes a one-to-one correspondence between assignments, f, with respect to S and assignments with respect to T.

6.5.1 Theorem *Let $\mathscr{L}$ be a formal second order language, let S and T be $\mathscr{L}$-structures, and let m be an isomorphism from S to T extended in the natural way to the functions and relations over the two structures. Then for every $\mathscr{L}$-formula, A, and every assignment, f, of values with respect to S to the free variables (of all sorts) of $\mathscr{L}$*

$$S \vDash A(f) \text{ iff } T \vDash A(m \circ f)$$

In particular, if A is a sentence (i.e. A has no free variables), then

$$S \vDash A \text{ iff } T \vDash A$$

Thus if m is an isomorphism from the $\mathscr{L}$-structure S to the $\mathscr{L}$-structure T, then S under the assignment f satisfies the same set of $\mathscr{L}$-formulas as does T under the assignment $m \circ f$. Note, by the way, that the isomorphism m does not effect a change of *notation* (as is sometimes carelessly said) but a change of *denotation*: the names remain the same, but what they denote is altered.

Theorem 6.5.1 gives a sharper formulation of what I have called the "fundamental dogma of the axiomatic method" – namely, that isomorphic structures share all their mathematically significant properties – in

6.5 Categoricity and the completeness of theories

which "mathematically significant property" is now interpreted to mean "property expressible by an $\mathscr{L}$-formula with parameters", where $\mathscr{L}$ is the formal language naturally associated with the structure in question. Stronger precise versions of this result can also be established (e.g. using logic of higher order).

Theorem 6.5.1 permits us to prove another important localisation theorem.

6.5.2 Theorem *(Localisation Theorem for Categorical Theories) Let $\mathscr{L}$ be a second order language, let Γ be a set of $\mathscr{L}$-sentences, let A be an $\mathscr{L}$-sentence, and let S be an $\mathscr{L}$-structure. Then from the hypotheses*

(i) *Γ is categorical*
(ii) *$S \vDash \Gamma$*
(iii) *$S \vDash A$*

we may infer that A is a logical consequence of Γ.

The sense in which this is a localisation theorem can be explained as follows. Suppose we have a categorical theory, Γ, and a model, S, of Γ. Then by virtue of Theorem 6.5.2, we are justified in taking the local proposition

$$S \vDash A$$

as an alternative characterisation of the sentence A's being a logical consequence of Γ. This assigns a definite truth value to all propositions of the form

A is a logical consequence of Γ

and thus allows us to negate such propositions and to combine them with other propositions using propositional connectives. We can also deduce that, for any such sentence A, either A is a logical consequence of Γ or $\neg A$ is a logical consequence of Γ, so that Γ is *logically complete*.

Thus a consistent and categorical set, Γ, of axioms logically determines a *definite mathematical theory*[33] in which every proposition expressible in the language in which Γ is couched has an objectively determined truth value which is logically independent of any *particular* model of Γ. Indeed, in some sense the theory, the set of logical consequences of Γ, is more fundamental than any of its particular models. (That the

---

[33] This is what Husserl calls a "manifold-form in the pregnant sense" (*Mannigfaltigkeitsform im prägnanten Sinn*), i.e. one that can be explained nomologically. See §31 of his *Formal and Transcendental Logic*.

logical consequences of Γ form a *set* and not just a semi-set follows from Theorem 6.5.2.)

Apply this to the categorical axiomatisations of the traditional theories – natural number arithmetic, real number arithmetic, and Euclidean geometry – and it seems as though those theories have somehow been absorbed into the logic of the terminology in which they are couched. We seem to have arrived at a kind of "linguistic" account of traditional mathematical objects. But it is not a *formalist* account, since it rests on *second order* logic, and therefore on the theory of sets.

Of course the fact that we have such a "linguistic" account of, say, Euclidean geometry, need not prevent the geometer from thinking of his points, lines, and planes in the traditional way as ideally exact objects to which he has direct access in his imagination, or in his understanding, or in some combination of both. He may even suppose that there *really are* such ideal entities.

But as long as he derives his theorems from the conventionally accepted axioms he does not have to *argue* for the existence of these peculiar "things" merely to defend the efficacy of his proofs or the legitimacy of his theory.

Paradoxically, it is precisely by adhering to the strict logical requirements of the axiomatic method that mathematicians are able to give free rein to their imaginations. They are free to postulate all sorts of systems of ideal or abstract objects, provided only that they can imagine or conceive them with sufficient precision to lay down axioms. For then the ontological status of those ideal or abstract objects becomes *logically* irrelevant.

In modern mathematics the "abstractness" or "ideality" of traditional mathematical objects reappears as an extra level of *generality* in the propositions. Thus the traditional proposition of real arithmetic that $\sin(\pi/2) = 1$ becomes the proposition that in every complete ordered field, if the function sin is defined by ... and π is defined by ... then $\sin(\pi/2) = 1$, where 1 is the multiplicative identity of the field in question and $2 = 1 + 1$.

Of course there is no harm *in mathematics* in referring to traditional mathematical objects such as real numbers: not to do so would result in an intolerable prolixity in our mathematical discourse. Indeed it seems likely that the axiomatic account of the traditional theories was so readily accepted by mathematicians precisely because it allowed them retain the old and familiar talk about the "objects" of their science. But the axiomatic method has rendered such talk *logically* unnecessary, and surely that represents a significant advance.

6.6 Mathematical objects

On the face of it, the basic axiomatic theories allow us only to *bypass* the ontological problems posed by the mathematical objects of tradition, not to *solve* them. The traditional philosophical problems (about the nature of the objects of traditional Euclidean geometry, for example) are still with us, and even if they no longer constitute a practical, logical problem for mainstream mathematics (as they did in the nineteenth century before the development of modern axiomatics), nevertheless they remain the subject of lively philosophical controversy.

Indeed, the attempt to confront and solve these problems inaugurated the modern, "analytic" or "linguistic" movement in philosophy. In particular, the now widespread idea that the solution to traditional problems in ontology and epistemology should be sought in logic and grammar originated in the attempt by mathematicians to solve the philosophical difficulties that surround the foundations of their science by means of mathematical logic.

The central figure here is, of course, Frege. Dummett tells us that "Frege's *Grundlagen* may justly be called the first work of analytic philosophy" on the grounds that in §62 of that work is to be found "the very first example of what has become known as the 'linguistic turn' in philosophy"[34]. Now if we turn to §62 of the *Grundlagen* we discover Frege in pursuit of an account of what natural numbers are. So if Dummett is right then the problem of the existence and status of natural numbers – in general, the problem of the existence and status of mathematical objects – is the very *fons et origo* of modern analytical philosophy.

The idea that there could be a purely logical or "linguistic" solution to this ancient ontological problem is still widely held, and, indeed, as we have just seen, there is a sense in which it can be justified. But that sense must be very heavily qualified, for we must count the whole of Cantorian set theory as logic if we are to maintain that there is a purely logical solution here.

Now that is not as absurd as it might at first appear. For one way of regarding the formalised version of Cantorian set theory is as a system of classical logic of absolutely infinite order, as a modern, Cantorian

---

[34] *Frege: Philosophy of Mathematics*, p.111.

version of Frege's conceptual notation (*Begriffsschrift*). But such a view, as I have repeatedly emphasised, requires us to recognise that there really are such things as sets (*arithmoi*). Of course in this respect it is rather like the point of view adopted by Frege in the *Grundgesetze*, where his new, mature semantics forces him to acknowledge value ranges (*Werthverläufe*) as genuine objects.

In fact, in passing from the *Grundlagen* to the *Grundgesetze*, Frege's views underwent a considerable change, requiring him, as he tells us[35], to discard a nearly completed manuscript; unfortunately for him, this change, he ruefully acknowledged[36], "stamped (his) views with an impress of paradox".

But his earlier views, the views of the *Grundlagen*, are, as Dummett has pointed out, the principal source of Frege's present day influence. Indeed, they have undergone something of a revival[37], and it will be useful here to contrast that earlier account of the nature of mathematical objects with the one that arises naturally from modern mathematical logic.

What was Frege's solution, in the *Grundlagen*, to the problem of mathematical objects? This question has occasioned considerable scholarly controversy, but one interpretation of that solution for the case of natural numbers might run as follows:

(1) Natural numbers, if they are anything at all, must be objects, because the terms we use for them function grammatically like object names.
(2) We cannot give the meanings of number names outside the context of a sentence. (This is the much discussed *context principle* as applied to such names.)
(3) We can, however, in principle, fix the meaning of every sentence (relevant to mathematics) in which such names occur, *and this is a necessary and sufficient condition for the existence of mathematical objects of the sort that numbers are.*

On this view to say that numbers, or, presumably, other sorts of basic "mathematical objects" exist is equivalent to saying that the mathematical sentences that contain "names" for these ostensible things have definite, objectively determined truth conditions. But how are we to understand this?

---

[35] *Grundgesetze der Arithmetik*, Introduction, p.ix.
[36] ibid., p.xi.
[37] In Dummett's *Frege: Philosophy of Mathematics*, for example. See also Crispin Wright's *Frege's Conception of Numbers as Objects*.

6.6 Mathematical objects

Does it mean that the objectivity of the truth conditions is *evidence* for the real existence of the objects in question? Or does it mean that the "existence" of the entities "named" in the propositions *just consists in* the circumstance that those propositions have objectively determined truth values?

Frege seems to have been in two minds about this. On the one hand, he speaks of the "self-subsistence" (*Selbständigkeit*) of the natural numbers as objects. Why "self-subsistence" and not "existence" here? Is Frege holding back from granting full-blown, *genuine* existence to natural numbers? Perhaps the explanation is to be found in §60 where he tells us that

> The self-subsistence that I am claiming for number is not to be taken to mean that a number word signifies something when removed from the context of a proposition, but only to preclude the use of such words as predicates or attributes, which appreciably alters their meanings.

This seems clear enough. There is no problem in supposing that ordinary naming expressions, like "the Eiffel Tower" or "Otto von Bismarck", succeed in signifying their *nominata* when removed from the context of a proposition, for the things they name really exist. But the "self-subsistence" attributed to numbers seems to be a matter of the grammar of number *words*.

What Frege says in the very next section (§61), however, seems to tell against this. For there he argues that numbers are objects even though they are not located in the external world:

> ... how is it possible for the number 4, which is objective, not to be anywhere? Now I contend that there is no contradiction in this whatever. It is a fact that the number 4 is exactly the same for everyone who deals with it; but that has nothing to do with being spatial. Not every objective entity (*objectiv Gegenstand*) has a place.

What are we to make of all this? Why isn't asking for the location of the number 4 of a piece with asking for the meaning of "4" outside the context of a sentence? Presumably the reason for enjoining us from asking for the meaning of "4" outside the context of a proposition is that "4" doesn't just name some *thing* in the straightforward way that "Otto von Bismarck" does. But surely if "4" does not simply function as a *name* in this way that can only be because there is no such *thing* as the number 4. And, of course, if there is no such *thing* (object – *Gegenstand*) as the number 4, then there is no question of where "it" is located.

All of these difficulties disappear – along with the context principle itself – in Frege's mature theory. There, as is well-known, the meaning (*Bedeutung*) of an expression is identified with the object that the expression names, so that the actual *meaning* of the expression "the wiliest Prussian statesman of the nineteenth century" wore side whiskers and plotted the invasion of France.

Moreover, in the mature conceptual notation of the *Grundgesetze* sentences are just names – of the truth values, *the True* and *the False* – and therefore can have no special role to play in the determination of meanings (*Bedeutungen*). Consequently, the context principle ceases to play a central role in Frege's theory. The mature Frege, it would seem, abandoned his earlier attempts to find "linguistic" solutions to the ontological problems of identity and existence[38].

But the Frege of the *Grundlagen* had not yet arrived at his mature position. And he was not the only mathematician of that era to entertain the idea that the existence of "mathematical objects" is either guaranteed by, or amounts to no more than, the objective determination of the truth or falsity of propositions expressed using names for the supposed objects. Indeed, it was a commonplace amongst geometers contemporary with Frege that the axioms of geometry were properly to be regarded, not as "given" truths about space, but rather as disguised definitions of the notions in terms of which they are expressed ("point", "line", "plane", etc.), so that by *stipulating* that the axioms, together with their logical consequences, are to be accounted true, the geometer "creates" the subject matter of his science.

The only check on this "creative" capacity, it would seem, is the Law of Contradiction. Hilbert, in his correspondence with Frege about the foundations of geometry, expressed just such a view of the nature of "mathematical" existence:

You write, "Axioms I call propositions ... From the fact that axioms are true it follows that they do not contradict one another." I was extremely interested to read just this proposition in your letter, because for as long as I have been thinking, writing, and lecturing about such things, I have always been saying the opposite: if the arbitrarily posited axioms together with all their consequences do not contradict one another, then they are true and the things defined by the axioms exist. For me, this is the criterion of truth and existence[39].

Surely the natural reading of Hilbert's remarks is that when he says that "the things defined by the axioms exist", the "things" in question are,

---

[38] The opening paragraphs of "Über Sinn und Bedeutung" seem strong evidence for this.
[39] E-H. W. Kluge, *Gottlob Frege on the Foundations of Geometry and Formal Theories of Arithmetic*, p.12.

6.6 Mathematical objects

for example, the points, lines, and planes referred to in the geometrical axioms. This is a muddle, for, as Frege pointed out in reply, it mistakes the *level* at which the definition takes place. But to make this objection clear requires a considerable logical and set-theoretical apparatus, and it is unlikely that Hilbert would have been prepared to countenance the use of such apparatus except in the context of yet another axiomatic theory – for sets, perhaps.

In any case, what does Hilbert mean when he says that if the axioms are consistent then they are *true*? True of what? Of the things they define? But what *are* those things?

The only hope of making sense of this farrago while, at the same time, observing Hilbert's scruples about the use of set-theoretical apparatus in logic, lies in adopting a strict version of formalism. Thus, for example, the geometer, when he is presenting his diagrams and arguments, is not, as we might naively suppose, trying to establish that triangles or circles have this or that property; nor is he trying to show that this or that property of triangles or circles holds in any model of his formal axioms: he is really trying to persuade his auditors that a proper, fully formalised proof of his proposition could be constructed, if only anyone could be bothered to do so. As Bourbaki tells us:

The *axiomatic method* is, strictly speaking, nothing but [the] art of drawing up texts whose formalisation is straightforward in principle[40].

Thus for Bourbaki formalisation is an end in itself, albeit an end which we habitually and deliberately abstain from pursuing. Unfortunately, however, he has neglected to explain to us what the *significance* of having a rigorously fomalised proof might be. And since, as we have seen, Bourbaki does not allow us to claim that our theorems are true, all we can conclude, if the geometer succeeds, is that he could abide by the strict rules of the game, if only he should decide it was worth his while to do so.

In any case, the Incompleteness Theorems disclose that this sort of formalism is untenable. Indeed, it cannot even give a coherent account of the very system of formal logic on which it so crucially depends. I have explained all this, in detail, in Section 6.5.

Perhaps Frege's *Grundlagen* account of mathematical objects[41] (or,

---

[40] *Theory of Sets*, p.8.
[41] Of course what Frege gave in the *Grundlagen* was not an account of "mathematical objects" in general, but rather of mathematical objects of a particular sort, namely, natural numbers. He does suggest, however, (e.g. in §60) that his methods have scope for wider application.

rather, the interpretation of it I gave in (1), (2) and (3) above) is not vulnerable to the same sort of criticism I have just given of the more radical views expressed in Hilbert's letter. But are Frege's own views, and those of his modern disciples, really immune to such a line of attack? The key question is how the context principle is to be applied.

Dummett, for example, tells us that

The context principle allows us to ascribe a reference to mathematical terms provided that we have fixed the truth-conditions of the sentences in which they occur[42].

Dummett is using "context principle" here not just for the doctrine that words have meanings only in the context of propositions (my condition (2) above), but for the further doctrine, inferred from the context principle proper, that we are allowed to ascribe a reference to *mathematical* terms (the qualification is surely essential) provided the truth-conditions for the relevant sentences have been fixed (condition (3)). But Dummett makes it clear that in such a case the fixing of the truth conditions cannot be carried out just anyhow.

The notion of reference, as applied to singular terms, is operative within a semantic theory, rather than semantically idle, just in case the identification of its referent is conceived as an ingredient in the process of determining the truth-value of a sentence in which it occurs. Hence the context principle, if it is to warrant an ascription of reference to a term, robustly understood, must include a further condition if it is to be valid. It is not enough that truth-conditions should have been assigned, in some manner or other, to all sentences containing the term: it is necessary also that they should have been specified in such a way as to admit a suitable notion of identifying the referent of the term as playing a role in the determination of the truth-value of a sentence containing it. With that further condition, the context principle ceases to be incoherent, and gains the cogency Frege took it to have ... [43]

But what cogency *did* Frege take this principle to have? The evidence is ambiguous. In §62 he says

How then are numbers to be given to us, if we cannot have any representations (*Vorstellungen*) or intuitions (*Anschauungen*) of them? Since it is only in the context of a proposition that words have any meaning, our problem becomes this: To define the sense of a proposition in which a number word occurs.

This is the very passage that Dummett identifies as the first example of the "linguistic turn" in philosophy. The meaning of the passage seems

---

[42] *Frege: Philosophy of Mathematics*, p.234.
[43] ibid., p.239.

straightforward enough: we just need to assign truth values to sentences containing number words. *That* is how numbers are given as objects.

Now we might very well baulk at accepting such an account. Is it really conceivable that the existence of numbers, or, indeed, of "mathematical objects" of any sort, could be guaranteed, or conferred, by such a means? Why should we grant the mathematician a licence we should not be willing to issue to the theologian[44]? We might even make a good case for the existence of Father Christmas or the Tooth Fairy on such grounds.

But, putting such scruples aside, how are we to reconcile this account of how numbers are "given" to us with the fact that in §68 Frege defines individual numbers outright:

My definition is therefore as follows: the number which applies to the concept F is the extension of the concept "equinumerous (*gleichzahlig*) with the concept F"

It is difficult to see in what sense the context principle continues to apply after this definition has been given. Can we not now say that the number 4 just *is*, for example, the extension of the concept "concept equinumerous with the concept *Galilean satellite of Jupiter*"? One might perhaps suppose there to be some difficulty involved in employing the notion of the extension of a concept here, but no, Frege tells us in a footnote to his definition of number that he assumes it to be known what the extension of a concept is.

Surely, if we do know what the extension of a concept is, then we know what "4" refers to, outside the context of any proposition whatsoever, just as we know what "the Eiffel Tower" and "Otto von Bismarck" refer to outside any such context.

This is all very puzzling. Perhaps Frege was in the process of rethinking his position while he was actually writing these sections of the *Grundlagen* – no one who has written on these matters will find such a supposition surprising – and thus what we have here is a juxtaposition of fragments of his earlier theory with the theory he hit upon in the course of writing the book.

He may well have persuaded himself that the earlier material should be retained, because the context principle, applied to "pre-Fregean" arithmetic, forestalls the obvious objection to his theory, namely, that his

---

[44] This is a point Frege himself made in his reply to Hilbert in the course of the correspondence referred to above. That, of course, was after he had abandoned his *Grundlagen* doctrines.

definition does not capture what mathematicians *in fact* think numbers to be.

To anyone raising such an objection Frege could say, "You are committing the error of asking for the meaning of number words outside the context of a proposition. What is scientifically essential here is that the objectively determined truth conditions for the propositions of arithmetic remain unaltered when my definition of number is adopted. And that they clearly do."

But let us leave these questions of exegesis aside. Dummett's revival of Frege's *Grundlagen* views depends, on his own account, on our finding some way of specifying the domains of our fundamental theories. Only if we can specify these domains in a suitable manner can we hope to set up a semantic theory which will satisfy the further condition that Dummett has imposed on the context principle in the passage I have quoted above. But this is, in Dummett's view, a difficult matter:

> The problem what constitutes a legitimate method of specifying the intended domain of a fundamental mathematical theory – one we do not treat as relating unproblematically to an already known totality of mathematical objects – remains intractable ... [45]

Since these intended domains will presumably be inhabited by the special "mathematical" objects (e.g. real numbers) that traditionally constitute the subject matter of these fundamental theories, the problem of specifying them seems intractable indeed.

In concentrating on these fundamental theories and their intended domains, Dummett unintentionally calls our attention to the uncomfortably close connection between the context principle, as he understands it, and the views of Hilbert I have quoted above. The crucial question remains the sense in which the context principle is to be understood: does the fixing of the truth conditions in which mathematical terms occur *guarantee* the existence (in the ordinary, "robust" sense) of mathematical objects as the references of those terms, or does the "existence" of those objects merely *consist in* the circumstance that those truth conditions are fixed?

Everything turns on which of these two interpretations of the context principle we adopt. Certainly if we adopt the second we cannot use the context principle as a stick with which to beat contemporary "nominalists". For the doctrine that "abstract objects" exist *only* in the sense that discourse in which "names" for the supposed objects occur can be given

---

[45] ibid., p. 235.

6.6 Mathematical objects

objective content by fixing truth-conditions for its sentences is surely *itself* a "nominalist" doctrine[46].

Dummett shows, by a shrewdly chosen example, that ordinary discourse is full of examples of reference to "abstract objects". He then goes on to remark that

> To deny to those things [countries, gases, political groups] the status of object, and to the corresponding expressions the function of referring to them, is to fall into the nominalist superstition, based ultimately on the myth of the unmediated presentation of genuine objects to the mind.

Dummett's example readily convinces us, if we needed convincing, of the crudity and untenability of much of contemporary "nominalism". But successfully to demolish such views does not sufffice, on its own, to establish the existence of the mathematical objects of tradition, as Dummett himself realises.

I am convinced that the essential difficulty here is that Dummett and others of his persuasion[47] have been too ready to plunge into the thickets of metaphysics and ontology without first allowing mathematics, and, in particular, mathematical logic, to carry them as far as it is capable of doing towards a resolution of these matters. We simply do not need this odd doctrine that language itself can somehow precipitate these peculiar "abstract objects" into existence. Indeed, we do not need those abstract objects themselves.

In the case of natural numbers the error originated with Frege and it is as much a historical error as a philosophical one. For if you are aware of the historical provenance of the natural numbers, you realise that you are not obliged to give an account of them *as objects* when you are laying down the foundations of mathematics. And you can give a logically coherent account of natural number arithimetic (or Euclidean geometry) using the axiomatic method, without referring to natural numbers (or points, lines, and planes) as abstract (or ideal) objects of a peculiar sort.

Ironically, the mathematical techniques made available to us by formal, mathematical logic permit us to carry out a mathematically rigorous version of the linguistic/metaphysical argument that Frege struggles to articulate in the *Grundlagen*. What Frege was missing was the technical notion of definition of a structure *up to isomorphism*.

---

[46] I think Dummett is aware of this difficulty, and that is why he has added his rider to the context principle which I have quoted above.

[47] I believe Husserl himself to be among their number. See his discussion of "definite manifolds" in §§ 9f and 9g of *The Crisis of European Sciences and Transcendental Phenomenology*.

We can specify which sentences of the formal language of second order arithmetic are true simply by laying down formalised second order versions of the five axioms of Peano Arithmetic given in Definition 5.1.7, and then stipulating that a sentence of that language is to be accounted true (i.e. in arithmetic) precisely when it is a logical consequence of those axioms. As we have seen in Section 6.5, the Consistency and Categoricity Theorems for that system of axioms guarantee that this stipulation determines the truth or falsity of every sentence of the language. There is, of course, no need to invoke the notion of an "intended domain" for the theory of natural numbers in establishing these key theorems.

Note that we have now arrived at something like the circumstances envisaged in *Grundlagen* §62. Our formal language contains terms intended to stand for "natural numbers". We have fixed the truth conditions for the sentences that contain those terms. Shall we now invoke the context principle? Are we to conclude from these facts that there are particular "abstract mathematical objects" called natural numbers of which all these sentences are true? Obviously not! Whatever would they be? Of what *use* would they be?

From the standpoint of our formal language, structures satisfying the Peano Axioms are absolutely indistinguishable. Indeed, any one of them would be absolutely indistinguishable from the "intended domain" for the theory of natural number arithmetic, since the latter, supposing there were such a thing, would also have to satisfy those axioms. The question what "the natural numbers" are is thus entirely pointless.

More than that, to posit such "objects" would simply be to clutter up mathematics with a lot of redundant junk. These observations become especially obvious when we recall the historical provenance of the natural numbers. For they did not even appear in mathematics until they were invented, along with the rational and real numbers, by the mathematicians of the sixteenth and seventeenth centuries.

Of course it might be objected that, whatever the logical foundations of arithmetic may be, and whatever may be the historical provenance of the natural numbers, nevertheless mathematicians think and speak of them as constituting a particular structure and of each of them as a particular mathematical object.

This may well be true, but it is utterly beside the point. After all, mathematicians also think of "the" Klein 4-group as a particular mathematical structure, even though its definition as "the" non-cyclic group of order 4 really only defines it up to isomorphism. But no one would think to ask what particular "mathematical object" its identity element

6.6 Mathematical objects

is[48]. The question what particular "mathematical object" the number 4 is should be seen as being equally senseless.

The point of view I have presented here is the common view of most mathematicians. Here, for example, is Spivak, commenting on his axiomatic definition of the real numbers in his classic introductory textbook on real analysis:

This theorem [i.e., the Categoricity Theorem for complete ordered fields] brings to an end our investigation of the real numbers, and resolves any doubts about them: There *is* a complete ordered field and, up to isomorphism, only one complete ordered field. It is an important part of a mathematical education to follow a construction of the real numbers in detail, but it is not necessary to refer ever again to this particular construction. It is utterly irrelevant that a real number happens to be a collection of rational numbers, and such a fact should never enter the proof of any important theorem about the real numbers. Reasonable proofs should use only the fact that the real numbers are a complete ordered field, because this property of the real numbers characterises them up to isomorphism, and any significant mathematical property of the real numbers will be true for all isomorphic fields. To be candid, I should admit that this last assertion is just a prejudice of the author, but it is one shared by almost all other mathematicians.

Exactly. I would only add the observation that what Spivak calls a "prejudice of the author" can be established as a theorem using the machinery of mathematical logic, as we have seen.

But what about set theory itself? How is its "intended domain" to be specified? The notion of set, or rather the notion of *arithmos* on which it is based, is fundamental in a sense that even the notions of natural number and real number are not. It is the oldest and most transparent mathematical notion we possess, and, accordingly, in laying the foundations of the theory of sets we are not confronted, as Frege was, with the problem of dealing with "objects" of whose nature we can form no clear conception.

There is nothing ineffable about the *arithmos* whose units are the books in my office bookcase. There *is* something puzzling about thinking of that *arithmos* as one single thing – that puzzled the Greeks too. But, after all, we can count *arithmoi* – we can take *arithmoi* as units in other

---

[48] The identity element is the only "identifiable" element of this group since given any two non-identity elements of a Klein 4-group, there is an automorphism of the group that interchanges them while leaving both other elements fixed. Does this mean that the "intended interpretation" you obtain from the categorical axiomatisation of "the" Klein 4-group using the context principle contains "indistinguishable" elements? Such are the quite unnecessary puzzles to which this version of the context principle gives rise.

arithmoi – and surely what we can count we can count (in the other sense of count, i.e. *take*) as being singular things.

No amount of argument can *prove* this, but then no amount of argument can refute it either. How, then, are these *arithmoi* "given" to us – how do we become aware of them? I have no answer to that question. But then I have no answer – no convincing answer – to the question how *anything* is "given" to us in the sense meant here. Neither, I suspect, has anyone else.

7
Axiomatic Set Theory

7.1 The Zermelo–Fraenkel axioms

The Zermelo–Fraenkel system of set theory is based on the notion of set that is the common property of all mathematicians. Indeed, my own exposition of set theory has been based on it. But now I want to consider that system as a formal axiomatic theory of the conventional sort. Of course it follows, *as a matter of logic*, that this theory, *qua* formal axiomatic theory, cannot serve directly as a foundation for mathematics; nevertheless its study is of central importance for set theory, properly so called, and therefore for the foundations of mathematics.

Zermelo's own version of the theory[1] was not formalised but was implicitly of second order. I shall consider both first and second order formal versions of his theory, ZF_1 and ZF_2, respectively. In his axioms Zermelo made provision for individuals that are not sets (*Urelementen*), but in modern treatments the formal theories deal only with pure sets, so that in models of them everything is "built up from the empty set". I shall follow the modern approach here.

Both of the formal theories ZF_1 and ZF_2 are based on the same primitive, non-logical vocabulary, a single binary predicate constant "$\in$" for membership, so that $\mathscr{L}(ZF_2)$ – the formal language of ZF_2 – differs from $\mathscr{L}(ZF_1)$ only in the presence of function and predicate variables of all orders. An interpretation for either of these languages is therefore provided by a structure of the form

$$M = (M, e)$$

where e is a binary relation on the universe of discourse M.

The formal axioms for ZF_2 constitute a formalised version of the

---
[1] See his "Über Grenzzahlen und Mengenbereiche".

axioms we have been considering all along, except that the Axioms of Pairing, Power Set, Union, and Replacement are given in existential form, and these axioms assert the existence only of supersets of the relevant sets: the existence of those sets themselves can then be derived from the Axiom of Comprehension.

I. *The Axiom of Extensionality*:

$$\forall v_0 \forall v_1 [[\forall v_2 [v_2 \in v_0 \supset v_2 \in v_1] \wedge \forall v_2 [v_2 \in v_1 \supset v_2 \in v_0]] \supset v_0 = v_1]$$

II. *The Axiom of Comprehension*:

$$\forall P \forall v_0 \exists v_1 \forall v_2 [v_2 \in v_1 \equiv [v_2 \in v_0 \wedge P(v_2)]]$$

III. *The Axiom of Pairing*:

$$\forall v_0 \forall v_1 \exists v_2 [v_0 \in v_2 \wedge v_1 \in v_2]$$

IV. *The Axiom of Power Set*:

$$\forall v_0 \exists v_1 \forall v_2 [\forall v_3 [v_3 \in v_2 \supset v_3 \in v_0] :\supset: v_2 \in v_1]$$

V. *The Axiom of Union*:

$$\forall v_0 \exists v_1 \forall v_2 [\exists v_3 [v_3 \in v_0 \wedge v_2 \in v_3]. \supset .v_2 \in v_1]]$$

VI. *The Axiom of Replacement*:

$$\forall f \forall v_0 \exists v_1 \forall v_2 [v_2 \in v_0. \supset .f(v_2) \in v_1]$$

VII. *The Axiom of Foundation*:

$$\forall P [\exists v_0 P(v_0). \supset .\exists v_0 [P(v_0) \wedge \forall v_1 [v_1 \in v_0 \supset \neg P(v_1)]]]$$

VIII. *Cantor's Axiom (the Axiom of Infinity)*:

$$\exists v_0 [\forall v_1 [\forall v_2 \neg [v_2 \in v_1] :\supset: v_1 \in v_0] :: \wedge :: \forall v_1 [v_1 \in v_0 \\ . :\supset: .\forall v_2 [\forall v_3 [v_3 \in v_2. \equiv .[v_3 \in v_1 \vee v_3 = v_1]] :\supset: v_2 \in v_0]]]$$

I shall denote by ZF_2 either the system of axioms just laid down or the conjunction of the eight sentences that compose it. The system Z_2 of *Zermelo set theory* is obtained from ZF_2 by deleting the Axiom of Replacement (Axiom VI)[2].

The only axiom in which a function variable is employed is the Axiom of Replacement. The axiom can be reformulated so as to employ only a predicate variable:

---

[2] This is, essentially, the theory Zermelo presented in his 1908 paper "Untersuchungen über die Grundlagen der Mengenlehre I".

7.1 The Zermelo–Fraenkel axioms

VI. *The Axiom of Replacement (alternative version)*:

$$\forall P \,[\forall v_0 \exists v_1 \forall v_2 [P(v_0, v_2) \equiv v_1 = v_2]$$
$$. \supset : .\forall v_0 \exists v_1 \forall v_2 \forall v_3 [v_2 \in v_0 \wedge P(v_2, v_3). \supset .v_3 \in v_1]]$$

Similarly, the Axiom of Foundation (Axiom VII), which is properly of second order, can be replaced by the following first order axiom:

IX. *The Axiom of Regularity*:

$$\forall v_0 [\exists v_1 [v_1 \in v_0]. \supset .\exists v_1 [v_1 \in v_0 \wedge \forall v_2 \neg [v_2 \in v_1 \wedge v_2 \in v_0]]]$$

The first order version, ZF_1, of Zermelo–Fraenkel set theory is the first order reduct of the second order version of the system just given, with Regularity (Axiom IX) replacing Foundation (Axiom VII), and augmented by the Axiom of Choice (suitably formulated).[3] This reduct is obtained by replacing each of the axioms properly of second order, namely the Axiom of Comprehension (Axiom II) and the Axiom of Replacement (Axiom VI), by the set of its first order substitution instances; for this purpose we must use the alternative version of Replacement.

Logicians have concentrated their attention on the first order version of this theory, but, as with the axiomatic theories of natural number arithmetic (the theory of simply infinite systems) and real analysis (the theory of complete ordered fields), the first order theory is natural only insofar as it is a reduct of the second order one[4].

To see why ZF_2 is natural we must first single out from among its models those which have the intended form, namely, those models $\mathbf{M} = (M, e)$ in which the members of M are sets, and e is the restriction of the membership relation, $\lambda xy[x \in y]$, to the set M. I shall call these intended interpretations of ZF_2 *normal domains*, following Zermelo.

7.1.1 Definition *Let $\mathbf{M} = (M, e)$ be an $\mathscr{L}(ZF_2)$-structure. Then, by definition, $\mathbf{M}$ is a normal domain if, and only if,*

(i) $(\forall x \in M)[x \text{ is a set}]$
(ii) $(\forall x \in M)(\forall y \in M)[(x, y) \in e \text{ .iff. } x \in y]$
(iii) $(M, e) \vDash ZF_2$

If we replace the last condition by

---

[3] The resulting theory is usually called ZFC. The Axiom of Choice is *not* required in ZF_2 as it is a (second order) consequence of the other axioms: this is because we account it true and have adopted it as an axiom for set theory proper.

[4] This is a point of considerable foundational significance and I shall discuss it in greater detail in the next section.

iii.) $(M,e) \models Z_2$

we obtain the notion of a *Zermelo domain*.

Let $M = (M, \in_M)$ be a normal domain. By a *class of M* (or *M-class*) I mean a subset of M; an *M-set* is a member of M; an M-class that is not an M-set is called a *proper M-class*. Note that by the Axiom of Union M is transitive ($(\forall x \in M)[x \subseteq M]$), and, by Comprehension, if $x \in M$ is an M-set and $y \subseteq x$ is a subset of x, then y is an M-set.

Normal domains have a special status among models of ZF_2: this follows from Theorems 7.1.3 and 7.1.4 below. Before stating these theorems I need to lay down an important definition.

7.1.2 Definition *Let* $M = (M, \in_M)$ *be a normal domain or a Zermelo domain. Then by definition*

$$Ord_M = \{x \in M : x \text{ is a von Neumann ordinal}\}$$

Ord_M *is called the* characteristic *of M.*

Since M is transitive, Ord_M is a von Neumann ordinal number, and is easily seen to be a proper M-class. In fact, Ord_M is a *strongly inaccessible number*, that is to say

(1) $(\forall f : Ord_M \to Ord_M)(\forall a < Ord_M)[sup(f``a) < Ord_M]$ (i.e. Ord_M is *regular*: every sequence of length $< Ord_M$ composed of ordinals $< Ord_M$ has a limit $< Ord_M$)
(2) $(\forall a < Ord_M)[||2^a|| < Ord_M]$

Condition (i) is imposed by the Axiom of Replacement, (ii) by the Axiom of Power Set.

Every normal domain is determined by its characteristic.

7.1.3 Theorem *Let* $M = (M, \in_M)$ *be a normal domain. Then* Ord_M *is strongly inaccessible and M is the set of all pure sets of rank $< Ord_M$, i.e.* $M = V_{Ord_M}$. *Conversely, if a is strongly inaccessible and $M = V_a$, then* $(M, \in_M)$ *is a normal domain.*

Thus a normal domain is an initial segment, V_a, of the cumulative hierarchy of pure sets, where a is a strongly inaccessible ordinal number.

Since the formal axioms of both ZF_1 and ZF_2 are formalised versions of the ordinary axioms of set theory, every proposition about pure sets that we can formulate in ordinary set theory has a formal counterpart in the language $\mathscr{L}(ZF_1)$ of ZF_1, and for every such proposition that we can prove in ordinary set theory, its formal counterpart is provable in ZF_1 and therefore is a logical consequence of ZF_2. In particular, the theory of von Neumann ordinal numbers can be seen to hold "internally" in models of ZF_2. Moreover, the Axiom of Foundation in ZF_2 guarantees that in any model $M = (M, e)$ of that theory (normal or not), the binary relation e on M is actually well-founded, so that the M-class of "von Neumann ordinals" in the model is well-ordered by e. By analogy, we can call the von Neumann ordinal which is the order type of that well-ordering the *characteristic* of M. These observations allow us to establish that ZF_2 is *quasi-categorical* in the sense of the following theorem.

7.1.4 Theorem *Let $M = (M, e)$ be a model of ZF_2, and let a be its characteristic. Then M is isomorphic (as an $\mathscr{L}(ZF_2)$-structure) to the normal domain $(V_a, \in_{V_a})$.*

Theorem 7.1.4 plays a role analogous to the categoricity theorems for Peano systems and complete ordered fields: it is our guarantee that the formal axiomatic theory ZF_2 is a natural one. Of course ZF_2 is only *quasi*-categorical because it will have non-isomorphic models if there is more than one strongly inaccessible von Neumann ordinal number, although given any two non-isomorphic models of ZF_2, one must be isomorphic to an initial segment of the other. (If there are no strongly inaccessible numbers ZF_2 is not even consistent.) This means that the analogy with the second order axiomatisations of natural number and real number arithmetic is imperfect.

In fact, Theorem 7.1.4 shows us that ZF_2 has, in the species of normal domains, a privileged species of natural models. In this respect ZF_2 stands in strong contrast with the theory of simply infinite systems and the theory of complete ordered fields, none of whose models enjoy a special status. This, of course, reflects the fundamental character of the notion of set on which the ZF_2 axioms are based.

Since ZF_2 axiomatises a species of initial segments of the cumulative hierarchy of pure sets, the study of ZF_2 as an axiomatic theory is obviously relevant to questions about the extent of that hierarchy. In fact the real reason for studying the formal axiomatic theories ZF_1, ZF_2, and their extensions is not because they deal directly with the universe of sets – Cantor's Absolute – but because they shed light on the structure of initial segments of that universe[5].

---

[5] This is a controversial matter to which I shall return in the next section.

None of the axioms considered in Chapters 4 and 5 suffices to establish the existence of even *one* normal domain. So if the study of ZF_2 as a formal axiomatic theory is to have any significance, we need to augment the basic principles of set theory with axioms which guarantee the existence of normal domains. Zermelo held, and in this I think most mathematicians who work in axiomatic set theory follow him, that it is reasonable to suppose that the species of normal domains is absolutely infinite: let us call this *Zermelo's Principle*.

I do not believe that the arguments that have been advanced for Zermelo's Principle are convincing. In particular, they do *not*, as both Zermelo and Gödel maintained, arise naturally out of the same considerations that lead us to the central principles – like Comprehension and Replacement – that underlie Cantorian set theory.[6] Nevertheless, it is useful to adopt it as a working hypothesis in discussing contemporary axiomatic set theory.

We can formulate Zermelo's Principle by introducing a global function, ζ, of one argument which, applied to an ordinal a, yields the first strongly inaccessible number beyond a.

7.1.5 Axiom (*Zermelo's Principle*) *Let a be a von Neumann ordinal. Then*

(i) $\zeta(a)$ *is strongly inaccessible.*

(ii) $a < \zeta(a)$.

(iii) $(\forall x < \zeta(a))[a < x : implies : x \text{ is not strongly inaccessible}]$.

Using transfinite induction we can now define a global function, γ, such that if a is an ordinal, γ_a is the ath strongly inaccessible number, counting ω as the 0th.

7.2 Axiomatic set theory and Brouwer's Principle

The axioms that I laid down in Chapters 4 and 5 for the Cantorian version of set theory, the version that underlies modern mathematical practice, would, if formalised (e.g. as in Appendix 1) constitute a weak subtheory of *first order* Zermelo–Fraenkel set theory, ZF_1. Thus the consistency even of ZF_1 cannot be proved from the axioms laid down so far, unless modern mathematics itself, with its Cantorian finitist assumptions, harbours a contradiction – this follows from Gödel's Second Incomplete-

---

[6] See the discussion in Section 5.7.

ness Theorem. Indeed, the consistency of ZF_1 is problematic, even from the standpoint of Cantorian finitism[7].

These facts might seem to suggest that I am recommending a weak version of conventional Zermelo–Fraenkel set theory as a foundation for mathematics. Nothing, however, could be further from my intention. On the contrary, my aim is to give a basic logical/set-theoretical framework for the axiomatic method. In particular, my aim is to make it possible to investigate axiomatic set theory itself while taking fully into account the Cantorian doctrine of the *absolute* infinity of the set-theoretical universe.

The system of Cantorian set theory presented in Chapters 3, 4 and 5 above should not, therefore, be regarded as a mere fragment of ZF_1, considered as a formal axiomatic theory. It is not a formal axiomatic theory at all, but is rather a partial description of the absolute universe of sets in which *all* conventional structures, including all models of both ZF_1 and ZF_2, are to be found.

The theory, as I have presented it, is not formalised. But, in the light of what I have said here, even the formalised version of the theory sketched in Appendix 1 ought not to be regarded as a particular theory formalised in (a fragment of) conventional first order logic: it is more useful to see the whole theory as itself constituting a formal logic of *absolutely infinite* order.

Of course the axioms and rules that I have actually laid down for that logic are incomplete – any axioms and rules that I *could* lay down would necessarily be incomplete: you cannot even lay down a complete system of proof procedures for *second* order logic. Indeed, if we include global quantifiers in our logic of absolutely infinite order it is not even clear what completeness might *mean*, since, as I have repeatedly emphasised, propositions expressed by means of those quantifiers do not have conventional truth values, and are not subject to conventional ("classical") laws.

The crucial feature of this ultimate system of higher order logic is Brouwer's Principle: *The conventional, classical laws of logic apply to a proposition only if the bound variable of each of its quantifiers ranges over the elements of a (Cantorian) finite species, that is to say, over the elements of a set.* However we might decide to augment our present supply of axioms and rules we must continue to acknowledge this principle, for

---

[7] I have discussed this matter at length in my articles "On the consistency problem for set theory" and "Global quantification in Zermelo–Fraenkel set theory".

to claim that our system is *ultimate* just *is* to acknowledge Brouwer's Principle.

In accordance with Brouwer's Principle there is only *one* notion of domain of discourse for interpreting any axiomatic theory, formal or informal, namely, the notion of a *set*, understood as a (Cantorian) finite plurality of objects: *there are no such domains of discourse outside the universe of set theory*. This means that any such domain has a power set, which, in turn, has a power set, and so on. Thus, for example, any structure can be equipped with a topology; and any structure suitable to interpret a first order language is a suitable interpretation for the second order language based on the same non-logical primitive vocabulary.

In this way Brouwer's Principle imposes a uniformity on the axiomatic method (formal and informal) which gives us free set-theoretical mobility, so to speak, in studying any particular structure. To accept Brouwer's Principle is thus radically to simplify both the theoretical foundations and the practice of axiomatics. In fact, the vast majority of mathematicians, most of whom are happily and complacently unaware of the foundational issues under discussion here, *do* accept Brouwer's Principle *de facto* simply by virtue of *taking it for granted* that conventional mathematical constructions, the ultimate justifications for which are to be found among the basic principles of set theory, can be applied to all the structures they consider.

Applying Brouwer's Principle to axiomatic set theory we are, of course, forced to accept that *neither ZF_1 nor ZF_2 can be regarded as applying to set theory as a whole*. This means that their models (if any) are ordinary structures, and therefore the "meta-theory" in which those models are studied is just set theory itself – which is another way of saying that the study of these theories is just like the sudy of any other formal axiomatic theory. But perhaps it is precisely this consequence of Brouwer's Principle that will discourage some from accepting it. We must consider this in more detail.

When I introduced Brouwer's Principle[8], I remarked that it shows us why we cannot apply the notion of *identity* to species and therefore why we cannot account them *objects*. It thus allows us to make a clear distinction between the finite and the infinite in Cantorian set theory, the *Cantorian* finite and the *absolutely* infinite. Of course, in any interpretation of ZF_2 in which the domain of discourse consists of sets, "the proper classes" of the domain are the analogues of absolutely

---

[8] In Section 3.5.

infinite species. Moreover, the "classical" nature of second order logic means that we are treating the totality of classes over the domain as itself a well-defined domain. Thus if we tried to interpret ZF_2 with the universe of sets as its underlying domain, not only would we be treating absolutely infinite "objects" as *values* of a quantified variable, we would be treating their (even more infinite) totality as a legitimate domain of discourse. Surely that would make it difficult to explain in what sense the Cantorian universe is *absolutely* infinite.

But perhaps some will argue that it is precisely here that we ought to draw the line, rejecting the universe of sets as a legitimate domain of interpretation for ZF_2, but allowing it for ZF_1. First order logic is applicable to that universe (so the argument might go) because the notion of "arbitrary set" is a clear one, but second order logic would not apply, since the notion of "arbitrary class (species, collection, ...)" of sets is not clear. This is, indeed, the position taken by Bernays in his textbook *Axiomatic Set Theory*[9].

There are, however, serious conceptual difficulties with this point of view. We have seen that introducing the apparatus of quantification subject to the conventional, classical laws of first order logic is formally equivalent to introducing an identity relation for infinite species and global functions (of both sorts) so that they can, in effect, be treated as objects.

Of course the first order language of ZF_1 does not even contain the vocabulary for talking about infinite species ("proper classes") or global functions. But we can introduce the machinery for forming class comprehension terms, and free variables over classes, as Bernays does[10], to obtain a conservative extension of ZF_1. In fact, we can add class quantifiers to the resulting theory conservatively, provided we restrict class comprehension to formulas without bound class variables (formulas with free class variables may used, however). The resulting theory VNB (von Neumann–Bernays set theory[11]) is equivalent to a finitely axiomatised theory BG (Bernays–Gödel[12]) (and both are conservative over ZF_1).

These *class–set theories*, whose elements are the extensions of properties defined by $\mathscr{L}(ZF_1)$-formulas with parameters from the universe of sets, are thus implicitly contained in ZF_1 itself, and, in particular, any conceptual difficulty we might experience in interpreting them as

---

[9] See his comments in §3 of Chapter I.
[10] op. cit.
[11] See Bernays' series of papers "A system of axiomatic set theory".
[12] See Gödel's monograph *The Consistency of the Continuum Hypothesis*.

applying to the universe of sets as a whole points to a cognate difficulty in interpreting ZF_1 in that way.

Now if we are to make sense of the axioms of ZF_1 as applying directly to Cantor's absolute universe of sets, we must make use of a prior, *general* notion of global function, both ordinary and propositional. Otherwise we cannot express the insights underlying the axiom schemata of Comprehension and Replacement. Since we see them *as schemata,* our conviction that a particular instance of Comprehension or Replacement holds cannot rest on a detailed insight into the logical structure of its principal formula: that formula will, in general, contain such a huge number of symbols that we could not even make a start at understanding its logical structure.

No, the insight that underlies Comprehension is that since a set is, by definition, a *finite, extensional plurality* (in the appropriate sense of "finite"), a sub*species* of a set whose defnining property is definite, that is to say, is given by a global propositional function – *any* global propositional function – forms a sub*set* of that set. Similarly, the insight that underlies Replacement is that the image of a set under a global function – *any* global function – is a set.

Thus the supposed insight that convinces us of the legitimacy of applying these formal, first order axiom schemata to the universe of sets must contain two components: first, a *set-theoretical* component, properly so called, which employs the prior, general notion of global function over that universe, and second, what we might call the *logical* component, namely, the further assumption that any suitable first order formula of $\mathscr{L}(ZF_1)$, interpreted over the universe of sets, can legitimately be used to define a global function of the sort under consideration. Of course it is only the *logical* component that must be called into question if we accept Brouwer's Principle.

But we encounter special additional difficulties if we accept one of the first order class–set systems *VNB* or *BG* as legitimate when applied to the whole universe of sets. For then we must assume the burden of explaining why the full, unrestricted versions of Comprehension and Replacement for formulas employing bound *class* variables should be rejected.

Why should the insight that leads us to suppose that the image of a set under a global function is a set apply *only* if that function is definable without using bound class variables? After all, since these class–set theories are formulated in conventional first order logic, the assumption that formulas involving bound class variables are *definite* in Zermelo's sense is built into their underlying logic antecedently, so to

7.2 Axiomatic set theory and Brouwer's Principle   247

speak. But if we allow formulas with bound class variables into our Class Comprehension schema, what we have, essentially, is ZF_2. I conjecture that this is why Bernays rejected the use of bound class variables in his formulation of the theory in his book, *Axiomatic Set Theory*[13].

Bernays sees classes as the "extensions of predicates" and therefore regards a class as "having the character of an ideal object"[14]. He contrasts such an *ideal* object with "a set, which as a collection, is a mathematical thing". In this way, perhaps, he thought he could provide a rationale for accepting the universe of sets as a legitimate domain of interpretation for $\mathscr{L}(ZF_1)$ while denying it such a role for $\mathscr{L}(ZF_2)$.

Of course, if we allow conventional, classical quantification over the domain of definition of the global propositional functions that provide the defining properties of these "ideal" objects, we can define extensional identity for them (as Bernays himself does), and it is difficult to invent a rationale for denying them status as "collections" and therefore as "mathematical things" (objects) of the very same sort that sets are.

But in any case, such considerations do not go to the heart of the matter. Surely, it is not the natures, or individuating peculiarities, of the inhabitants of a domain, *considered severally and individually*, that determine the suitability or otherwise of proposed laws governing quantification over that domain. Surely all that is relevant to such questions is the *extent* of the domain – its "size", and the sharpness of the boundary that seperates its members from its non-members.

From the standpoint of Brouwer's Principle, conventional, classical quantification over the universe of sets is disallowed, not because of any peculiarity or irregularity in the natures of the inhabitants of that universe *considered each on its own*, but rather because *together* they are absolutely infinite in multitude. What is relevant here is the difference between a finite domain and an absolutely infinite one: *that* is what must be reflected in the logic of first order quantification over the domain.

This point of view gains plausibility from the following fact: if we attempt to describe the universe of sets using a suitable version of ZF_1 based on an underlying *intuitionistic* first order logic, inserting an antecedent clause in the Comprehension schema requiring decidability (excluded middle) for the principal formula, the resulting theory is con-

---

[13] In his article "Zur Frage der Unendlichkeitsschemata in der axiomatischen Mengenlehre", Bernays adopted a full second order version of set theory, incorporating a strong second order Reflection Principle.

[14] *Axiomatic Set Theory*, Chapter I, §1.

servative over that part of the theory which is free of global quantifiers. This is, of course, not true if the underlying logic is classical, as in ZF_1[15].

To suppose, as Bernays does, that there could be a structure interpreting $\mathscr{L}(ZF_1)$ that cannot interpret $\mathscr{L}(ZF_2)$ is, from the standpoint of "ordinary" mathematics, to assume that some sets (classes, collections, domains ... – call them what you will) have power sets (power classes, power collections, power domains ...) while others do not. But if the logician is allowed to consider the universe of sets as the underlying domain of a structure for interpreting the first order language of set theory, why can't the group theorist use it as the underlying domain for an interpretation of the first order language for the theory of groups? Indeed, why can't the group theorist form the free group over that universe?

But then that group has no automorphism group, for its underlying domain has no power set! Is it necessary, then, in group theory to distinguish between those groups that have automorphism groups and those which do not? More seriously, if the elements of a group form a set without a power set, how do we speak of its normal subgroups? Are we allowed to acknowledge only those we can define in the formal language of set theory based on $\in$? Why don't mathematicians at large make this fundamental distinction between those structures whose domains have power sets and those whose domains do not?

All this may seem an extravagant fantasy, but analogous difficulties have actually arisen in category theory[16]. The fact is that any attempt to treat absolutely infinite species as legitimate domains of discourse for conventional mathematical structures will inevitably be intellectually unstable. For it will require us to make a fundamental distinction between "small" structures, which inhabit the universe of sets, and "large" structures, which do not, and of which the only natural examples we possess are the universe of sets itself, and its absolutely infinite subspecies. But the basis for this distinction will be forgotten or become obscured, and there will be controversy over what "constructions" are permitted on "large" structures. Gradually mathematicians will come to apply *all* the standard set-theoretical concepts and constructions to these "large" structures (to form functor categories over "large" categories, for exam-

---

[15] This point is discussed in my article "Global quantification in Zermelo–Fraenkel set theory".

[16] See Mac Lane's dicussion of these matters in his *Categories for the Working Mathematician*, especially in Section 6 of Chapter I. I have considered the issues raised there at some length in my article "On the consistency problem for set theory" (Part I).

7.2 Axiomatic set theory and Brouwer's Principle

ple) and the rationale, such as it is, for the large–small distinction will fade from consciousness.

In a letter to Dedekind[17], Cantor attempted to explain his doctrine of the absolutely infinite:

> a multiplicity can be such that the assumption that *all* of its elements 'are together' leads to a contradiction, so that it is impossible to conceive of the multiplicity [*Vielheit*] as a unity [*Einheit*], as 'one complete thing'. Such multiplicities I call *absolutely* infinite or *inconsistent multiplicities*.

So many objects satisfy the defining property of an absolutely infinite species that they cannot be consistently said even to *compose* a totality (as the presence of the indefinite article – *a* totality – indicates). Indeed, on Cantor's account here the very expression "universe of set theory" is misconceived on etymological grounds (*unum*, one + *versus*, past participle of *vertere*, to become). It would seem that to say anything on this difficult matter is to stretch the resources of language to the breaking point. Perhaps this is what led Cantor to say

> [The Absolute] surpasses human power of comprehension, as it were, and, in particular, eludes mathematical determination[18].

To adopt Brouwer's Principle is to acknowledge this. It is to acknowledge that outside "the" universe of sets there is nothing, certainly not a mathematical structure called "the universe of sets", talk about "which" is really only a *façon de parler*. Sets and individuals are all that there are.

If, on the other hand, we reject Brouwer's Principle and follow the examples of Gödel[19] and Kreisel[20] in positing the existence of structures interpreting standard first and second order languages whose domains of discourse are not members of the universe of sets, we are immediately confronted with all sorts of questions. What is the nature of those domains? Which of them can occur as elements in further such domains? Do the subdomains of such a domain themselves form a domain? If not, why not? If so, then why can't we regard the universe of pure sets itself as just another normal domain? And why can't we use the well-orderings of the proper classes of that domain to extend the cumulative hierarchy? What closure conditions apply to the new "ordinals" in such extensions? And what content can we now give to the assertion that the universe of sets is *absolutely* infinite? Why should we not regard the union of those

---

[17] A translation of this letter is included in van Heijenoort's *From Frege to Gödel*.
[18] *Gesammelte Abhandlungen.*, p. 405
[19] "What is Cantor's continuum problem?" (Section 3).
[20] "Informal rigour and completeness proofs" (Section 1).

extensions as the *real* universe of set theory? But then how are we to treat *that* universe? Does Brouwer's Principle apply *there*? If not, the whole process must start all over again, and so on .... We find ourselves pursuing a mirage, a *contradictio in adjecto*, a Kantian *Idea of Reason* in the form of a completion of what is essentially incomplete.

Such problems inevitably arise when we do not treat Cantor's Absolute as an *absolute*. But these difficulties about extending the cumulative hierarchy are not the only problems that confront us. Presumably there could, in principle, be all sorts of domains lying outside the cumulative hierarchy, however extravagantly that hierarchy is extended beyond the Absolute, and that raises still further questions. Is there a universal domain, a ("classical") domain which contains *all* the domains that can serve as universes of discourse for interpretations of standard predicate languages? If not, why not? If so it must be a member of itself, and that raises the question of the subdomain of the universal domain consisting of all those domains that are not included among their own members.

Such are the difficulties that confront us if we fail, or refuse, to acknowledge Brouwer's Principle. They have, I'm afraid, a familiar ring. All the old problems that preceded, and, indeed, inspired Zermelo's analysis of the notion of set return to plague us. It is as if that analysis had been in vain.

Surely this provides us with a strong incentive to adopt Brouwer's Principle, maintaining in doing so a sharp distinction between the (Cantorian) finite and the (absolutely) infinite. In the next section we shall see that this provides us with a coherent standpoint from which to address difficult problems in the global semantics of second order logic.

There is one final issue that I need to confront. It might be objected that Brouwer's Principle imposes a curb on the study of strong axioms of infinity. This, however, is not the case There is nothing essential in the current study of these axioms that is incompatible with it. The principal bearing that the acknowledgement of Brouwer's Principle has on these investigations lies in the way that we regard them – in what we take their significance to be.

The way these principles are, in fact, studied is by adding them as new axioms to the *first order* theory ZF_1[21]. This is an eminently practical

---

[21] This, I believe, explains why we find it natural to call them "axioms". They are, after all, "axioms" in the technical terminology of mathematical logic. But that terminology is inappropriate, as I have already pointed out in Section 5.6, since they cannot be accounted "axioms" in the traditional sense of being uncontroversial starting points for proofs. Such are the confusions we land in if we try to treat our foundational theory as a conventional axiomatic theory.

7.3 The localisation problem for second order logic

and sensible way to go about about things. After all, we want to investigate what we can and, equally important, what we *cannot*, prove from them. The first order formulation puts all the model-theoretic and proof-theoretic techniques of modern logic at the disposal of the investigator.

But from the perspective of Brouwer's Principle these theories are interesting primarily insofar as they describe initial segments of the cumulative hierarchy and therefore insofar as they are *second order consistent*. In other words, *if T is such a theory, then $T \cup \{ZF_2\}$ is the theory we are really interested in*. Put another way, from a foundational standpoint we are primarily interested in *standard models* of these theories in which the domain of the model is a normal domain in Zermelo's sense.

Some of these strong infinity axioms can be naturally formulated directly in set theory itself – the axioms of Mahlo and Zermelo, for example. Some, on the other hand, can *only* be formulated relative to first order axiomatic theories – axioms that postulate non-trivial elementary embeddings of the universe of the theory into class sized submodels of itself, for example. But even in the latter cases we can suppose that the theories in question have *standard models* in which the universe of the model is a normal domain and therefore is a model of ZF_2.

On a naive level when you are dealing with theories that are extensions of ZF_1 by large cardinal axioms, you tend to think, informally, that the intended model is the universe of sets, so that, for example power set "means" the real power set. But if the intended model is an initial segment of the cumulative hierarchy – a normal domain – these intuitions can remain in place, and adopting Brouwer's Principle doesn't interfere with them.

Note that from this perspective there is a natural measure of the strength of these theories: T_1 is stronger than T_2 if the smallest standard model of the latter is an initial segment of the smallest standard model of the former. Thus we should not be surprised that these first order theories are linearly ordered with repect to interpretability. Indeed it would be interesting to find pairs of strong infinity axioms (or axiom schemata) that, as far as we can tell, are each *separately* compatible with the basic set-theoretical axioms, but which are either mutually inconsistent, or for which we can prove that neither implies the other.

7.3 The localisation problem for second order logic

I want now to return to the problem of localisation for second order logic briefly touched on in Section 6.4. The problem, you will recall, arises because the definitions of the basic logical notions of consistency (satisfiability), universal validity, and logical consequence all require global quantification for their formulation. This means, *prima facie*, that, in accordance with Brouwer's Principle, these key notions are not subject to the ordinary ("classical") laws of logic; and this, in turn, means that, for example, the species of subsets, Γ, of the set, *Sent*, of all $\mathscr{L}(ZF_2)$-sentences such that Γ is satisifiable is a candidate for a natural example of what Vopenka calls a *semi-set* – a sub*species* of a set (in this case the set $P(Sent)$) which does not form a sub*set* of that set. The localisation problem is to find characterisations of these key notions that do not require global quantification for their formulation.

From the standpoint of axiomatic set theory, the approach to this problem that immediately suggests itself is to extend the axioms of ZF_2 to a theory whose natural models encompass sufficiently large initial segments of the cumulative hierarchy to include models witnessing the consistency of all consistent second order theories. There is, after all, only a continuum of theories, consistent and inconsistent.

The internal definition of satisfaction in a Zermelo domain is *absolute*, that is to say, if $(M, \in_M)$ is a Zermelo domain, then for every $X \in M$ and every $\Delta \subseteq Sent$[22]

$$(M, \in_M) \vDash \ulcorner [v_1 \vDash v_0] \urcorner (\Delta/v_0, X/v_1) \ .iff.\ X \vDash \Delta$$

Hence the internal definition of second order logical consistency in any Zermelo domain, *a fortiori* in any normal domain, is a partial characterisation of the *absolute* notion of second order logical consistency Thus for Zermelo domains $(M, \in_M)$

$$(M, \in_M) \vDash \ulcorner \exists v_1 [v_1 \vDash v_0] \urcorner (\Delta/v_0) \ .implies.\ (\exists X \in M)[X \vDash \Delta]$$

Of course it need not be the case that the internal definition of second order logical consistency in a Zermelo domain actually *coincides* with the absolute definition. Indeed, if there are any normal domains, there must necessarily be normal domains whose internal definition of second order logical consistency does not coincide with the absolute one.

The question thus naturally arises whether we can extend ZF_2 to a theory Γ with $ZF_2 \in \Gamma \subseteq Sent(\mathscr{L}(ZF_2))$ such that in normal domains satisfying Γ the internal definition of second order logical consistency actually coincides with the absolute notion.

---

[22] Notice that I use Quine's corners "$\ulcorner$" and "$\urcorner$" to distinguish formulas in $\mathscr{L}(ZF_2)$ from propositions in the meta-theory.

7.3 The localisation problem for second order logic

I need first to give a precise formulation of this question and of the corresponding questions for universal validity and logical consequence in second order logic.

I. Γ *defines satisfiability for* $\mathscr{L}(ZF_2)$-*sentences absolutely:*

$\forall XY [\ulcorner ZF_2 \urcorner \in \Gamma \subseteq Sent\ .:and:.$
$\quad [(X, \in_X) \vDash \Gamma\ .implies.(\forall z \in Sent)[Y \vDash z\ implies$
$\quad\quad (X, \in_X) \vDash \ulcorner \exists v_1 [v_1 \vDash v_0] \urcorner (z/\ulcorner v_0 \urcorner)]]\,]$

II. Γ *defines logical validity for* $\mathscr{L}(ZF_2)$-*sentences absolutely:*

$\forall XY [\ulcorner ZF_2 \urcorner \in \Gamma \subseteq Sent\ .:and:.$
$\quad [(X, \in_X) \vDash \Gamma\ : implies :\ (\forall z \in Sent)[(X, \in_X) \vDash$
$\quad\quad \ulcorner \forall v_1 [v_1\ is\ an\ \mathscr{L}(ZF_2)\text{-}structure\ \supset v_1 \vDash v_0] \urcorner (z/\ulcorner v_0 \urcorner)$
$\quad\quad .implies.\ [Y\ is\ an\ \mathscr{L}(ZF_2)\ structure\ implies\ Y \vDash z]\,]]\,]$

III. Γ *defines satisfiability for sets of* $\mathscr{L}(ZF_2)$-*sentences absolutely:*

$\forall XY [\ulcorner ZF_2 \urcorner \in \Gamma \subseteq Sent\ .:and:.$
$\quad [(X, \in_X) \vDash \Gamma\ .implies.$
$\quad\quad (\forall \Delta \subseteq Sent)[Y \vDash \Delta\ implies\ (X, \in_X) \vDash$
$\quad\quad\quad \ulcorner \exists v_1 [v_1 \vDash v_0] \urcorner (\Delta/\ulcorner v_0 \urcorner)]]\,]$

IV. Γ *defines logical consequence for* $\mathscr{L}(ZF_2)$ *absolutely:*

$\forall XY [\ulcorner ZF_2 \urcorner \in \Gamma \subseteq Sent\ .:and:.\ [(X, \in_X) \vDash \Gamma\ : implies :$
$\quad (\forall \Delta \subseteq Sent)(\forall u \in Sent)[(X, \in_X) \vDash$
$\quad\quad \ulcorner \forall v_2 [v_2\ is\ an\ \mathscr{L}(ZF_2)\text{-}structure\ \supset$
$\quad\quad\quad [v_2 \vDash v_0 \supset v_2 \vDash v_1]] \urcorner (\Delta/\ulcorner v_0 \urcorner, u/\ulcorner v_1 \urcorner)\ .implies.$
$\quad\quad [Y \vDash \Delta\ implies\ Y \vDash u]]\,]]$

Note that these definitions are all expressed by Π_1 propositions. To simplify the logic of the argument I have formulated them so that they are all satisfied vacuously by inconsistent sets, Γ, of $\mathscr{L}(ZF_2)$-sentences. A straightforward argument establishes that I and II entail one another, as do III and IV.

I can now prove that no consistent set of $\mathscr{L}(ZF_2)$-sentences defines satisfiability for sets of $\mathscr{L}(ZF2)$-sentences absolutely.

7.3.1 Theorem *The hypothesis*

Γ *defines satisfiability for sets of* $\mathscr{L}(ZF_2)$-*sentences absolutely*

entails the conclusion

$$\forall X\ not[(X, \in_X) \vDash \Gamma]$$

Proof Suppose as hypothesis that Γ defines satisfiability for sets of $\mathscr{L}(ZF_2)$-sentences absolutely:

$\forall X Y [\ulcorner ZF_2 \urcorner \in \Gamma \subseteq Sent$.:and:. $[(X, \in_X) \vDash \Gamma$ .implies.
$(\forall \Delta \subseteq Sent)[Y \vDash \Delta$ implies $(X, \in_X) \vDash \ulcorner \exists v_1 [v_1 \vDash v_0] \urcorner (\Delta / \ulcorner v_0 \urcorner)]]]$

Then, instantiating $\forall X$ with M, $\forall Y$ with $(M, \in_M)$, $(\forall \Delta \subseteq Sent)$ with Γ, and simplifying, we obtain

(1) $\ulcorner ZF_2 \urcorner \in \Gamma \subseteq Sent$
(2) $(M, \in_M) \vDash \Gamma$ implies $(M, \in_M) \vDash \ulcorner \exists v_1 [v_1 \vDash v_0] \urcorner (\Gamma / v_0)$

Now suppose, by way of contradiction, that

(3) $(M, \in_M) \vDash \Gamma$

We may assume, without loss of generality, that $(M, \in_M)$ is the smallest normal domain that satisfies Γ:

(4) $(M, \in_M) \vDash \ulcorner \forall v_1 \neg [(v_1, \in_{v_1}) \vDash v_0] \urcorner (\Gamma / v_0)$

From (2) and (3) we may deduce

(5) $(M, \in_M) \vDash \ulcorner \exists v_1 [v_1 \vDash v_0] \urcorner (\Gamma / v_0)$

But since $\ulcorner ZF_2 \urcorner \in \Gamma$ (by (1)), any model of Γ is isomorphic to a normal domain. Moreover, the translation of the proposition expressing this fact into $\mathscr{L}(ZF_2)$ is a logical consequence of ZF_2. Hence by (1) and (5)

(6) $(M, \in_M) \vDash \ulcorner \exists v_1 [(v_1, \in_{v_1}) \vDash v_0] \urcorner (\Gamma / v_0)$

Together (4) and (6) yield

(7) $\forall X$ not $[(X, \in_X) \vDash \Gamma]$

as required. □

As a corollary of Theorem 7.3.1 we immediately obtain

7.3.2 Corollary *The hypothesis*

Γ *defines logical consequence for* $\mathscr{L}(ZF_2)$ *absolutely*

entails the conclusion

$\forall X$ not$[(X, \in_X) \vDash \Gamma]$

Note that in 7.3.1 and 7.3.2 there is no restriction placed on Γ in respect of definability: no extension whatsover of ZF_2 in $\mathscr{L}(ZF_2)$ captures these key notions of second order global semantics absolutely. *A fortiori* no such extension that we could actually employ – no recursively enumerable extension, for example – could do the required job.

Thus in this regard the notions of second order logical consistency and

7.3 The localisation problem for second order logic

second order logical consequence lie outside the scope of the axiomatic method, and therefore are *essentially* global in character. Theorems 7.3.1 and 7.3.2 therefore tell us that *the global semantics of second order logic transcends ordinary mathematical practice.*

Let us now turn our attention to the problem of second order logical consistency (satisfiability) for individual sentences. We can easily establish an analogue of Theorem 7.3.1, this time for *Euclidean finite* extensions of ZF_2.

7.3.3 Theorem *The hypotheses*

(i) Γ *is Euclidean finite*
(ii) Γ *defines satisfiability for $\mathscr{L}(ZF_2)$-sentences absolutely*

together entail the conclusion

$$\forall X\, not[(X, \in_X) \vDash \Gamma]$$

The proof here is essentially the same as that for Theorem 7.3.1. As an immediate corollary of Theorem 7.3.3 we obtain

7.3.4 Corollary *The hypotheses*

(i) Γ *is Euclidean finite*
(ii) Γ *defines universal validity for $\mathscr{L}(ZF_2)$-sentences absolutely*

together entail the conclusion

$$\forall X\, not[\,(X, \in_X) \vDash \Gamma]$$

Now let us drop the assumption that Γ is Euclidean finite and turn to the general problem of finding an extension, Γ, of ZF_2 which defines satisfiability for individual $\mathscr{L}(ZF_2)$-sentences. Whether we can eliminate all such Γ is not clear, but, as I shall show in Theorem 7.3.6, we can eliminate all such Γ that are "absolutely definable" in the sense of the following definition.

7.3.5 Definition *Let Γ be a set of $\mathscr{L}(ZF_2)$-sentences, $A(v_0)$ an $\mathscr{L}(ZF_2)$-formula whose only free variable is v_0 and Δ a finite extension of Z_2. Then to say that $A(v_0)$ defines Γ absolutely in Δ is to say that*

$$\forall X[(X, \in_X) \vDash \Delta \text{ .implies.}$$
$$(\forall y \in Sent)[(X, \in_X) \vDash \ulcorner A(v_0)\urcorner(y/\ulcorner v_0 \urcorner) \text{ iff } y \in \Gamma]]$$

7.3.6 Theorem Let Γ be a set of $\mathscr{L}(ZF_2)$-sentences, $A(v_0)$ an $\mathscr{L}(ZF_2)$-formula whose only free variable is v_0, Δ a finite extension of Z_2 and M a set. Then the hypotheses

(i) $A(v_0)$ defines Γ absolutely in Δ
(ii) Γ defines satisfiability for $\mathscr{L}(ZF_2)$-sentences absolutely
(iii) $(M, \in_M) \models \Delta$

together entail the conclusion

$$\forall X\ not[(X, \in_X) \models \Gamma]$$

Proof Assume (i) – (iii) as hypotheses, and suppose, by way of contradiction, that

(1) $(X, \in_X) \models \Gamma$

We may assume, without of loss generality, that $(X, \in_X)$ is the smallest normal domain with this property.

We can now establish that

(2) $(X, \in_X) \models \ulcorner \exists v_1 [(v_1, \in_{v_1}) \models \ulcorner \Delta \urcorner] \urcorner$

Pf. Since Γ defines satisfiability for $\mathscr{L}(ZF_2)$-sentences absolutely (hypothesis ii) and Δ is satisfiable (hypothesis iii), $(X, \in_X) \models \ulcorner \exists v_1 [v_1 \models \ulcorner \Delta \urcorner] \urcorner$. Since $(X, \in_X) \models ZF_2$ (1 and hypothesis ii), and since in any normal domain the existence of a model of Δ implies the existence of a standard model of Δ (since $Z_2 \in \Delta$), 2 follows.

(3) $(X, \in_X) \models \ulcorner \exists v_1 [(v_1, \in_{v_1}) \models v_0] \urcorner (\Gamma / \ulcorner v_0 \urcorner)$

Pf. Let B be the $\mathscr{L}(ZF_2)$-sentence $Z_2 \wedge \exists v_1 [(v_1, \in_{v_1}) \models \ulcorner \Delta \urcorner] \wedge B_1$, where B_1 is the sentence

$$\exists v_1 [\forall v_2 [(v_2, \in_{v_2}) \models \ulcorner \Delta \urcorner . \supset .\\ \forall v_3 [(v_2, \in_{v_2}) \models \ulcorner A(v_0) \urcorner (v_3 / \ulcorner v_0 \urcorner) \supset (v_1, \in_{v_1}) \models v_3]]]$$

The sentence B is satisfiable. In fact, if $a = rank(X) + \omega$, then

$$(V_a, \in_{V_a}) \models B$$

For since $A(v_0)$ defines Γ absolutely in Δ (hypothesis i), and since satisfaction for $\mathscr{L}(ZF_2)$-sentences is absolute in Zermelo domains,

$$(V_a, \in_{V_a}) \models \ulcorner \forall v_1 \forall v_2 [(v_2, \in_{v_2}) \models \ulcorner \Delta \urcorner :\supset:\\ [\forall v_3 [(v_2, \in_{v_2}) \models \ulcorner A(v_0) \urcorner (v_3 / \ulcorner v_0 \urcorner) \supset (v_1, \in_{v_1}) \models v_3]. \equiv .\\ \forall v_3 [v_3 \in v_0 \supset (v_1, \in_{v_1}) \models v_0]]] \urcorner (\Gamma / \ulcorner v_0 \urcorner)$$

Since we clearly have $(V_a, \in_{V_a}) \models \ulcorner \exists v_1 [(v_1, \in_{v_1}) \models v_0] \urcorner (\Gamma / \ulcorner v_0 \urcorner)$, we may conclude that $(V_a, \in_{V_a}) \models \ulcorner B_1 \urcorner$ and therefore $(V_a, \in_{V_a}) \models \ulcorner B \urcorner$ as claimed.

7.3 The localisation problem for second order logic

Since Γ defines satisfiability for $\mathscr{L}(ZF_2)$-sentences absolutely (hypothesis (ii)) and B extends Z_2, we may conclude that

$$(X, \in_X) \vDash \ulcorner \exists v_1 [(v_1, \in_{v_1}) \vDash \ulcorner B \urcorner] \urcorner$$

Hence there is a Y in X such that $(Y, \in_Y) \vDash \ulcorner B \urcorner$ and therefore, since $A(v_0)$ defines Γ absolutely in Δ (hypothesis (i))

$$(Y, \in_Y) \vDash \ulcorner \exists v_1 [(v_1, \in_{v_1}) \vDash v_0] \urcorner (\Gamma / \ulcorner v_0 \urcorner)$$

Hence there is a $Z \in Y \subseteq X$ such that $(Z, \in_Z) \vDash \Gamma$, and (3) follows.

We have now arrived at our desired contradiction, since (3) is clearly incompatible with the minimality of $(X, \in_X)$ as a model of Γ. □

Theorem 7.3.6 immediately yields the following corollaries.

7.3.7 Corollary *Let Γ be a set of $\mathscr{L}(ZF_2)$-sentences, $A(v_0)$ an $\mathscr{L}(ZF_2)$-formula whose only free variable is v_0, Δ a finite extension of Z_2 and M a set. Then the hypotheses*

  (i) *$A(v_0)$ defines Γ absolutely in Δ*
  (ii) *Γ defines universal validity for $\mathscr{L}(ZF_2)$-sentences absolutely,*
  (iii) *$(M, \in_M) \vDash \Delta$*

together entail the conclusion

$$\forall X \; not[(X, \in_X) \vDash \Gamma]$$

Thus no extension of ZF_2 definable absolutely in the sense of Definition 7.3.5 can define satisfiability (or, equivalently, universal validity) for $\mathscr{L}(ZF_2)$-sentences. In particular, no recursively enumerable extension of ZF_2 can do this, nor can any extension definable in arithmetic, or in analysis, or ...

7.3.8 Corollary *Let Γ be a recursively enumerable set of $\mathscr{L}(ZF_2)$-sentences. Then the hypothesis*

  Γ defines satisfiability for $\mathscr{L}(ZF_2)$-sentences absolutely

entails the conclusion

$$\forall X [\; not(X, \in_X) \vDash \Gamma]$$

Proof Since Γ is recursively enumerable, we can find a suitable $\mathscr{L}(ZF_2)$-formula $A(v_0)$ that defines Γ absolutely in Z_2. Thus in this particular case hypotheses (i) and (iii) of Theorem 7.3.6 can be proved outright. □

Of course we immediately obtain

7.3.9 Corollary *Let Γ be a recursively enumerable set of $\mathscr{L}(ZF_2)$-sentences. Then the hypothesis*

Γ defines universal validity for $\mathscr{L}(ZF_2)$-sentences absolutely

entails the conclusion

$$\forall X \; not[(X, \in_X) \vDash \Gamma]$$

Does the species of all satisfiable subsets of the set, *Sent*, of all $\mathscr{L}(ZF_2)$-sentences form a sub*set* of that set? Theorem 7.3.1 is surely *compatible* with the supposition that it does not, and since Brouwer's Principle gives us reason *prima facie* to suppose this, doesn't 7.3.1 provide strong evidence that this species provides a natural example of an *indefinite* species that is not absolutely *infinite*, i.e. of a semi-set?

Theorem 7.3.6 and its corollaries raise similar questions regarding the notions of *satisfiability* and *universal validity* for individual $\mathscr{L}(ZF_2)$-sentences. Here, where we are dealing with the semantics of *individual* sentences, we cannot establish the kind *absolute* results that we can for the semantics of *sets* of sentences. But in this important case we can show that *no extension of ZF_2 that we can describe absolutely can define second order logical validity or second order satisfiability absolutely.* It follows immediately that no local formula of our theory can define second order satisfiability or second order validity, for such formulas are absolute in models of Z_2.

In fact, I believe that the evidence here strongly suggests that in all these cases we have natural examples of *semi-sets*, and that *the global semantics of second order logic cannot be localised..*

Part Four
Euclidean Set Theory

There is perhaps something mysterious in the fact that we seem to know instinctively what the natural numbers actually are. For as children (or adults) we are provided with just a comparatively small number of descriptions as to what "zero", "one", "two", "three", etc., mean ("three oranges", "one banana", etc.); yet we can grasp the entire concept despite this inadequacy. In some Platonic sense, the natural numbers have an absolute conceptual existence independent of ourselves.

Roger Penrose, *Shadows of the Mind*

What is essential is to regard the natural numbers as mental constructions, generated in determinate manner by repeated application of the successor operation to zero. Considered as an infinite structure, the totality N of natural numbers is uniquely determined: there cannot be non-isomorphic structures each with an equally good claim to represent N.

Michael Dummett, *Elements of Intuitionism*

8
Euclidean Finitism

8.1 The serpent in Cantor's paradise

The foundations of modern mathematics, based, as I have explained, on Cantor's extension of the classical conception of the finite, are obscured by a miasma of irrelevancies and misconceptions. But the plain fact is there is nothing *obviously* wrong with Cantorian set theory. When Cantor first proposed that species previously regarded by everyone as genuinely infinite – the natural numbers, the real numbers, the points in space, etc. – could be assigned definite, and indeed different, sizes, he was greeted with scepticism, a perfectly reasonable scepticism given the radical nature of his proposal. Who could have said, at the outset, that such an attempt would not come to grief on what Bolzano called the "paradoxes of the infinite"?

In fact, as we now know, that did not happen. What Cantor saw only dimly and incompletely in the early 1870s had become a clear and powerful vision by the end of the 1880s. The light shed by Cantor's doctrines and theorems on the foundations of mathematical analysis so impressed his younger contemporaries that by the end of the century his approach to analysis had carried all before it. Even the discovery of the so-called "paradoxes of set theory" at the beginning of the twentieth century did no more than briefly slow down the steady advance of Cantorian ideas.

Cantor himself regarded these "paradoxical" discoveries with relative equanimity, and quite rightly so; for the facts that underlie the arguments of Burali-Forti, Russell *et al.* seem paradoxical only if one fails to distinguish between the *arithmetical* notion of *set* and the *logical* notion of *class* or *species*. The essence of this distinction is already present in Cantor's doctrine of the Absolute, which dates from the early 1880s.

If we examine the most common objections to Cantor's ideas, from their very first appearance right down to the present day, we discover that they invariably beg the very question at issue. For all of these objections boil down to the observation that we ought not to treat infinite pluralities on the same basis as finite ones; to which the Cantorian reply is: "Of course we cannot treat the finite and the infinite on the same basis; but what do you mean by 'finite' and 'infinite'?"

Historically, the anti-Cantorians have drawn the distinction between finite and infinite in operationalist terms. They have maintained, for example, that the natural numbers must be constructed, one at a time, in a process of generation: each such number is finite precisely because it can be generated; but the totality of such numbers is infinite because the process of generation can never be completed. But such an "explanation" can only be understood metaphorically. Taken literally it raises more questions that it answers.

Is the process of generation an actual one, a kind of Cosmic Clock that has been ticking away since the Big Bang? Surely one must be speaking of a *possible* process or processes here. But what is the possibility grounded on? Physical laws? Our actual capacities as human beings? Or on some idealised extrapolation of those laws or of those capacities? If the latter, what grounds the possibility of the extrapolation? Even supposing that these questions can be answered – and I do not believe that they can be – further, equally difficult questions present themselves.

Are the numbers not yet generated predestined to have the mathematical properties they will have after being generated? Surely it is impossible to conceive that the numbers with which we are already acquainted could have had properties other than those they are now perceived to possess: this one is prime, that perfect, this, yet again, is expressible as the sum of two cubes in two different ways. How could it ever have been otherwise? But if they are predestined for those properties, why can we not speak of them as if they had already been generated?

Surely the distinction between what is actual and what is merely potential is difficult to maintain when we are speaking of abstract things like numbers. A potential house, however minutely specified may be its blueprint, is a quite different matter from an actual house: one can't live in it, for example (except, potentially, perhaps). But what is there to a number apart from its blueprint, so to speak? And what is it that corresponds to the building of the house in the case of a number? How does one actualise potential numbers? By generation? How does one "generate" even a single natural number? By writing it down? By calling out its name? What notations, what words, is one permitted to use on such occasions? What happens when one runs out of breath, or out of ink and paper? More seriously, what happens when one runs out of space and time?

Perhaps one need only *think* of a number, perform a "mental construction", in order to "generate" it, to bring it into being. But thinking, as we human beings do it, is, or is accompanied by, a physiological process. Are the principles of mathematics to be discovered in the laws of physiology then? Or in the laws of physics on which they depend? But how can we formulate or understand those laws without mathematics?

The questions accumulate; the answers are not forthcoming. But even if (*per impossible*, I say) these questions could be answered, and the formidable difficulties they raise be laid to rest, that would still not constitute an argument against Cantorian finitism. It would be merely to present an alternative: one does not refute a man simply by pointing out to him that his position is in opposition to one's own. And the Cantorian position has weighty *practical* arguments in its favour.

Si monumentum requiris, circumspice

The greatest testimony to the power, the profundity, and the soundness of Cantor's ideas comes from modern mathematics itself. Those ideas are constantly employed by mathematicians in universities and research institutes all over the world. Every year, in every major university, apprentice mathematicians are introduced to Cantor's approach to mathematical analysis; thousands of mathematicians engaged in research, many of them highly gifted, the overwhelming majority at least competent, put his ideas to work in novel circumstances. No one has ever discovered the merest hint of a contradiction lurking in his central assumptions. If Cantor was mistaken in those assumptions, it was an extremely subtle mistake, not a gross and obvious one.

Where should we be without the concepts and techniques Cantor and his successors have provided us? We should have to begin afresh, think through again all those principles that we now take for granted in our teaching and in our research. That is a daunting prospect! It is not surprising that Hilbert should have been moved to assert that "We shall not allow anyone to drive us from the paradise that Cantor has created for us". Surely every mathematician will understand the strength of Hilbert's determination; there is, indeed, a great deal at stake.

Are we, then, to be driven from Cantor's paradise? The issue comes to this: must we accept that for a plurality to be finite, or, equivalently, to have a size, it must be larger than any of its proper subpluralities; or may we suppose, with Cantor, that the species of natural numbers (however defined) is finite, despite the fact that it can be placed in one-to-one correlation with most[1] of its proper subpluralities?

Cantor's Axiom, the so called Axiom of Infinity, does represent a bold extrapolation beyond our ordinary experience. But everyone must concede, pro and anti-Cantorians alike, that there is nothing *obviously* wrong with it. If it contains a contradiction, then that contradiction has eluded the notice of the best mathematicians for the past one hundred and twenty-five years.

Such negative evidence is not, of course, decisive, however impressive it may seem to us. For we may very well be inclined to overestimate our own collective capacity. Perhaps there is a contradiction buried in Cantor's ideas which we have simply not had the wit to discover.

But there is other, positive evidence that can be adduced here. If Cantor's fundamental finiteness assumption is just wrong, how could it have given rise to the profound, beautiful, and coherent theory of transfinite cardinal and ordinal numbers, whose development was the culmination of his achievement? That such a theory could arise by accident, so to speak, as the result of a mere blunder or misconception, is surely a hypothesis that beggars belief. Indeed, it is clearly incumbent on anyone proposing to reject Cantor's point of view to explain why it *seems* so plausible and so coherent.

Despite these weighty considerations, however, there is evidence that something may be amiss in Cantor's paradise, evidence that it may, in fact, harbour a serpent. This evidence comes from two sources – from mathematics itself, and from physics. It is not decisive, but must be disturbing to a Cantorian.

In assessing this evidence we must continually bear in mind the *nature* of Cantor's radical innovation in the interpretation of finiteness. Surely it is correct to say that if we did not know the outcome of that innovation, if we did not know how profoundly it simplifies foundations and frees the mathematical imagination, then we should be inclined to view Cantor's proposed extension of the concept of finiteness with scepticism, as, indeed, his contemporaries did view it. For there can be no question but that it represents a considerable extrapolation beyond our immediate experience of the finite; and this might suggest, if not the likelihood of out and out contradiction, at least the possibility that in laying down the assumption

---

[1] There are only $\aleph_0$ subsets that are strictly smaller, but $2^{\aleph_0}$ subsets of the same size.

8.1 The serpent in Cantor's paradise

that the species of natural numbers is really finite, in the root sense of that word, and in trying to determine the precise sizes of the resulting new species of finite plurality, whose existence follows from that assumption, we should be dealing with merely *ideal* notions, notions corresponding to nothing in reality, and that, in consequence, we should be led to raise questions about them which we cannot answer, questions, indeed, which *have* no answers, despite having every appearance of being natural and fundamental.

These considerations must give even the most committed Cantorian pause when he reflects on the seemingly intractable difficulties that surround the most fundamental question in Cantorian set theory, namely, the question of the size of the linear continuum, the so-called *continuum problem*. We know that the continuum is not countable, and that its cardinal number cannot be expressed as the limit of a countable number of smaller cardinals. Beyond that we know virtually nothing: we do not even know whether there are uncountable subsets of the continuum which are not of the size of the continuum.

Cantor himself devoted much of his career to this problem: his work on the point-set structure of the line, especially his work on perfect sets, was undertaken in an attempt to settle it, as was his work on the second number class (the countable transfinite ordinals). Many of his most distinguished successors – including even Hilbert himself – worked on this problem, all to no avail.

Indeed, we now know that they were all predestined to fail. For by the work of Gödel in the late 1930s and of Cohen in the early 1960s, we now know that the Continuum Hypothesis – the hypothesis that there is no size intermediate between the size of the set of natural numbers and that of the set of real numbers – is logically undecided by the first order axioms of Zermelo–Fraenkel set theory, even if we include the Axiom of Choice among the latter. And any proof in set theory which commands general assent is virtually certain to be formalisable in that theory.

In any case, the consistency and independence results of Gödel and Cohen are remarkably stable. They continue to apply even when the standard axioms of Zermelo–Fraenkel are augmented by powerful axioms of strong infinity which posit the existence of transfinite cardinals of quite unimaginable size. Indeed, I think it is fair to say that no one has ever put forward, as a serious candidate for a natural extension of the Zermelo–Fraenkel system, any axiom which decides the Continuum Hypothesis.

The Continuum Hypothesis is not the only celebrated, "classical" problem to have been shown independent of the first order Zermelo–Fraenkel axioms. The Souslin Hypothesis, the Kurepa Conjecture, even certain "mainstream" problems in algebra have been shown to be independent of those axioms, and of any reasonable extension of them.

From the standpoint of foundations, these results, and the difficulties to which they give rise, are quite unprecedented. One often hears the facile comparison made between the independence results in set theory and the nineteenth century discovery that the Axiom of Parallels – Euclid's Fifth Postulate – is independent of the other geometrical axioms. There is a lot of speculation about "non-Cantorian set theory" (in which the Continuum Hypothesis is false) on the analogy of non-Euclidean geometry (in which Euclid's Axiom of Parallels is false).

But the analogy is without substance, as Kreisel has carefully and, one would have thought, definitively explained[2]. The essential point is this: the Axiom of Parallels is *second order* independent of the other geometrical axioms, whereas the Continuum Hypothesis is decided by the second order axioms of Zermelo–Fraenkel; indeed, it is decided by the second order axioms for analysis (the axioms for a complete ordered field). The set-theoretical independence results are all relative to *first order* formulations of Zermelo–Fraenkel, and depend essentially on the pathologies of first order logic, the Löwenheim-Skolem Theorem in particular. All of the models produced in proving these results are radically non-standard, and consequently provide no insight into the true facts of the case.

Of course we cannot abandon the idea that there are true facts of the case here without abandoning the essential principles of Cantorian finitism: we must abandon either the assumption that the natural numbers form a set, or the assumption that every set, including the set of natural numbers, has a power set. Indeed, if we abandon Power Set it is not clear that we do not thereby abandon the natural numbers as well, for it is difficult to see how we could then obtain the general theory of simply infinite systems.

In any case, abandoning either of these principles would entail a radical change in our whole approach to mathematics, something not to be undertaken lightly or for essentially frivolous reasons. And the idea that the fact of a proposition's being formally independent of formal, first order Zermelo–Fraenkel set theory is sufficient, *on its own*, to establish that the proposition in question fails to have an objectively determined truth value is essentially frivolous. For it fails to take into account the

---

[2] In his paper "Informal rigour and completeness proofs".

8.1 The serpent in Cantor's paradise

fact that the general theory of first order logic, both its semantics and its syntax, depends essentially on Cantorian finitist assumptions. Without those assumptions there is no first order logic as we now understand it.

No, it is not the simple *fact* of the independence results, but their apparent *stability* under natural extensions of the theory that is disquieting. Of course this situation could change overnight: someone might produce a new axiom which everyone could agree was plausible and natural and which decided, say, the Continuum Hypothesis. But I believe that those in the know – set theorists themselves – are sceptical about such a possibility.

Now there is a Cantorian finitist explanation, or, perhaps, excuse, for these phenomena: by a kind of metaphysical accident, but, in any case, as a contingent matter of fact, we humans are capable of directly apprehending only such totalities as are finite, and only such configurations as are finitely articulated, where "finite" here must be understood in the traditional, pre-Cantorian sense of Euclid[3]. The configurations of signs that express our propositions and proofs must be finite in this older sense of "finite". That is why countable, first order languages capture the formal aspects of our reasoning so well.

But if we could directly apprehend totalities and configurations manifesting the new species of the *Cantorian* finite, then we might actually incorporate uncountably many names, c_x, for the elements, x, of an uncountable set a in our language, and employ generalized inference rules

From the (uncountable) set $\{\Phi(c_x) : x \in a\}$ of premises infer $(\forall x \in a)\Phi(x)$ as conclusion

which would enable us to decide questions like the Continuum Hypothesis. The impossibility of our doing this is not, on the Cantorian finitist view, a *logical* impossibility; it does not rest on the nature of the finite (e.g. the power set of the natural numbers is finite, on the Cantorian view) but rather on contingent facts about our nature and the nature of the world in which we live[4].

---

[3] Euclidean finiteness is a necessary but, of course, not sufficient condition here. It is obvious, for example, that we cannot directly apprehend totalities with 10^{30} elements, to pick an easy upper bound.

[4] This is essentially the point Russell was making when he remarked that our inability to count through all the natural numbers in a finite time was a "medical" limitation, not a logical one.

This sort of explanation does save appearances. It explains, or explains away, the mathematical difficulties in Cantorian finitism which the results of Gödel and Cohen seem to raise. But the explanation appeals to supposed contingencies in the natural constitution of human beings, and in the physical world they inhabit, which themselves constitute an embarrassment to the Cantorian finitist standpoint. For the new species of the finite that the Cantorian view posits are not directly manifested in the physical world. The Cantorian transfinite is "there", so to speak, but like the luminiferous aether, is not detectable by any means available to us.

Here we have come to the second source of disquiet about Cantor's "paradise". Modern physics seems to have banished what our ancestors would have called "infinities" from the physical world. The Newtonian idea that the universe is a three-dimensional, Euclidean space, unlimited in all directions, which has endured through an unlimited past and will endure through an unlimited future has been cast in doubt by the general theory of relativity, which can accommodate a universe that is strictly limited in extent, both in space and in time (or, more correctly, in space-time).

In a sense, these ideas are not new: many of Newton's predecessors (e.g. Aristotle) held that space was limited in extent. The Aristotelian universe is a sphere outside whose surface there is nothing (not even space)[5], and the Christian Aristotelians held that time, as well, had a beginning and will have an end.

But if general relativity tells us that space and time need not be unlimited in *extent*, quantum theory tells us that they need not be unlimited in *divisibility* either. To look into a small region of space we must illuminate that region with electromagnetic radiation[6] whose wavelength, λ, is small relative to the dimensions of that region. But electromagnetic radiation is emitted and absorbed in quanta whose energy is inversely proportional to its wavelength. The energy, E, of a quantum of electromagnetic radiation of wavelength λ is given by

$$E = h/\lambda$$

where h is Planck's constant. As the wavelength decreases without bound,

---

[5] See *De Caelo* (279a11).
[6] We can also use particles of matter, as in an electron microscope. The argument for that case is essentially the same as the one I am presenting here.

8.1 The serpent in Cantor's paradise

so the energy of each quantum increases without bound. This clearly suggests that there is a lower limit to sizes that have any physical significance. In a finite universe with a finite energy content there would be an absolute physical bound on smallness.

Of course, the mathematics in which general relativity and quantum theory are expounded is based on Cantorian finitist principles. But this simply reflects the fact that *all* our mathematics is based on such principles. Indeed, it may well be that the "infinities" which turn up in these physical theories (e.g. singularities in general relativity) are simply artefacts of the mathematics in which they are expressed. This alone should supply us with sufficient incentive to search for Euclidean finitary alternatives to conventional real analysis. The methods of non-standard analysis and synthetic differential geometry may provide us with useful clues on how to proceed here.

But in pursuing such speculations I stray from the central point. Cantorian finitism posits a vast new realm of possibilities for the sizes of finite totalities. None of them, however, is realised in nature, as far as we can now determine. In mathematics we have a precise scale for measuring these new sizes – the absolutely infinite sequence of initial ordinals, the $\aleph$'s – but we cannot determine the size of the linear continuum on that scale. Nothing that we now know is incompatible with the possibility that we may have run up against an absolute barrier here, that we can never know the size of the continuum, in the sense of deriving it from some natural and plausible extension of the generally accepted principles of Cantorian set theory.

We ask simple and straightforward questions which, on Cantorian finitist grounds, must have definite answers; and yet nature, or whatever it is that corresponds to nature here, seems to conspire to prevent our discovering those answers, just as to physicists at the end of the nineteenth century, nature seemed to conspire to prevent their determining the velocity of the earth with respect to the luminiferous aether.

Of course none of this is decisive against Cantor's views. New discoveries may shed an entirely new light on these circumstances. But as things stand now, the facts are exactly as we should expect them to be if Cantor's new species of the finite were mere phantasms, corresponding to nothing in reality, and prompting us to raise questions which seem significant, but which do not have objectively determined answers. This is the serpent in Cantor's paradise.

8.2 The problem of non-Cantorian foundations

What is the alternative to Cantorian orthodoxy? If we are to be driven from Cantor's paradise, or if we decide to leave it voluntarily, what could be put in the place of those familiar techniques and principles that now guide our mathematical practice? How could we lay the foundations of a non-Cantorian mathematics?

The first prominent anti-Cantorian was the German algebraist and number theorist Leopold Kronecker (1823–1891). Kronecker was a bitter opponent of Cantor's innovations from the outset: "God made the natural numbers", he said, "everything else is the work of man". To a modern mathematician a consequence of that slogan is that the basic principles of natural number arithmetic – the principle of proof by mathematical induction and the principle of definition by recursion – stand on their own as fundamental principles, and do not require justification, or formulation, in the theory of sets. This point of view has been embraced by the whole anti-Cantorian movement in modern mathematics: Poincaré, Weyl, Brouwer and the Dutch school of intuitionists, and, more recently, Bishop, have all followed Kronecker in this regard.

But the idea that natural number arithmetic is a self-contained, autonomous discipline, logically independent of, and perhaps in some sense logically prior to, the theory of sets, is by no means confined to anti-Cantorians: it is widespread even among mathematicians who are otherwise quite well-disposed to Cantor's ideas. It arises naturally out of what is, at bottom, an *operationalist* conception of natural number. For the natural numbers are, on this conception, what we get when we start with zero and *iterate* the successor operation, $x \mapsto x + 1$, *ad infinitum*.

To be sure, this way of "defining" the natural numbers does at least appear to carry with it an implicit justification of the principles of proof by induction and definition by recursion. But the appearance is illusory, and such a definition of the natural numbers is completely spurious. Indeed, the belief that the concept of natural number can be defined in this operationalist way constitutes a serious obstacle to clear thinking about alternatives to Cantorian finitism, the current mathematical orthodoxy.

Kronecker, in fact, got it wrong: it was not *God* who made the natural numbers; on the contrary, they were invented by the mathematicians of the seventeenth century, along with negative, rational, and real numbers. We must disabuse ourselves of the illusion that our concept of natural number is fundamental. This is why I have already devoted a whole chapter, Chapter 2, to this crucial issue.

8.2 The problem of non-Cantorian foundations

There I presented a series of arguments, all pointing to the conclusion that the ancient notion of *arithmos* is superior to the modern notion of natural number as a means of explaining the familiar facts of simple arithmetic. Those arguments may not have seemed, at the time, to have direct relevance to the *practice* of mathematics. But now that we have come to consider alternatives to the orthodox Cantorian standpoint that underlies modern mathematics, the issues raised in Chapter 2 can be seen to have direct, *practical* significance, for reasons that I am about to adduce.

The danger that always confronts us when we discuss these really fundamental matters is that we may be tempted to rush forward offering "philosophical" explanations and justifications for our ideas when what is really required is that we subject them to a careful *mathematical* analysis. The danger is especially acute when the foundations of natural number arithmetic are under consideration.

We can see the truth of this if we compare Wittgenstein's operationalist account of natural number arithmetic in the *Tractatus* with the set-theoretical account of that theory given in Dedekind's monograph, *Was sind und was sollen die Zahlen?* Dedekind's work, although published thirty years before Wittgenstein's, exposes, when compared with the latter, the futility of propounding a philosophical account of number that does not rest upon a careful and thorough mathematical investigation[7].

The precision and subtlety of Dedekind's analysis, and the difficulty we experience in following its details, call our attention to what it is that is problematic about the natural numbers. The surprisingly difficult proof of the theorem on definition by recursion (§126) is especially enlightening in this regard.

The content of the theorem seems quite straightforward – after all, from an operationalist perspective it seems merely to say that we can iterate a function as often as we like – but its proof is quite subtle, and as we reflect on that subtlety we begin to realise that the theorem itself is not at all as straightforward as we at first supposed.

---

[7] Indeed, Dedekind's work constitutes an anticipatory refutation of the series of laconic pronouncements that constitute Wittgenstein's account of mathematics in the *Tractatus*: "A number is an exponent of an operation" (6.021 – the basis for this remark is the bogus "definition" of natural number given in 6.02); "The theory of classes is completely superfluous in mathematics" (6.031); "The propositions of mathematics are equations and therefore pseudo propositions" (6.2). Nowhere in this farrago do we find a discussion of either mathematical induction or recursion, although we are told (5.2523) "The concept of successive applications of an operation is equivalent to the concept 'and so on'".

Of course, one hears it said, even today, that Dedekind's analysis of the concept of natural number is *unnecessarily* subtle, that it merely results in the "reduction" of a clearly understood theory (that of the arithmetic of natural numbers) to an obscurely understood one (the theory of sets), that the familiar axioms for natural number arithmetic are "intuitively obvious" truths, etc., etc.

Dedekind himself was sensitive to criticisms of this sort and, in a letter to Dr. Hans Keferstein of Hamburg[8], presented a masterly rebuttal of all such objections. Dedekind's argument is so much to the point here that I feel justified in quoting it at some length.

> I have shown..., however, that these facts[9] are still far from being adequate for completely characterising the nature of the number sequence N. All these facts would hold also for every system S that, besides the number sequence N, contained a set T, of arbitrary additional elements t, to which the mapping ϕ [i.e. the successor function for N] could always be extended while remaining $1-1$ and satisfying $\phi(T) = T$. But such a system S is obviously something quite different from our number sequence N, and I could so choose it that scarcely a single theorem of arithmetic would be preserved in it. What, then, must we add to the facts above in order to cleanse our system S of such alien intruders t as disturb every vestige of order and restrict it to N? *This was one of the most difficult points of my analysis and its mastery required lengthy reflection* [emphasis added]. If one presupposes knowledge of the sequence N of natural numbers and, accordingly, allows oneself the use of the language of arithmetic, then, of course, one has an easy time of it. One need only say: an element n belongs to the sequence N, if and only if, starting with the element 1, and counting on and on steadfastly, that is, going through a finite number of iterations of the mapping ϕ (see the end of article 131 in my essay), I actually reach the element n at some time; by this procedure, however, I shall never reach an element t outside the sequence N. But this way of characterising the distinction between the elements t that are to be ejected from S and those elements n that alone are to remain is surely quite useless for our purpose; it would, after all, contain the most pernicious and obvious kind of vicious circle. The mere words "finally get there at some time", of course, will not do either; they would be of no more use than, say, the words *"karam sipo tatura"*, which I invent at this instant without giving them any clearly defined meaning.

Can we not detect a note of irritation slipping in here? To anyone who

---

[8] The letter is dated 27 February 1890. The passage below is quoted from a translation of the letter by Hao Wang and Stefan Bauer-Mengelberg, and is to be found in van Heijenoort's *From Frege to Gödel*.

[9] The facts Dedekind is referring to here are, roughly speaking, those expressed in the Dedekind–Peano axioms for natural number arithmetic, *apart from the Induction Axiom*. (See Section 5.1.)

8.2 The problem of non-Cantorian foundations 273

has thought about these matters, the point Dedekind is making is surely obvious. In any case, he continues

Thus how can I, without presupposing any arithmetic knowledge, give an unambiguous conceptual foundation to the distinction between the elements n and the elements t? Merely through consideration of the *chains* (articles 37 and 44 of my essay), and yet by means of these completely! If I wanted to avoid my technical expression "chain" I should say: an element n of S belongs to the sequence N if, and only if, n is an element of every part [i.e. subset] K of S that possesses the following two properties: (i) the element 1 belongs to K and (ii) the image $\phi(K)$ is part of K [i.e. is a subset of K]. In my technical language: N is the intersection 1_0 or $\phi_0(1)$ of all those chains (in S) to which the element 1 belongs. Only now is the sequence N characterised completely.

Dedekind then adds, significantly

In passing I should like to make the following remark on this point. Frege's *Begriffsschrift* and *Grundlagen der Arithmetik* came into my possession for the first time for a brief period last summer [i.e. 1889], and I noted with pleasure that his way of defining the non-immediate succession of an element upon another in a sequence agrees in *essence* with my notion of chain (articles 37 and 44); only one must not be put off by his somewhat inconvenient terminology.

Surely it is significant that both Dedekind and Frege embarked upon their investigations into the concept of natural numbers determined to avoid any logical appeal to our supposed capacity to generate the number sequence by iteration of the successor function.

Indeed, Dedekind's analysis, which is deeper mathematically, contains (in §§126–131) a precise investigation into what is *meant* by "iteration". It provides us with a rigorous, mathematical *definition* of that symbol generated concept, and in so doing grounds, and justifies, the operationalist chatter with which we surround our discussions of natural number arithmetic.

But as it stands, *Dedekind's theory rests upon Cantorian finitist assumptions*[10], *and therefore cannot justify either proof by induction or definition by recursion in the absence of such assumptions.*

That these two principles stand in need of justification, that the operationalist's incantation of the syllables "*karam sipo tatura*" will not suffice here, is a conviction that will impose itself on anyone who makes a serious effort to follow either Frege's or Dedekind's analysis.

In any case, it is a perfectly natural *mathematical* question whether, or to what extent, Dedekind's analysis of natural number can be carried

---

[10] In particular, there must be a set that can be placed in one-to-one correspondence with one of its proper subsets, and that set must have a power set.

out without Cantorian finitist assumptions. This is the question to which Zermelo turned his attention in his 1909 article "*Sur les ensembles finis et le principe de l'induction complète*". Zermelo was able to establish various set-theoretical versions of the principle of mathematical induction without appealing to Cantor's Axiom. But he stopped short of considering the central result in Dedekind's essay, namely, the theorem that justifies the definition of a function by recursion along a simply infinite system (§126 of the essay). Unfortunately, this is the very point in Dedekind's exposition where the need for Cantorian finitist assumptions seems greatest.

Tarski also carried out extensive investigations along these lines[11], though he, too, ignored the problem of justifying recursive definition. Moreover, neither Zermelo nor Tarski employed the assumption that all sets are Euclidean finite: rather they merely avoided explicit appeal to Cantorian finitist assumptions. This actually makes things more difficult, and both Zermelo and Tarski were forced to adopt technical definitions of Euclidean finiteness which are more complicated than the straightforward definition according to which a set is Euclidean finite if it cannot be placed in one-to-one correspondence with any of its proper subsets.

In any case, this sort of fence sitting won't do; for if you carry out a thorough investigation of the principles of induction and recursion, you will eventually come up against the question of what logical principles govern quantification over the species of natural numbers, and when this happens, you must choose between the Cantorian and the Euclidean conceptions of finiteness.

But all this brings us back to the question I asked at the beginning of this section: what is the alternative to Cantorian orthodoxy? The obvious alternative, I submit, is simply to deny that Cantor was justified in extending the notion of finiteness in the way that he did, and to return to the older, Euclidean conception of finiteness in accordance with which a finite plurality cannot be placed in one-to-one correspondence with any of its proper subpluralities. On that conception of the finite, if you take away elements from a finite plurality, even just one element, the elements that remain together form a *smaller* plurality.

This means, of course, that we must abandon Cantor's Axiom, the misnamed "Axiom of Infinity". But which of the other finiteness principles that constitute the orthodox view should we retain? Surely the only

---

[11] In his article "Sur les ensembles finis".

8.2 The problem of non-Cantorian foundations

candidates for sceptical doubt here are the axioms of Power Set and Union.

Of these axioms Union has the more ancient pedigree: it is what underlies and sanctions the idea of multiplication. The Axiom of Power Set is of more recent provenance. It is only since the Renaissance that the idea of exponentiation as an arithmetical operation has had wide currency, and even now classical number theory is largely confined to the additive and multiplicative properties of numbers.

But, to question the validity of Power Set is to suppose that there could be a plurality which, though finite, was so large that it contained an infinity of subpluralities. This seems plausible only on the supposition that the finite is to be identified with what we can actually count out. Maybe even the U.S. national debt would turn out to be infinite on that criterion.

Still, it would be of mathematical interest to investigate what could or could not be proved in the absence of either, or both, of these two principles. Nevertheless I shall not pursue such investigations here, for it seems to me that, in Euclidean set theory, both of them have reasonable claims to self-evidence.

But what I have left unsaid is really the principal point. In abandoning Cantorian finitism we must not abandon the idea that the foundations of mathematics are to be found in the theory of sets. On the contrary, to abandon set theory as a foundational theory would be a retrograde step: it would be to abandon the ideal of objectivity in mathematics. The concept of set, in its ancient guise of *arithmos*, is the oldest, simplest, and clearest mathematical concept that we possess.

As I argued at length in Chapter 2, we cannot regard sets as merely subjective artefacts of our thinking. Whatever things there are, and in whatever sense they may be said to be, there are finite pluralities – sets – composed of those things. The facts of simple arithmetic are facts about those sets. They are objective facts: they do not depend on our conventions of notation, or on our thought processes, or on anything of that sort, though, of course, those facts must be *expressed* in language, and language does depend upon our conventions and mental processes.

It follows that if we abandon Cantorian finitism, we cannot then simply take the natural numbers as "given" and the basic principles of natural number arithmetic, the principles of proof by induction and of definition by recursion, as self-evidently valid. On the contrary, the first task that confronts non-Cantorian mathematics is to reconstitute natural number arithmetic, insofar as such reconstitution is possible. Induction and recursion are not to be taken as self-evident first principles, but as problems to be investigated. We must determine whether, and to what extent, we can establish those principles on a sound, *set-theoretical* basis. This is the project upon which I am about to embark.

By following Dedekind in insisting upon a strictly set-theoretical account of induction and recursion, I shall not allow myself any appeal to the supposed "intuition" that would permit us to postulate that any mathematical operation can be "iterated" any finite number of times. This means that my arguments and definitions will be constructive in a very strong and concrete sense.

For example, I shall count what is conventionally seen as a function, "f" from "natural numbers" to "natural numbers" as well defined only if it is possible to exhibit a well-defined set-theoretical operation which, when applied to any set of size n, yields a set of size $f(n)$[12].

I shall also be forced to pay careful attention to the distinction between local propositions, to which the conventional ("classical") laws of logic apply, and global propositions, which must be governed by different laws.

8.3 The Axiom of Euclidean Finiteness

The classical, or, perhaps better, *neo*-classical, Euclidean alternative to modern, Cantorian set theory rests on the traditional assumption, included in Euclid's *Elements* as Common Notion (Axiom) 5, that *the whole is greater than the part*. The Common Notions (*koinai ennoiai*) were seen by the mathematicians of ancient Greece as applying to all kinds of quantity (lines, surfaces, solids, *arithmoi*, ...) and as having a different meaning in each such application.

In arithmetic Common Notion 5 means that if one or more units are removed from an *arithmos* the remainder of its units form a smaller *arithmos*. On the Euclidean view, then, a plurality has a size – is *finite* – only if it grows smaller when any of its elements are deleted. This was, of course, the "common sense" view of finiteness that prevailed before Cantor.

Thus to obtain Euclidean set theory from basic set theory we must add to it the following alternative to Cantor's Axiom

---

[12] A precise explanation of this is given in the discussion in Section 9.1 that accompanies Definition 9.1.1.

8.3 The Axiom of Euclidean Finiteness

8.3.1 Axiom *(The Axiom of Euclidean Finiteness)*

$$\forall x Y [x : Y \to Y \text{ and } x \text{ is } 1-1 : \text{implies} : x \text{ is onto}]$$

It follows that if S is a set and T a proper subset of S then $S \not\cong_c T$ and therefore $T <_c S$.

As an immediate consequence of Euclid's axiom we obtain a set-theoretical version of the principle of mathematical induction.

8.3.2 Theorem *(The Principle of One Point Extension Induction – Local Version) Let S be any set and T be any set of subsets of S satisfying the following two conditions:*

(i) *The empty set is a member of T.*
(ii) *For every set, X, that is a member of T and every object, y, that is a member of S, $X \cup \{y\}$ is a member of T.*

Then S itself is a member of T.

Proof Suppose, by way of contradiction, that S and T, satisfying the hypotheses of the theorem, are given, but that, contrary to the conclusion, S is not itself a member of T. We notice immediately that for every set X which is a member of T, there must be an element y of S such that y is not a member of X. The condition (ii) placed on T then guarantees that $X \cup \{y\}$ must also lie in T. Thus T must contain sets of infinitely many different sizes. In fact, we can obtain the required contradiction by considering the set Q of all equivalence classes of T under the equivalence relation *sameness of size*. Q can be shown to violate the Axiom of Euclidean Finiteness since we can define an injective function

$$f : Q \to Q - \{\{\emptyset\}\}$$

The required f is defined by stipulating that, for $X \in Q$, $f`X$ is the set of all elements of T that are *one point extensions*[13] of elements of X.

The argument here is straightforward, but of considerable foundational importance, so let us formulate it carefully. For each $X \in T$, let $\hat{X}$ be defined by

$$\hat{X} = \{Y \in T : Y \cong_c X\}$$

and let

$$Q = \{\hat{X} : X \in T\}$$

---

[13] A *one point extension* of the set X is any set of the form $X \cup \{a\}$, where $a \notin X$.

278   Euclidean Finitism

Since $\emptyset \in T$ and T consists exclusively of sets (subsets of S, in fact)

$$\hat{\emptyset} = \{\emptyset\} \in Q.$$

For each $Z \in Q$, define

$$Z^+ = \{X \in T : (\exists Y \in Z)(\exists y \in S - Y)[X \cong_c Y \cup \{y\}]\}$$

Then for each $Z \in Q$, $Z^+ \in Q$.

Thus if we define

$$f = \{(X, X^+) : X \in Q\}$$

we have established that

$$f : Q \to Q - \{\{\emptyset\}\}$$

since $\{\emptyset\} \neq Z^+$ for any $Z \in Q$. It only remains to show that f is injective. But that follows immediately from the fact that if $X \cup \{x\}$ and $Y \cup \{y\}$ are proper one point extensions of X and Y, respectively, and $X \cup \{x\} \cong_c Y \cup \{y\}$, then $X \cong_c Y$.

Hence f is 1–1 but not onto from Q to Q, contradicting the Axiom of Euclidean Finiteness. It follows that the assumption that $S \notin T$ is false and the theorem is established. □

As an immediate corollary of Theorem 8.3.2 we obtain

8.3.3 Theorem (*The Principle of One Point Extension Induction – Global Version*) *Let Φ be any first level global propositional function of one argument. Then the hypotheses*

  (i) $\Phi(\emptyset)$
  (ii) $\forall xy[\Phi(x) \text{ .implies. } \Phi(x \cup \{y\})]$

entail the conclusion

$$\forall x[x \text{ is a set .implies. } \Phi(x)]^{14}$$

Proof Assume (i) and (ii) as hypotheses, and let S be any set. Define

$$T = \{X \in P(S) : \Phi(X)\}.$$

---

[14] The proposition asserted in this theorem is a global entailment. Its formal expression in the conceptual notation would be

$$\Phi(\emptyset), \forall xy[\Phi(x) \text{ implies } \Phi(x \cup \{y\})] \vdash \forall x[x \text{ is a set .implies. } \Phi(x)]$$

(See section A1.1 of Appendix 1).

8.3 The Axiom of Euclidean Finiteness

Then S and T, so defined, satisfy the hypotheses of 8.3.2. It follows that $S \in T$, so $\Phi(S)$ as required. □

These two theorems (8.3.2 and 8.3.3) raise an important methodological issue. For it is not difficult to see that if we laid down either of them as an axiom we could then derive the Axiom of Euclidean Finiteness as a theorem.

Of course the mere fact that the Principle of One Point Extension Induction in either of its forms is *logically* capable of standing in for the axiom adopted should not, of itself, recommend it to us *as an axiom*. For an *axiom* ought to be self-evidently true; it should be capable of standing at the beginning of an argument, without support from any other proposition.

No doubt since it is an induction principle that is in question, however, there will be those who would argue that it *does* have the character of an axiom, that the principle is plainly valid on the face of it, that, indeed, it merely anticipates and expresses the outcomes of an infinite number of sorites, each one of which is obviously valid: for if Φ satisfies the hypotheses of 8.3.2, then Φ is true of $\emptyset$, and therefore of $\emptyset \cup \{a\} = \{a\}$, and hence of $\{a\} \cup \{b\} = \{a,b\}$ and hence of $\{a,b\} \cup \{c\} = \{a,b,c\}$, and so on. And since, the argument continues, *every Euclidean finite set can be built up from the empty set by successive addition of elements one at a time*, there must be such a sorites to establish $\Phi(S)$ for every Euclidean finite set S.

It is this last part of the argument that contains the fatal fallacy, namely, the assumption, to put it another way, that the global function $\lambda xy(x \cup \{y\})$ somehow *generates* all sets in the Euclidean version of set theory.

Such a view of the meaning of Euclidean finiteness is not so much false *simpliciter* as unintelligible and therefore false in a profounder sense. What does it *mean* to "build up" a set one element at a time, or, indeed, to "build up" a set in any way whatsoever?

We can assemble the elements of a set of tennis balls before us on a table; but those particular balls composed the set before they were so assembled. Do we have to *list* the elements of a set – to produce a *name* for it? But then what we have built up is not the thing itself; it is only a name for that thing. Besides, what happens when we run out of paper or ink?

Perhaps we merely have to *think* of the elements one at a time to generate the set. But how much time is to be allotted for such a procedure?

Surely we don't want to tie down finitude to our actual capacities. But then what *ideal* capacities shall we posit? The capacity "in principle" to perform any finite task, including the "task" of thinking of each of the elements of a set one a time? But it is precisely "finiteness" in this Euclidean sense that we are trying to analyse and explain. How does it help to say that a *finite* plurality is one that we could ("in principle" of course!) build up element by element in a *finite* time, even supposing that we knew what it was to "build up" a plurality? The supposed *explicans* contains the *explicandum*. Indeed, as with all such operationalist "explanations" the notion supposedly explained not only turns up in the explanation, but turns up encumbered with all sorts of irrelevant associations (finite *time*, etc.).

In fact, the proof of 8.3.2 actually *explains* why One Point Extension Induction, in this form, is a valid principle. The proof is perfectly rigorous and set-theoretical; it makes no appeal to our capacities actual or idealised, no references to temporal processes or imagined "constructions" of unspecified sorts. The alternative operationalist explanation explains nothing.

One Point Extension Induction provides us with the means actually to prove, in Euclidean set theory, a fundamental principle which in the Cantorian version of the theory must be laid down as an axiom: I mean the *principle* (or *axiom*) *of choice*. Recall (Definition 5.3.1) that a choice function for a set S is a local function $f : P(S) - \{\emptyset\} \to S$ defined on the non-empty subsets of S with values in S that assigns to each such subset one of its own members.

$$f`X \in X \text{ for all } X \in P(S) - \{\emptyset\}$$

Speaking metaphorically we may say that the function "chooses" an element from each non-empty subset.

8.3.4 Theorem *(The Principle of Choice) Every set has a choice function.*

Proof Let Φ be the global propositional function of one argument which takes the value *true* precisely at those sets which have choice functions. Then Φ is true of the empty set vacuously. Now suppose that S is any set that has a choice function and a is any object. If f is a choice function for S, we may define a choice function

$$g : P(S \cup \{a\}) - \{\emptyset\} \to S \cup \{a\}$$

8.3 The Axiom of Euclidean Finiteness

by setting, for each non-empty subset, X, of $S \cup \{a\}$

$$g`X = \begin{cases} f`X \text{ if } a \notin X \\ a \text{ if } a \in X \end{cases}$$

We have thus established the hypotheses of the global form of One Point Extension Induction, so we may conclude that Φ is true of every set, i.e. that every set has a choice function as required. □

The proof of the Principle of Choice is completely routine. I have included it here in order to call attention to the fact that the principle of choice is a straightforward consequence of the Axiom of Euclidean Finiteness, so that we are not required to appeal to the supposed possibility of carrying out any finite number of "acts of choice" in order to justify the existence of choice functions. Like the Principle of One Point Extension Induction, the Principle of Choice does not rest upon operationalist assumptions.

One Point Extension Induction can also be used to establish a Euclidean version of Zorn's Lemma.

8.3.5 Theorem (*Zorn's Lemma*) *Every non-empty partial ordering has both maximal and minimal elements.*

Proof Let $\leq$ be a partial ordering with field S, and let T be the set of all subsets of S that either are empty or contain both maximal and minimal elements with respect to the restriction of $\leq$ to the elements of the given subset. Then T contains the empty set and is closed under one point extensions from S.

Applying the local version of One Point Extension Induction (8.3.2) we conclude that S itself lies in T, i.e. S has both maximal and minimal elements. □

Zorn's Lemma affords us the means to prove the following important theorem.

8.3.6 Theorem *All sets can be compared as to size. More precisely, given any sets S and T*

$$S \leq_c T \text{ or } T \leq_c S$$

Proof Consider the partial ordering whose field is the set of all 1-1 functions g for which $Domain(g) \subseteq S$ and $Range(g) \subseteq T$, and where the ordering is just set-theoretical inclusion restricted to $P(S \times T)$. (The ordering is non-empty since the 1 − 1 onto function $\emptyset : \emptyset \to \emptyset$ lies in its field). By Zorn's Lemma this partial ordering must contain a maximal element, say f. Then either f bears witness to $S \leq_c T$ or f^{-1} bears witness to $T \leq_c S$. □

8.4 Linear orderings and simple recursion

Recall[15] that a linear ordering is a local relation that is reflexive, antisymmetric, transitive and total.

8.4.1 Theorem *Let S be any set. Then there is a subset r of $S \times S$ such that r is a linear ordering and $Field(r) = S$.*

Proof Let S be a set and let $T \subseteq P(S \times S)$ be the set of all linear orderings whose fields are subsets of S. Then T is partially ordered by inclusion and therefore by the Euclidean version of Zorn's Lemma (8.3.5), T has maximal elements with respect to inclusion. Any such maximal element is a linear ordering of S. □

Thus every set can be linearly ordered. Recall[16] that a linear ordering is a *well-ordering* if every non-empty subset of its field has a least element.

8.4.2 Definition *Let r be any set. Then by definition r is a double wellordering if, and only if, r and r^{-1} are well-orderings.*

Thus a linear ordering r is a double well-ordering if, and only if, every non-empty subset of the field of r has both a greatest and a least element.

The Euclidean version of Zorn's Lemma (8.3.5) immediately yields the following theorem.

8.4.3 Theorem *Every linear ordering is a double well-ordering.* □

Theorems 8.4.1 and 8.4.3 imply that every set can be doubly well ordered. Moreover by 8.4.3, every linear ordering has both a first and a last element. Furthermore it is a consequence of 8.4.3 that given any linear ordering r, we can define a unique successor function for that ordering which sends each element of its field, except the last, to the next succeeding element in the ordering. These observations lead to the following definition.

---

[15] Definition 4.9.1.
[16] ibid.

8.4 Linear orderings and simple recursion

8.4.4 Definition *Let r be any linear ordering. Then by definition*

(i) $First(r) = \iota(\{x \in Field(r) : (\forall y \in Field(r))[x \leq_r y]\}, \emptyset)$
(ii) $Last(r) = \iota(\{x \in Field(r) : (\forall y \in Field(r))[y \leq_r x]\}, \emptyset)$
(iii) $Next(r) = \{x \in r : 1^{st}(x) <_r 2^{nd}(x)$ and
$(\forall y \in Field(r))[1^{st}(x) <_r y$ implies $2^{nd}(x) \leq_r y]\}$ □

In Euclidean set theory we may be tempted to say, speaking metaphorically, that a linear ordering r is "generated" from its first element $First(r)$ by the appropriate number of applications of the successor operation

$$x \mapsto Next(r)`x$$

But this is only a metaphor: the facts that make it apt can easily be established without appeal to the operationalist notion of generation.

8.4.5 Theorem *Let r be a linear ordering. Then*

(i) *For every x in the field of r, either $x = First(r)$ or there is a y in the field of r such that $x = Next(r)`y$.*
(ii) *For every x in the field of r, if $x \neq Last(r)$ then $Next(r)`x \neq First(r)$.*
(iii) *For all x and y in the field of r, if neither x nor y is the last element of r and if $Next(r)`x = Next(r)`y$ then $x = y$.*
(iv) $Next(r) : Field(r) - \{Last(r)\} \to Field(r) - \{First(r)\}$ *is $1-1$ and onto.*

A Euclidean version of mathematical induction for linear orderings is also straightforward to establish.

8.4.6 Theorem *(The Principle of Induction for Linear Orderings) Let r be a non-empty linear ordering and let S be a subset of its field such that*

(i) *The first element of r lies in S.*
(ii) *For every x in S, if x is not the last element of r, then $Next(r)`x$ lies in S.*

Then S is the whole of the field of r.

The principle of induction for linear orderings can be used to establish a principle of definition by simple recursion along such orderings, as I shall now show, but first some matters of notation. Recall (Definition 4.8.3) that

$$r_{\leq a} = \{x \in r : 2^{nd}(x) \leq_r a\}$$

so that $r_{\leq_a}$ is the initial segment of r determined by a. Clearly if r is a non-empty linear ordering then

$$r = r_{\leq Last(r)}$$

and, in any case,

$$r = \bigcup \{r_{\leq_x} : x \in Field(r)\}$$

Moreover, if a is in the field of r,

$$First(r_{\leq_a}) = First(r) \text{ and } Last(r_{\leq_a}) = a$$

These facts will be of use in the following proof.

8.4.7 Theorem (*The Principle of Definition by Simple Recursion along a Linear Ordering*) *Let r be a non-empty linear ordering and let S be a set, a an element of S and $g : S \to S$ a local function mapping S to itself. Then there is exactly one local function $f : Field(r) \to S$ that satisfies the following two conditions (called the recursion equations for f):*

(i) $f\text{'}First(r) = a$.
(ii) $f\text{'}Next(r)\text{'}x = g\text{'}f\text{'}x$, for all x in the field of r except for $Last(r)$.

Proof We can establish this by induction along r. Thus let T be the set of all elements, x, in the field of r such that there exists a local function $f_x : Field(r_{\leq_x}) \to S$ such that

(i)$_x$ $f_x\text{'}First(r) = a$.
(ii)$_x$ $f_x\text{'}Next(r)\text{'}y = g\text{'}f\text{'}y$ for all y in the field of $r_{\leq_x}$ except for x.

Clearly $First(r) \in T$ since we may set $f_{First(r)} = \{(a, a)\}$. Now suppose $x \in T$ so that (i)$_x$ and (ii)$_x$ hold for some local function f_x. If $x \neq Last(r)$, define $f_{Next(r)\text{'}x}$ by setting

$$f_{Next(r)\text{'}x} = f_x \cup \{(Next(r)\text{'}x, g\text{'}f_x\text{'}x)\}$$

Then $f_{Next(r)\text{'}x} : Field(r_{Next(r)\text{'}x}) \to S$ and satisfies (i)$_{Next(r)\text{'}x}$ and (ii)$_{Next(r)\text{'}x}$. Hence $Next(r)\text{'}x \in T$. By induction along r, T is the whole of the field of r. In particular, $Last(r) \in T$, so that there is a function

$$f_{Last(r)} : Field(r_{\leq Last(r)}) \to S$$

satisfying the appropriate recursion equations. But $r_{\leq Last(r)}$ is just r itself. Hence we may set

$$f = f_{Last(r)}$$

8.4 Linear orderings and simple recursion 285

to obtain a function satisfying (i) and (ii). That f so defined is the only such function follows by induction along r. □

Theorem 8.4.7 is an important result, both for its technical utility and for the light it and its proof throw on the whole enterprise of providing non-Cantorian foundations for mathematics.

Some might be tempted to explain that proof naively as saying that we may "iterate" the function $g : S \to S$ as many times as there are terms in the linear ordering r. But the temptation to do so must be resisted, for such an "explanation", in fact, explains nothing.

We do not need the sanction of 8.4.7 in order to write down an expression of the form

$$g`g`\ldots g`a$$

using as many g's as we like: that permission is implicit in our notational conventions. However *nothing of any mathematical significance turns on such implicit permission*.

Indeed, if we trace back through the sequence of definitions and theorems that connect both the statement and proof of 8.4.7 with the set-theoretical elements and principles that underlie them, we see that the idea of iteration, understood literally in the operationalist sense, simply has no role to play whatsoever.

Theorem 8.4.7 does not, in fact, tell us that a function may be iterated as often as a linear ordering has terms; on the contrary it provides us with a precisely defined *objective correlative* to that vague, symbol generated idea.

The force of these observations is brought home to us when we examine the proof of 8.4.7 in detail. For our notational conventions permit us to "iterate" not only *local* functions

$$g`g`\ldots g`a$$

but also *global* functions as well

$$\sigma(\sigma(\ldots(\sigma(a))\ldots))$$

But that the function "iterated" in 8.4.7 be *local* is essential to the efficacy of the proof. Let us see why.

In the proof of 8.4.7 we must first define the set T to be the subset of $Field(r)$ whose elements x satisfy the condition

$$(\exists y \in S^{Field(r_{\leq x})})[\, y \text{ satisfies the recursion equations for } r_{\leq x}]$$

286 *Euclidean Finitism*

and then prove, by induction along r, that $T = Field(r)$. But this definition depends upon the fact that $g : S \to S$ is a *local* function; for that means we can confine the values of the "approximating functions" f_x to the set S; and that, in turn, means that the initial existential quantifier in the definining condition for T is *local*.

If we were to substitute a global function σ for the local function g, the corresponding T would have to be the *species* of all x in the field of r that satisfy the defining condition

$$\exists y [\, y \text{ is a function .and. } Domain(y) = Field(r_{\leq x})$$
$$\text{.and. } y \text{ satisfies the recursion equations for } r_{\leq x}]$$

But since this defining condition employs a *global* existential quantifier[17] for its expression, we cannot infer that the species T of which it is the defining condition forms a set. True, T cannot be genuinely *infinite* since it is a subspecies of a set; but it may be *indefinite*: it may be what Vopenka calls a *semi-set*.

But the defining condition under consideration here is of the logically simplest type among those employing global quantifiers, for it is Σ_1. Induction for such properties seems to constitute a special case. Let us, therefore, examine the matter in some detail.

As we have seen, if we try to adapt the proof of 8.4.7 to accommodate the case in which the local function g is replaced by a global function σ, then the defining condition for T takes the Σ_1 form

$$\exists y \Phi(\sigma, r, s, y)$$

where Φ is a global propositional function with one first level function and three objects as arguments; moreover the function argument, σ, and the object argument, r, remain fixed, as parameters, throughout the entire proof. By inspection of the proof of 8.4.7 we see that we can establish

$$\Phi(\sigma, r, First(r), \{(First(r), a)\}) \qquad (1)$$

Moreover, we can explicitly define a global function, τ_σ (uniformly in σ), for which we can establish

$$\forall xy [x \in (Field(r) - \{Last(r)\}) :implies:$$
$$[\Phi(\sigma, r, x, y) \text{ implies } \Phi(\sigma, r, Next(r)`x, \tau_\sigma(y))]] \qquad (2)$$

---

[17] This corresponds to the familiar fact that in natural number arithmetic Σ_1 induction is required in order to establish the existence of primitive recursive functions.

8.4 Linear orderings and simple recursion

Thus if b witnesses the validity of the existential assertion

$$\exists y \Phi(\sigma, r, s, y)$$

where s lies in the field of r but $s \neq Last(r)$, then $\tau_\sigma(b)$ witnesses the validity of the existential assertion

$$\exists y \Phi(\sigma, r, Next(r)\text{'}s, y)$$

Thus (1) and (2) are the basis and induction steps of a traditional induction argument. Moreover, (2) gives a "constructive" meaning to the passage from the proposition for s to the proposition for $Next(r)$'s.

But to show that $T = Field(r)$ without actually *invoking* Σ_1-induction, we need to show that for all x in the field of r

$$\exists y \Phi(\sigma, r, x, y)$$

and to do this we need to define a global function γ_σ and then to prove that

$$(\forall x \in Field(r))[\Phi(\sigma, r, x, \gamma_\sigma(x))]$$

However, *the only way to define such a γ_σ is by recusion along r with respect to the global function τ_σ*. So the argument is transparently circular: to establish recursion with respect to σ we must apply recursion to τ_σ.

I have considered this matter in so much detail because it is a straightforward example of the Σ_1 *barrier* to induction and recursion that we shall encounter again and again in these investigations.

The fact that recursion with respect to global functions cannot be established will make it impossible for us to prove that all simply infinite systems are isomorphic, as we shall see in Chapters 10 and 11. This means that *there is no way to define a unique natural number sequence in Euclidean set theory*.

In the absence, then, of a "given" natural number system, we shall have to use linear orderings as surrogate initial segments of "the" natural numbers. Accordingly, I shall sometimes use expressions like

$$[1_r, \ldots, n_r] \text{ or } [1, \ldots, n]_r$$

or if I wish to start with zero

$$[0_r, \ldots, n_r] \text{ or } [0, \ldots, n]_r$$

to designate a linear ordering r. If the ordering in question is clear from the context, I shall just write

$$[1, \ldots, n] \text{ or } [0, \ldots, n]$$

suppressing the subscript r. Needless to say, *this is merely a notation*, and the numerals and letters employed are *not* to be taken as designating "natural numbers". The field of r may then be designated by

$$\{1_r, \ldots, n_r\} \text{ or } \{1, \ldots, n\}_r$$

and, again, the subscript r may be suppressed if doing so is unlikely to cause confusion.

The terms of a linear ordering must, of course, be distinct. Accordingly, I must introduce the notion of a *sequence*.

8.4.8 Definition *Let r be a linear ordering and S and T be sets. Then*

(i) *T is an r-sequence on S if, and only if, T is an ordered pair, (f, r), where $f : Field(r) \to S$.*
(ii) $Seq(r, S) = \{x \in S^{Field(r)} \times \{r\} : r \text{ is a linear ordering}\}$.

Thus if r is a linear ordering, $Seq(r, S)$ is the set of all r-sequences on S. We are forced to define sequences as ordered pairs of this form since there are, in the general case, many distinct linear orderings whose field is the same as the field of r. The linear ordering r is the *length* of the r-sequence (f, r).

8.4.9 Definition *Let $S = (f, r)$ and $T = (g, s)$ be sequences, and let $a \in Field(r)$. Then by definition*

(i) $lh(S) = r \, (= 2^{nd}(S))$.
(ii) $(S)_a = f`a \, (= 1^{st}(S)`a)$.
(iii) $S * T = (h, r +_o s)$, where h is defined by setting

$$h = \{((x, 0), f`x) : x \in Field(r)\} \cup \{((y, 1), g`y) : y \in Field(s)\}$$

Thus $lh(S)$ is the *length* of S, $(S)_a$ is the *ath term* of S, and $S * T$ is the *concatenation* of S and T (in that order).

In conformity with the notation introduced for linear orderings I shall use expressions like

$$\langle a_{1_r}, \ldots, a_{n_r} \rangle \text{ or } \langle a_1, \ldots, a_n \rangle_r$$

to designate r-sequences. As before the subscripted r may be omitted if circumstances warrant.

We need to establish a more general version of recursion (Theorem 8.4.7).

8.4 Linear orderings and simple recursion

8.4.10 Theorem (*The Principle of Definition by Primitive Recursion along a Linear Ordering*) *Let $r = [0,\ldots,n]$ be a linear ordering, S a non-empty set, $a \in S$ a distinguished element of S, and $h : \text{Field}(r) \times S \to S$. Then there is exactly one local function $f : \text{Field}(r) \to S$ that satisfies the following equations for local primitive recursion:*

(i) $f`0 = a$
(ii) $f`(k+1) = h`(k, f`k)$ *for all* $k \in \{0,\ldots,n-1\}$

Proof Let $g : \text{Field}(r) \times S \to \text{Field}(r) \times S$ be defined, for all $x \in S$, by

$$g`(k, x) = \begin{cases} (k+1, h`(k, x)) & \text{if } k \in \{0,\ldots,n-1\} \\ (n, a) & \text{if } k = n \end{cases}$$

Theorem 8.4.7 now guarantees the existence of a unique

$$f_0 : \text{Field}(r) \to \text{Field}(r) \times S$$

defined by recursion along r such that

$$f_0`k = \begin{cases} (0, a) & \text{if } k = 0 \\ g`f_0`(k-1) & \text{if } k \in \{1,\ldots,n\} \end{cases}$$

Then for each $k \in \{0,\ldots,n\}$, $f_0`k = (k, x)$ for some $x \in S$. This is true for $k = 0$, and if it is true for $k \in \{0,\ldots,n-1\}$, so that $f_0`k = (k, x)$ for $x \in S$, then

$$f_0`(k+1) = g`f_0`k = g`(k, x) = (k+1, h`(k, x))$$

and it is true for $k + 1$ as well.

It follows that $f_0``\{0,\ldots,n\} : \{0,\ldots,n\} \to S$. In fact

$$f = f_0``\{0,\ldots,n\}$$

is the function required by the theorem. For clearly $f`0 = a$, and for all $k \in \{0,\ldots,n-1\}$

$$\begin{aligned} f`(k+1) &= 2^{nd}(f_0`(k+1)) \\ &= 2^{nd}(g`(f_0`k)) \\ &= 2^{nd}(g`(k, f`k)) \\ &= 2^{nd}((k+1, h`(k, f`k))) \\ &= h`(k, f`k) \end{aligned}$$

The uniqueness of f follows by a straightforward induction along r. □

The idea of definition by simple recursion has an important connection with the idea of a full order morphism[18], as the following theorem discloses.

8.4.11 Theorem *Let r_1 and r_2 be non-empty linear orderings, and let $m : Field(r_1) \to Field(r_2)$. Then m is a full order morphism if, and only if, m satisfies the recursion equations*

(i) $m\text{`}First(r_1) = First(r_2)$
(ii) $m\text{`}Next(r_1)\text{`}x = Next(r_2)\text{`}m\text{`}x$ *for all* $x \in (Field(r_1) - \{Last(r_1)\})$

Proof (*only if*) Suppose that m is a full order morphism from r_1 to r_2. Since m is full, $m\text{`}First(r_1) = First(r_2)$.

Now suppose, by way of contradiction, that there is an exception to (ii) and let x_0 be the $\leq_{r_1}$ least such. Setting $m\text{`}x_0 = y_0$ we have

$$m\text{`}x_0 = y_0 <_{r_2} Next(r_2)\text{`}y_0 <_{r_2} m\text{`}Next(r_1)\text{`}x_0$$

So, since m is full, there is an x_1 such that $m\text{`}x_1 = Next(r_2)\text{`}y_0$. Clearly $x_1 \neq x_0$. But if $x_1 <_{r_1} x_0$, then $m\text{`}x_1 = Next(r_2)\text{`}y_0 <_{r_2} m\text{`}x_0 = y_0$ which is impossible. And if $x_0 <_{r_1} x_1$, then $Next(r_1)\text{`}x_0 \leq_{r_1} x_1$ but

$$m\text{`}x_1 = Next(r_2)\text{`}y_0 <_{r_2} m\text{`}Next(r_1)\text{`}x_0$$

which is also impossible. The required conclusion now follows.

(*if*) Suppose that m satisfies the recursion equations. We must show that it is both full and order preserving. To that end, for each $x \in Field(r_1)$ let $m_x : Field((r_1)_{\leq x}) \to Field(r_2)$ be the restriction of m to the initial segment of r_1 consisting of the elements $\leq x$. A straightforward induction over r_1 suffices to show that m_x is full and order preserving for all $x \in Field(r_1)$. For this is vacuously the case for $x = First(r_1)$, and the argument for the induction step goes through on the observation that

$$m_{Next(r_1)\text{`}x} = m_x \cup \{(Next(r_1)\text{`}x, Next(r_2)\text{`}m\text{`}x)\}$$

Since $m = m_{Last(r_1)}$, the desired conclusion follows.

□

Since functions defined by simple recursion are unique, Theorem 8.4.11 yields, as an immediate corollary, the result, already established in Theorem 4.9.7, that there is at most one full order morphism from one linear ordering to another. In particular, order isomorphisms are unique. We

---

[18] Definition 4.9.6.

also obtain an alternative, Euclidean, proof that any two linear orderings are comparable in length.[19]

As a corollary of these results we obtain, using the Axiom of Euclidean Finiteness (8.3.1), the following corollary.

8.4.12 Corollary *Let r_1 and r_2 be linear orderings. Then*
(i) $r_1 \cong_o r_2$:iff: $Field(r_1) \cong_c Field(r_2)$
(ii) $r_1 <_o r_2$:iff: $Field(r_1) <_c Field(r_2)$
In particular, any two linear orderings of the same set are equal in length.

The relation "$<_o$" among linear orderings is "doubly well-founded" locally, as the following theorem shows.

8.4.13 Theorem *Let S be a non-empty set of linear orderings. Then S contains elements of maximal length and elements of minimal length.*

Proof Let T be the set of equivalence classes of S under the equivalence relation induced on S by $\cong_o$, and let r be the order relation whose field is T and which is defined by stipulating that for $U, V \in T$, $(U, V) \in r$ if, and only if, some (and therefore every) element of U is less than or equal to some (and therefore every) element of V in length. It is obvious that r so defined is a partial ordering; that it is also linear follows from Theorem 8.3.6. Since the elements of T are equivalence classes, neither $First(r)$ nor $Last(r)$ is empty. All elements of $First(r)$ are elements of S having minimal length in S; all elements of $Last(r)$ have maximal length in S. □

Theorem 8.4.13 cannot be extended to apply to arbitrary non-empty species of linear orderings. Of course we could not expect an infinite species to contain maximal length elements in general, but we cannot even show that they contain minimal length ones in the general case.

8.5 Local cardinals and ordinals

Corollary 8.4.12 tells us that the "lengths" or "order types" of the linear orderings of a set are all the same and are determined by the cardinality of the set. Thus the order types of linear orderings, like the cardinalities of sets, correspond to "natural numbers", and the latter might be regarded, naively, as "abstract entities" associated to – or

---

[19] Theorem 4.9.10 (viii).

perhaps somehow extracted from – equivalence classes of linear orderings under $\cong_o$.

This is a traditional way of "creating" mathematical entities, but it is completely inadequate for our purposes. For we can form no clear idea of what such abstract entities might be, or, more importantly, of what "abstract operations" we could perform on them. We cannot simply identify them with the equivalence classes themselves, since the latter, as infinite species, are not objects. The problem of defining these order types, or their associated cardinalities ("natural numbers") *as objects*, therefore, comes down to the problem of assigning to each linear ordering, or to each set, a particular "canonical" object representative of its equivalence class (under ordinal or cardinal equivalence, as appropriate). It is not difficult to see that this would amount to selecting, in a canonical manner, a unique element from each such equivalence class. As we shall see in Chapter 11, this is by no means a straightforward task.

In the case of linear orderings a variant of this problem already arises locally. For there are many ways of linearly ordering a set with two or more elements, and none of these orderings is, in the general case, in any way privileged or natural. It is useful, therefore, to devise a uniform method whereby, given any set, we can define a linear ordering which represents the mutually equivalent linear orderings of that set. There is an obvious way to do this that is suggested by Lemma 8.5.2 below. But first we need a definition.

8.5.1 Definition *Let S be any set. Then by definition*

$$\chi(S) = \{\{y \in P(S) : y \cong_c x\} : x \in P(S) - \{S\}\}$$

Thus given a set S, $\chi(S)$ is the set of equivalence classes of all proper subsets of S under cardinal equivalence ($\cong_c$).

8.5.2 Lemma *Let S be any set. Then*

$$S \cong_c \chi(S)$$

Proof The proof proceeds by global one-point-extension induction with respect to the global propositional function $\lambda x[x \cong_c \chi(x)]$.

Basis case ($x = \emptyset$) This case follows immediately since $\chi(\emptyset) = \emptyset$.

Induction Suppose, as hypothesis of induction, that $S \cong_c \chi(S)$, and let a be given such that $a \notin S$. We must show that $S \cup \{a\} \cong_c \chi(S \cup \{a\})$.

8.5 Local cardinals and ordinals

To that end let $f : S \to \chi(S)$ be a bijection witnessing $S \cong_c \chi(S)$. Define $g : \chi(S) \to \chi(S \cup \{a\})$ by setting

$$g`X = X \cup \bigcup\{\{(z - \{y\}) \cup \{a\} : y \in z\} : z \in X\}$$

for $X \in \chi(S)$. Then $g`X \in \chi(S \cup \{a\})$ and $(\forall y \in X)(\forall z \in g`X)[y \cong_c z]$. Now define $h : S \cup \{a\} \to \chi(S \cup \{a\})$ by setting

$$h`x = \begin{cases} \{(S - \{x\}) \cup \{a\} : x \in S\} \cup \{S\} & \text{if } x = a \\ g`f`x & \text{if } x \in S \end{cases}$$

Then h is bijective and $S \cup \{a\} \cong_c \chi(S \cup \{a\})$, as required.

The lemma now follows by induction. □

In the light of this lemma we could define the "local ordinal" of a set, S, to be the ordering of $\chi(S)$ which reflects the cardinalities of the members of the sets belonging to $\chi(S)$. I shall adopt a rather more complicated definition, however, since I also want to define a *local cardinal*, card(S), for each set S in such a manner that $S \cong_c$ card(S), and, in addition, card(S) has exactly one member of each cardinality smaller than that of S. The *local ordinal*, ord(S), can then be defined to be the restriction of $\leq_c$ to card(S).

8.5.3 Definition *Let S be any object. Then by definition*

(i) $\text{ord}_0(S) = \{x \in \chi(S) \times \chi(S) : (\forall y \in 1^{st}(x))(\forall z \in 2^{nd}(x))[y \leq_c z]\}$
(ii) $\text{card}(S) = \{\emptyset\} \cup \{Field((\text{ord}_0(S))_{\leq_x}) : x \in Field(\text{ord}_0(S)) - \{Last(\text{ord}_0(S))\}\}$
(iii) $\text{ord}(S) = \{x \in \text{card}(S) \times \text{card}(S) : 1^{st}(x) \leq_c 2^{nd}(x)\}$

Thus if $S = \{a, b, c\}$, then

$$\chi(S) = \{A, B, C\}$$

where

$$A = \{\emptyset\}$$
$$B = \{\{a\}, \{b\}, \{c\}\}$$
$$C = \{\{a,b\}, \{a,c\}, \{b,c\}\}$$

and therefore

$$\text{ord}_0(S) = [A, B, C]$$
$$\text{card}(S) = \{\emptyset, \{A\}, \{A, B\}\}$$
$$\text{ord}(S) = [\emptyset, \{A\}, \{A, B\}]$$

294 Euclidean Finitism

The properties of the global function $\lambda x[\text{card}(x)]$ that make it useful are presented in the following theorem.

8.5.4 Theorem *Every set is of the same size as its local cardinal (i.e. $S \cong_c \text{card}(S)$). Moreover, if S is any set and T is any smaller set (i.e. $T <_c S$), then there is exactly one element of $\text{card}(S)$ which is of the same size as T.*

Proof Let the local function m be defined for $x \in \text{Field}(\text{ord}_0(S))$ by

$$m^\iota x = \text{Field}((\text{ord}_0(S))_{<x})$$

Then m is an isomorphism $\text{ord}_0(S) \to \text{ord}(S)$. Moreover if $X \in \text{card}(S)$ then $X \equiv_c Y$ for every $Y \in m^\iota X$. Since the field of $\text{ord}_0(S)$ consists of cardinal equivalence classes of $P(S) - S$ the theorem follows.

□

To summarise: if S is a set then $S \cong_c \text{card}(S)$, where $\text{card}(S)$ contains exactly one set of each cardinality less than that of S, and $\text{ord}(S)$ is the linear ordering of $\text{card}(S)$ which arranges its elements in order of increasing size.

8.6 Epsilon chains and the Euclidean Axiom of Foundation

In Section 4.11 I discussed the consequences of the extensional character of sets for the analysis of the relation of set membership. The treatment there was rather sketchy, and I explained that although the Cantorian and Euclidean versions of set theory shared the Principle of Regularity, which is a consequence of the extensional character of sets, nevertheless these differing versions of set theory lead to analyses of the membership relation that differ significantly in detail. Having given the Cantorian analysis in Section 5.4, I shall now present the Euclidean analysis. The reader will note a certain overlapping of terminology, although the definitions differ in significant ways.

In Euclidean set theory, as in Cantorian, I call the objects that go, directly or indirectly, to make up a set the *constituents* of that set. Thus the members of a set, S, are constituents of S; and if S contains sets among its members, then their members, in turn, are also constituents of S, likewise *their* members as well, and so on. In general, any constituent of a constituent of a set is also itself a constituent of that set. But how to make this naive idea mathematically precise?

8.6 Epsilon chains and the Euclidean Axiom of Foundation

As in the Cantorian case, the constituents of a set together compose what is usually called the *transitive closure* of that set, the smallest transitive set that includes the given set as a subset. Thus T is the transitive closure of S if it satisfies the following three conditions:

(1) S is a subset of T. (Members of S are constituents of S.)
(2) T is *transitive*, that is to say, every set that is a member of T is also a subset of T. (Members of constituents of S are constituents of S.)
(3) The only transitive subset of T of which S is a subset is T itself. (The set of constituents of S is the smallest set satisfying (i) and (ii))

But how do we know that every set *has* a transitive closure? Well, the constituents of a set are precisely those objects the existence of which as well-defined objects is presupposed by the existence of that set itself as a well-defined object. Therefore, if the constituents of a set were indeterminate in number that would constitute an indeterminacy in the set itself, which, surely, would violate our conception of a set as a *finite* entity, that is to say, an *object*.

It is thus reasonable to suppose that the constituents of a set are finite in number and therefore together compose a set. The argument here is akin to the argument for the existence of a power set for each set: surely it is implausible, in each of these cases, that the absolutely infinite should be *rooted* in a finite object. How could a set, which by definition is a *finite* plurality, have an absolute infinity of either subsets or constituents?

These considerations would seem to justify our introducing a new primitive global function to represent the first level global function given by the correspondence

$$S \mapsto \text{the transitive closure of } S$$

and laying down an axiom that uses (i), (ii) and (iii) above to characterise this new function.

Now such an axiom would clearly incorporate a finiteness principle, and in that respect would resemble all the other principal axioms of the theory. But it would not take the quasi-definitional form those other axioms all share. For (i), (ii) and (iii) do not simply lay down necessary and sufficient conditions for an object to be a constituent of a set. Moreover, the natural form for condition (iii) would be the global generalisation

$$\forall X [X \text{ is a transitive set and } S \subseteq X \; : implies : \; T \subseteq X]$$

which actually asserts (rather than merely implies) the minimality of T among transitive supersets of S. Of course this global generalisation would follow from (iii) since the intersection of transitive sets is transitive; but it seems undesirable that an axiom should express its content indirectly, so to speak.

Accordingly, I shall approach these matters from a different direction, defining the notion of *constituent* in terms of the notion of an *epsilon chain*. An epsilon chain beginning with a given object is a non-empty linear ordering whose first term is the given object and each of whose non-initial terms (if any) is a member of the immediately preceding term[20]. Thus if $[a, b, c]$ is an epsilon chain, then

$$c \in b \in a$$

On so important a matter it is perhaps best to lay down an official definition.

8.6.1 Definition *Let a be any set or individual and r a linear ordering. Then by definition*

$$r \text{ is an epsilon chain from } a$$

if, and only if

 (i) $r \neq \emptyset$.
 (ii) $First(r) = a$.
 (iii) $(\forall x \in Field(r))[x \neq Last(r) \text{ implies } Next(r)`x \in x]$.

Surely the epsilon chains beginning with a given set must be finite in multitude, for the same reason that the constituents of a set must be finite in multitude. The totality of such chains beginning with a given set corresponds to a complete analysis of that set as an *extensional* plurality, and it is surely reasonable to suppose that such an analysis should itself be finite. Surely we should not expect to encounter the absolutely infinite in the course of analysing the essential definition of an *object*, which in accordance with our conventions and terminology is a *determinate* entitiy.

It is reasonable, therefore, to assume that the species of all epsilon chains beginning with a given object, a, is finite, and, in consequence, forms a set called the *epsilon fan of a*, which I will denote by

$$\epsilon(a)$$

---

[20] The notion of an epsilon chain thus corresponds to the notion of an analysing sequence in Cantorian set theory. See Definition 5.4.1.

8.6 Epsilon chains and the Euclidean Axiom of Foundation

The correspondence

$$a \mapsto \epsilon(a)$$

thus defines a first level global function of one argument.

In accordance with the general pattern established for incorporating finiteness assumptions in our theory we may now lay down the following axiom.

8.6.2 Axiom *(The Epsilon Fan Axiom) Let a be any set or individual. Then*

(i) *$\epsilon(a)$ is a set.*

(ii) *$\forall x[x \in \epsilon(a)$ iff x is an epsilon chain from a].*

The notion of *constituent* can now be defined.

8.6.3 Definition *Let a and b be any sets or individuals.*

$$a \text{ is a constituent of } b \; :\text{iff}: \; (\exists x \in b)(\exists y \in \epsilon(x))[a \in Field(y)]$$

Thus the constituents of a set S are those objects that appear as terms in epsilon chains beginning with elements of S. If $a \in S$ and a is not a set, then $\epsilon(a) = \{[a]\}$, so the only constituent a contributes to S is a itself. Individuals and the empty set have no constituents.

8.6.4 Definition *Let a be any set or individual. Then by definition*

$$Constituents(a) = \bigcup \bigcup \{\{Field(x) : x \in \epsilon(y)\} : y \in a\}$$

The following theorem is now easily established.

8.6.5 Theorem *Let a be any set or individual. Then*

(i) $a \subseteq Constituents(a)$

(ii) $(\forall x \in Constituents(a))[x \subseteq Constituents(a)]$
    (i.e. $Constituents(a)$ is transitive)

(iii) $(\forall X \subseteq Constituents(a))[a \subseteq X$ and $(\forall y \in X)[y \subseteq X] \; :$ implies $:$
    $X = Constituents(a)]$

Proof (i) and (ii) are immediately obvious. To establish (iii), let $[b,\ldots,c]$ be an epsilon chain where $b \in a$, and let $S \subseteq Constituents(a)$, where $a \subseteq S$ and $(\forall y \in S)[y \subseteq S]$. A straightforward induction along the linear ordering $[b,\ldots,c]$ establishes that each of its terms lies in S. □

As I have said, the constituents of a set constitute precisely those sets and individuals whose status as well-defined, legitimate objects is presupposed by the existence of that set itself as a well-defined *extensional* plurality. For the extensional nature of sets means that what a set *is*, its essence, so to speak, is determined completely and solely by what its elements are; and if there are sets among the elements of S what each of them is, in turn, is determined by what *its* elements are, and so on. Thus what S *is* – its very *being*, so to speak – is determined, ultimately, by what each of its constituents is.

Thus from the *extensional* character of sets it follows that no set can be a constituent of itself or, indeed, of any of its constituents, for that would constitute an obvious *circulus in definiendo*. Accordingly, we must now lay down the following axiom.

8.6.6 Axiom *(The Euclidean Axiom of Foundation)*

$$\forall x \forall y [x \in Constituents(y) \text{ implies } y \notin Constituents(x)]$$

From this axiom we can immediately deduce that $S \notin S$ for every set S. In fact, we can establish the Principle of Regularity.

8.6.7 Theorem *(The Principle of Regularity)* *Let S be any non-empty set. Then S has an element which has no element in common with S.*

Proof Let T be the set of all epsilon chains beginning with S, all of whose non-initial terms belong to S. Then T, as a non-empty set of linear orderings, has elements of maximal length. Let $[S, \ldots, s]$ be an element of T of maximal length. Then s is the required element of S having no elements in common with S. For if t were an element of both S and s, then the Euclidean Axiom of Foundation would guarantee that t did not already lie in the field of $[S, \ldots, s]$, and therefore that $[S, \ldots, s, t]$ lay in T. That would contradict the maximality of the length of $[S, \ldots, s]$. □

If we define the *ultimate constituents* of a set to be those among its constituents (if any there be) which have no constituents themselves (i.e. are either individuals or the empty set), the we obtain the following as a corollary of Regularity.

8.6.8 Corollary *Every non-empty set has ultimate constituents.*

Proof Any member of the set of all constituents of S which has no

8.6 Epsilon chains and the Euclidean Axiom of Foundation 299

elements in common with that set of constituents must be an ultimate constituent of S. □

We can use the Principle of Regularity to establish an important form of induction.

8.6.9 Theorem *(The Principle of Epsilon Induction) Let Φ be any first level global propositional function of one argument. Then the hypothesis*

$$\forall X[(\forall y \in X)\Phi(y) \text{ implies } \Phi(X)]$$

entails the conclusion

$$\forall X \Phi(X)$$

Proof The hypothesis being granted, the empty set and all individuals satisfy Φ vacuously. Now let S be any non-empty set and let T be the set consisting of all those constituents of S that do *not* satisfy Φ. I claim that T must be empty. For otherwise T must, by Regularity, have an element $X \in T$ such that $X \cap T = \emptyset$. Since every member (if any) of X is a constituent of S it follows that every member of X, as a member of $Constituents(S) - T$, must satisfy Φ. Thus, by the hypothesis of the theorem, X must satisfy Φ. Since this contradicts the definition of T, T must be empty as claimed, and every constituent of S satisfies Φ. In particular, every *element* of S satisfies Φ and therefore, by the hypothesis of the theorem, S itself satisfies Φ, as required. □

9
The Euclidean Theory of Cardinality

9.1 Arithmetical functions and relations

Two ways of regarding natural numbers naturally arise: on the one hand, they name, or correspond to, the various species of Euclidean finite cardinality; on the other, they are what we use to count out the elements of a set. Correspondingly, there are two approaches to natural number arithmetic that readily suggest themselves. The first is to regard that arithmetic as the general theory of the relation of *cardinal equivalence* or *sameness of size*; the second is to regard the natural numbers as the successive terms in a simply infinite sequence

$$0, 1, 2, \ldots, n, n+1, \ldots$$

In Cantorian set theory, these two approaches lead to essentially the same theory, so that what is Euclidean finite is what can be counted out using the terms of a simply infinite system. But, as we shall see, Euclidean set theory holds out the prospect that simply infinite systems may differ in length and, indeed, that there may be no such system sufficiently long to count out the elements of every (Euclidean finite) set.

In this chapter I shall pursue the first alternative, in which the "natural" numbers are taken to be, not particular *objects*, but rather *infinite species*, namely, equivalence classes of the global relation, $\lambda xy[x \cong_c y]$, of cardinal equivalence. On this approach, sets correspond to the numbers (*arithmoi*) of Aristotle and Euclid, except that we allow a *zero* (the empty set) and *ones* (singletons) among our "numbers".

Doing natural number arithmetic in this way has considerable advantages. By dealing directly with sets (the modern *arithmoi*), rather than with "natural" numbers considered as particular "abstract objects", we cut our thinking free of its customary dependence upon imaginary cal-

9.1 Arithmetical functions and relations

culating processes: at the deepest level of our theory we are confronted with questions of what sets there are, rather than with questions of what calculations are possible with natural numbers.

This shift of attention away from possibilities of calculation is essential if we are to make a clear distinction between what is self-evident and what is problematic in the foundations of arithmetic. For, in any case, problems about what calculations are possible "in theory" must necessarily reduce, ultimately, to questions about finiteness, and those questions, when properly formulated, are questions about what sets there are.

The idea behind the approach to natural number arithmetic taken in this chapter is thus that the subject matter of arithmetic is sets, but the propositions of natural number arithmetic, properly so called, treat of sets only insofar as they are examples of their sizes. This can be made precise by means of the notions of *arithmetical function* and *arithmetical propositional function* or *arithmetical relation*, which I shall now define.

9.1.1 Definition

(i) *Let σ be a global function of one argument. Then to say that σ is arithmetical is to say that*

$$\forall xy[x \cong_c y \text{ .implies. } \sigma(x) \cong_c \sigma(y)]$$

(ii) *Let Φ be a global propositional function of one argument. Then to say that Φ is arithmetical is to say that*

$$\forall xy[x \cong_c y \text{ .implies. } [\Phi(x) \text{ iff } \Phi(y)]]$$

These definitions can be extended in the obvious way to global functions, of both types, having more than one argument.

A proposition to the effect that such and such a global function (propositional or ordinary) is arithmetical is a Π_1 proposition, and, as such, cannot be combined with other propositions, global or ordinary, to form compound propositions. Such propositions can be employed in two ways only: they can be asserted, or they can be laid down as hypotheses[1].

In this chapter, then, natural number arithmetic is the study of arithmetical functions and relations defined on sets.

---

[1] See Section A1.2 of Appendix 1.

The Euclidean Theory of Cardinality

9.1.2 Theorem *The following global functions (propositional and ordinary) are arithmetical:*

(i) *The power set function,* $\lambda x P(x)$
(ii) *The Cartesian product,* $\lambda xy(x \times y)$
(iii) *The exponential function,* $\lambda xy(x^y)$
(iv) *The relation of cardinal equivalence (equality of size),* $\lambda xy[x \cong_c y]$
(v) *The cardinal less than relation,* $\lambda xy[x <_c y]$

Proof Suppose that a, b, c and d are given such that $a \cong_c c$ and $b \cong_c d$, where $f : a \to c$ and $g : b \to d$ are bijections witnessing these equivalences.

(1) Define $h : P(a) \to P(c)$ $(1-1$ onto$)$ by setting

$$h`X = \{f`x : x \in X\}$$

for each set $X \subseteq a$.

(2) Define $h : a \times b \to c \times d$ $(1-1$ onto$)$ by setting

$$h`(x, y) = (f`x, g`y)$$

for each $(x, y) \in a \times b$.

(3) Define $h : a^b \to c^d$ $(1-1$ onto$)$ by setting

$$h`x = \{(f`u, g`v) : (u, v) \in x\}$$

for each $x \in a^b$.

(4) It follows immediately from the symmetry and transitivity of $\cong_c$ that $a \cong_c b$ if, and only if, $c \cong_c d$.

(5) If $h : a \to b$ $(1-1$ but not onto$)$, then $g \circ h \circ f^{-1} : c \to d$ $1-1$ but not onto.

$\square$

In arithmetical contexts I shall sometimes write "$s \times_c t$" instead of "$s \times t$" and "$\exp_c(s, t)$" instead of "s^t".

Arithmetical propositional functions are closed under propositional combination and arithmetical bounded quantification.

9.1.3 Theorem *Let Φ and Ψ be arithmetical propositional functions. Then so are*

(i) $\lambda x_1 x_2 x_3 y_1 y_2 y_3 [\Phi(x_1, x_2, x_3)$ and $\Psi(y_1, y_2, y_3)]$
(ii) $\lambda x_1 x_2 x_3 y_1 y_2 y_3 [\Phi(x_1, x_2, x_3)$ or $\Psi(y_1, y_2, y_3)]$
(iii) $\lambda x_1 x_2 x_3 y_1 y_2 y_3 [\Phi(x_1, x_2, x_3)$ implies $\Psi(y_1, y_2, y_3)]$

9.1 Arithmetical functions and relations 303

(iv) $\lambda x_1 x_2 x_3 y_1 y_2 y_3 [\Phi(x_1, x_2, x_3) \text{ iff } \Psi(y_1, y_2, y_3)]$
(v) $\lambda x_1 x_2 x_3 \text{ not } \Phi(x_1, x_2, x_3)$
(vi) $\lambda x_1 x_2 x_3 (\forall z \in \text{card}(x_3)) \, \Phi(x_1, x_2, z)$
(vii) $\lambda x_1 x_2 x_3 (\exists z \in \text{card}(x_3)) \, \Phi(x_1, x_2, z)$

This theorem can, of course, be extended, in obvious ways, to functions with different numbers of arguments. In arithmetical contexts I shall sometimes use the notation "$(\forall x <_c S)$" and "$(\exists x <_c S)$" for "$(\forall x \in \text{card}(S))$" and "$(\exists x \in \text{card}(S))$", respectively.

The standard functions of conventional natural number arithmetic are (or correspond to) global arithmetical functions. By 9.1.2 we already have multiplication and exponentiation. We can easily define the other basic arithmetical functions.

9.1.4 Definition

(i) *Let a be any set. Then by definition*

$$a' = a \cup \{a\}$$

(ii) *Let a and b be any sets. Then by definition*

$$a +_c b = (a \times \{\emptyset\}) \cup (b \times \{\{\emptyset\}\}) \quad (= (a \times \{0\}) \cup (b \times \{1\}))$$

(iii) *Let a and b be any sets. Then by definition*

$$a -_c b = \text{card}(a) - \{x \in \text{card}(a) : (\exists y \in \text{card}(b))[x \cong_c y]\}$$

(iv) *Let a and b be sets. Then by definition*

$$\log_b(a) = \begin{cases} 0 & \text{if } b \leq_c 1 \\ 1 & \text{if } 1 <_c b \text{ and } a \cong_c 0 \\ \max(\{x \in \text{card}(a) : b^x \leq_c a\}) +_c 1 & \text{if } 0 <_c a \end{cases}$$

Here, of course, $0 = \emptyset$, $1 = 0' = \{0\}$, $2 = 1' = \{0, 1\}$ (and similarly for $3, 4, \ldots$). $a +_c b$ is sometimes called the *disjoint union* of a and b. The function $\lambda xy(x -_c y)$ is the analogue of *modified subtraction* in recursion theory: $a -_c b$ corresponds to the ordinary difference if $b <_c a$ and is $0 (= \emptyset)^2$ if $a \leq_c b$. The function $\lambda x \log_b(x)$ can perhaps best be regarded

---

[2] In arithmetical contexts I shall usually write "0" instead of "$\emptyset$" and "1" instead of "$\{\emptyset\}$".

(naively) as giving the number of digits in the card(b)-ary numeral designating the cardinality of its argument[3].

9.1.5 Theorem

(i) $\lambda x(x')$ is arithmetical.
(ii) $\lambda x(x +_c y)$ is arithmetical.
(iii) $\lambda x(x -_c y)$ is arithmetical.
(iv) $\lambda xy(\log_x(y))$ is arithmetical.

Proof Suppose that a, b, c and d are given such that $a \cong_c c$ and $b \cong_c d$, where $f : a \to c$ and $g : b \to d$ are bijections witnessing these equivalences.

(i) Define $h : a' \to c'$ $(1-1$ onto$)$ by setting $h = f \cup \{(a, c)\}$.

(ii) Define $h : a +_c b \to c +_c d$ $(1-1$ onto$)$ by setting

$$h = \{((x, 0), (y, 0)) : (x, y) \in f\} \cup \{((x, 1), (y, 1)) : (x, y) \in g\}$$

(iii) Define $h : a -_c b \to c -_c d$ $(1-1$ onto$)$ by setting

$$h`x = \iota\{y \in c -_c d : x \cong_c y\}$$

for each $x \in a -_c b$.

(iv) If $0 <_c a$, then clearly

$$\max(\{x \in \text{card}(a) : b^x \leq_c a\}) \cong_c \max(\{x \in \text{card}(c) : b^x \leq_c c\})$$

If $0 \cong_c a \cong_c c$, then $\log_b(a) = 1 = \log_b(c)$. $\square$

Addition, multiplication, and exponentiation can be shown to satisfy the familiar recursion equations[4] that in conventional natural number arithmetic are often taken as their definitions.

9.1.6 Theorem

(i) $a +_c 0 \cong_c a$ and $a +_c b' \cong_c (a +_c b)'$
(ii) $a \times_c 0 \cong_c 0$ and $a \times_c b' \cong_c (a \times_c b) +_c a$
(iii) $a^0 \cong_c 1$ and $a^{(b')} \cong_c (a^b) \times_c a$

---

[3] Of course we haven't yet defined those numerals with mathematical precision – this will be done in Section 9.3.
[4] Notice that I am using "$\cong_c$" in place of "$=$" for those equations in 9.1.6.

9.1 Arithmetical functions and relations

Proof (i) To establish the second conjunct define $f : a +_c b' \to (a +_c b)'$ (1 – 1 onto) by setting $f = id(a +_c b) \cup \{((b, 1), a +_c b)\}$.

(ii) To establish the second conjunct define $f : a \times_c b' \to (a \times_c b) + a$ (1–1 onto) by setting $f = \{(x, (x, 0)) : x \in a \times_c b\} \cup \{((x, b), (x, 1)) : x \in a\}$.

(iii) To establish the second conjunct define the function

$$f : a^{b'} \to (a^b) \times_c a \ (1\text{–}1 \text{ onto})$$

by stipulating that, for $y : b \cup \{b\} \to a$,

$$f`y = (y - \{(b, y`b)\}, y`b)$$

□

The basic properties of these familiar functions can also be established by direct calculation in this way.

9.1.7 Theorem *Let a, b and c be any sets. Then*

(i) $(a +_c b) +_c c \cong_c a +_c (b +_c c)$
(ii) $a +_c b \cong_c b +_c a$
(iii) $(a \times_c b) \times_c c \cong_c a \times_c (b \times_c c)$
(iv) $a \times_c b \cong_c b \times_c a$
(v) $a \times_c (b +_c c) \cong_c (a \times_c b) +_c (a \times_c c)$
(vi) $\exp_c(a, b +_c c) \cong_c \exp_c(a, b) \times_c \exp_c(a, c)$
(vii) $\exp_c(a, b \times_c c) \cong_c \exp_c(\exp_c(a, b), c)$

The basic properties of the global function log can also be established in this way.

9.1.8 Theorem *Let $1 <_c a$ and p and q be sets with $0 <_c p, q$.*

(1) $p <_c q$ implies $\log_a(p) \leq_c \log_a(q)$
(2) $a^{(\log_a(p) -_c 1)} \leq_c q \leq_c a^{\log_a(p))} -_c 1$: implies : $\log_a(p) = \log_a(q)$
(3) $(\log_a(p) +_c \log_a(q)) -_c 1 \leq_c \log_a(p \times_c q) \leq_c \log_a(p) +_c \log_a(q)$
(4) $q \times_c (\log_a(p) -_c 1) +_c 1 \leq_c \log_a(p^q) \leq_c q \times_c \log_a(p) +_c 1$

Proof (i) and (ii) follow immediately from the definition of log. To prove (iii) let p and q be given with $\log_a(p) \cong_c m +_c 1$ and $\log_a(q)) \cong_c n +_c 1$. Then

$$\log_a(p \times_c q) \geq_c \log_a(a^m \times_c a^n)$$
$$\cong_c \log_a(a^{m+_c n})$$
$$\cong_c (m +_c n) +_c 1$$
$$\cong_c (\log_a(p) +_c \log_a(q)) -_c 1$$

Conversely, since $p \leq_c a^{m+_c 1} -_c 1$ and $q \leq_c a^{n+_c 1} -_c 1$,

$$\begin{aligned}
\log_a(p \times_c q) &\leq_c \log_a((a^{m+_c 1} -_c 1) \times_c (a^{n+_c 1} -_c 1)) \\
&\cong_c \log_a(a^{m+_c n+_c 2} -_c a^{m+_c 1} -_c a^{n+_c 1} +_c 1) \\
&\leq_c \log_a(a^{m+_c n+_c 2} -_c 1) \\
&\cong_c m +_c n +_c 2 \\
&\cong_c \log_a(p) +_c \log_a(q)
\end{aligned}$$

It only remains to prove (iv). To that end let $\log_a(p) \cong_c m +_c 1$. Then $a^m \leq_c p$ and therefore

$$\begin{aligned}
\log_a(p^q) &\geq_c \log_a((a^m)^q) \\
&\cong_c \log_a(a^{m \times_c q}) \\
&\cong_c \log_a(a^{q \times_c m}) \\
&\cong_c q \times_c m +_c 1 \\
&\cong_c q \times_c (\log_a(p) -_c 1) +_c 1
\end{aligned}$$

Conversely

$$\begin{aligned}
\log_a(p^q) &\leq_c \log_a((a^{m+_c 1})^q) \\
&\cong_c \log_a(a^{(m+_c 1) \times_c q}) \\
&\cong_c \log_a(a^{q \times_c (m+_c 1)}) \\
&\cong_c q \times_c (m +_c 1) +_c 1 \\
&\cong_c q \times_c \log_a(p) +_c 1
\end{aligned}$$

□

9.1.9 Definition *Let σ be a first level global function of one argument, and let a be any object. Then by definition*

(i) $\sum_{x \in a} \sigma(x) = \bigcup \{\sigma(x) \times \{x\} : x \in a\}$
(ii) $\sum_{x < a} \sigma(x) = \sum_{x \in \text{card}(a)} \sigma(x)$
(iii) $\prod_{x \in a} \sigma(x) = \{y \in (\bigcup \{\sigma(x) : x \in a\})^a : (\forall x \in a)[y`x \in \sigma(x)]\}$
(iv) $\prod_{x < a} \sigma(x) = \prod_{x \in \text{card}(a)} \sigma(x)$

Note that on this definition $\sum_{x<0} \sigma(x) = 0$ and $\prod_{x<0} \sigma(x) = 1$.

The *bounded sum* and *bounded product* functions, defined in 9.1.9 (ii) and (iv), respectively, are second level arithmetical functions in the sense given by the following theorem.

9.1 Arithmetical functions and relations

9.1.10 Theorem *Let σ be a first level global function of one argument. Then the hypothesis*

$$\sigma \text{ is arithmetical}$$

entails the conclusion

$$\lambda y \sum_{x<y} \sigma(x) \text{ is arithmetical}$$

and also the conclusion

$$\lambda y \prod_{x<y} \sigma(x) \text{ is arithmetical}$$

Proof Assume as hypothesis that

$$\sigma \text{ is arithmetical}$$

and let a and b be sets with $a \cong_c b$. We need to show that

$$\sum_{x<a} \sigma(x) \cong_c \sum_{x<b} \sigma(x) \text{ and } \prod_{x<a} \sigma(x) \cong_c \prod_{x<b} \sigma(x)$$

The key to defining the bijections that bear witness to the truth of these assumptions is to be found in two obvious facts:

(1) Since $\text{card}(a) \cong_c \text{card}(b)$ there is a (in fact, exactly one) bijective function $f : \text{card}(a) \to \text{card}(b)$ such that

$$(\forall x \in \text{card}(a))[x \cong_c f`x]$$

(2) Since σ is arithmetical

$$(\forall x \in \text{card}(a))[\sigma(x) \cong_c \sigma(f`x)]$$

where f is the unique function given in (1).

Hence, by the Principle of Choice, there is a function h defined on $\text{card}(a)$ such that

$$(\forall x \in \text{card}(a))[h`x : \sigma(x) \to \sigma(f`x) \, (1\text{--}1, \text{ onto})]$$

We can now use these two functions, f and h, to construct the witnessing bijections we require.

First, to show that $\sum_{x<a} \sigma(x) \cong_c \sum_{x<b} \sigma(x)$, define g on $\text{card}(a)$ by stipulating that, for $x \in \text{card}(a)$ and $y \in \sigma(x)$,

$$(g`x)`(y, x) = ((h`x)`y, f`x)$$

Then

$$(\forall x \in \text{card}(a))[g`x : \sigma(x) \times \{x\} \to \sigma(f`x) \times \{f`x\}, (1\text{–}1, \text{ onto })]$$

Since $(\sigma(x) \times \{x\}) \cap (\sigma(y) \times \{y\}) = \emptyset$ whenever $x \neq y$, the bijections $g`x$ for $x \in \text{card}(a)$ have disjoint domains and ranges. It follows that if we set $G = \bigcup\{g`x : x \in \text{card}(a)\}$ then

$$G : \bigcup\{\sigma(x) \times \{x\} : x \in \text{card}(a)\} \to \bigcup\{\sigma(f`x) \times \{f`x\} : x \in \text{card}(a)\}$$

and G is bijective. But since f is bijective

$$\bigcup\{\sigma(f`x) \times \{f`x\} : x \in \text{card}(a)\} = \bigcup\{\sigma(y) \times \{y\} : y \in \text{card}(b)\}$$

Hence the bijection G verifies that $\sum_{x<a} \sigma(x) \cong_c \sum_{x<b} \sigma(x)$.

Now to show that $\prod_{x<a} \sigma(x) \cong_c \prod_{x<b} \sigma(x)$ define a function

$$j : \prod_{x<a} \sigma(x) \to \prod_{x<b} \sigma(x)$$

by stipulating that, for each $y \in \prod_{x<a} \sigma(x)$,

$$j`y : \text{card}(b) \to \bigcup\{\sigma(x) : x \in \text{card}(b)\}$$

is defined, for each $z \in \text{card}(b)$, by

$$(j`y)`z = [(h \circ f^{-1})`z]`[(y \circ f^{-1})`z]$$

Since for each $y \in \prod_{x<a} \sigma(x)$ and each $z \in \text{card}(b)$

$$(y \circ f^{-1})`z \in \sigma((f^{-1})`z)$$

and

$$(h \circ f^{-1})`z : \sigma((f^{-1})`z) \to \sigma(f`(f^{-1})`z) \,(= \sigma(z))$$

it follows that j does, indeed, map $\prod_{x<a} \sigma(x)$ to $\prod_{x<b} \sigma(x)$.

That j is $1-1$ follows from the fact that if $u, v \in \prod_{x<a} \sigma(x)$ and $x \in \text{card}(a)$ with $u`x \neq v`x$, then

$$(j`u)`(f`x) \neq (j`v)`(f`x)$$

as a straightforward calculation will disclose.

That j is onto follows from the fact that if $u \in \prod_{x<b} \sigma(x)$, and we define y by stipulating that for $x \in \text{card}(a)$

$$y`x = [(h \circ x)^{-1}]`[(u \circ f)`x]$$

then, again, a straightforward calculation shows that $y \in \prod_{x<a} \sigma(x)$ and $j`y = u$.

9.1 Arithmetical functions and relations

Thus j is a bijection and $\prod_{x<a} \sigma(x) \cong_c \prod_{x<b} \sigma(x)$, as required.

□

It is not difficult to show that the bounded sum and product functions satisfy the expected recursion equations.

9.1.11 Theorem *Let σ be an arithmetical function of one argument. Then the functions $\lambda y \sum_{x<y} \sigma(x)$ and $\lambda y \prod_{x<y} \sigma(x)$ satisfy the following recursion equations:*

$$\sum_{x<0} \sigma(x) \cong_c 0 \quad \sum_{x<a'} \sigma(x) \cong_c \left(\sum_{x<a} \sigma(x)\right) +_c \sigma(a)$$

$$\prod_{x<0} \sigma(x) \cong_c 1 \quad \prod_{x<a'} \sigma(x) \cong_c \left(\prod_{x<a} \sigma(x)\right) \times_c \sigma(a)$$

Now that the basic properties of bounded sums and products have been established it follows that we can parallel the definitions of functions in Grzegorczyk class $\mathscr{E}^3$ by definitions for corresponding arithmetical functions in Euclidean set theory[5].

I now need to introduce a series of definitions for arithmetical functions which I will call *logarithmic exponentials*. These will be needed in Chapter 11 when we take up the study of simply infinite systems in Euclidean set theory.

9.1.12 Definition *Let a, b, and c be any sets. Then by definition*

(i) $\text{lexp}_1(a, b, c) = \exp(b, \log_a(c))$
(ii) *Suppose that the function lexp_n has been defined. Then*

$$\text{lexp}_{n+1}(a, b, c) = \exp(b, \text{lexp}_n(a, \log_a(c), \log_a(c)))$$

Definition 9.1.12 is not so much a definition as a set of instructions for carrying out a series of definitions. For the bold face letter "**n**" cannot logically be construed as ranging over "natural numbers" here. In fact it goes proxy for a concrete numeral.

The generality indicated here is not *theoretical* generality with respect to "natural numbers" – we have not yet determined what such "things" might be – but rather *practical* generality with respect to actual inscriptions and the actual laying down of definitions of particular functions.

---

[5] See H. E. Rose, *Subrecursion: Functions and Hierarchies*, p. 3 for the definition of this class in conventional natural number arithmetic.

The question now naturally arises as to the status of the principles of proof by induction and of definition by recursion in arithmetic done in this neo-Euclidean style. In the case of proof by induction, there is a quite straightforward answer, and it is given by the next theorem.

9.1.13 Theorem (*The Principle of Mathematical Induction*) *Let Φ be a first level global propositional function of one argument. Then the hypotheses*

  (i) *Φ is arithmetical*
  (ii) *$\Phi(0)$*
  (iii) *$\forall x[\Phi(x)$ implies $\Phi(x')]$*

together entail the conclusion $\forall x \Phi(x)$.

Proof Since Φ is arithmetical, hypothesis (iii) entails

$$\forall xy[\Phi(x) \text{ implies } \Phi(x \cup \{y\})]$$

The required result now follows by one-point-extension induction (Theorem 8.3.3). □

The problem of definition by recursion is much less straightforward. I shall take it up in the next section.

9.2 Limited recursion

Let us suppose that γ and χ are arithmetical functions of **n** and **n + 2** arguments[6], and consider the global recursion equations

$$\varphi(\emptyset, \vec{b}) \cong_c \gamma(\vec{b})$$
$$\varphi(a \cup \{a\}, \vec{b}) \cong_c \chi(a, \varphi(a, \vec{b}), \vec{b})$$

These are just the familiar equations for *primitive recursion* expressed in terms suitable to Euclidean arithmetic. In these equations the function γ is the *initial function* and the function χ is the *recursion function*.

The question we must now address is whether, and in what circumstances, these equations can be "solved" for a global function φ. The most general answer that can be given is that they can be solved for φ when the equations are *bounded locally* by a previously given global function. Just what this means I shall now explain.

---

[6] In the equations, I use the notation "$\vec{b}$" to stand for a list $b_1, \ldots, b_n$, of free variables, where "**n**" goes proxy for a concrete numeral.

9.2 Limited recursion

9.2.1 Definition *Let γ, χ, and θ be arithmetical functions of $\mathbf{n}$, $\mathbf{n}+2$, and $\mathbf{n}+1$ arguments respectively. Then*

(i) [7]*f is said to satisfy the recursion equations determined by γ (as initial function) and χ (as recursion function) on a at the arguments $b_1, \ldots, b_n$ $(= \vec{b})$ if, and only if, f is a function defined on $\operatorname{card}(a)$ and*

   (a) $f`\emptyset = \gamma(\vec{b})$.
   (b) $(\forall u, v \in \operatorname{card}(a))[v \cong_c u +_c 1 \text{ implies } f`v = \chi(u, f`u, \vec{b})]$.

(ii) [8]*θ is said to bound the recursion equations determined by γ (as initial function) and χ (as recursion function) locally if, and only if, for all x, $y_1, \ldots, y_n$ $(= \vec{y})$, and z, if z satisfies the recursion equations determined by γ (as initial function) and χ (as recursion function) on x at the arguments $y_1, \ldots, y_n$, then*

   (a) $\gamma(\vec{y}) <_c \theta(\emptyset, \vec{y})$.
   (b) $(\forall u \in \operatorname{card}(x))[\chi(u, z`u, \vec{y}) <_c \theta(u, \vec{y})]$.

When a system of recursion equations is bounded in this way it can be construed as defining an arithmetical function which is unique up to cardinal equivalence. That is the import of the following theorem.

9.2.2 Theorem *(The Principle of Definition by Limited Recursion) Let γ and χ be arithmetical functions of $\mathbf{n}$ and $\mathbf{n}+2$ arguments respectively, and let θ be an arithmetical function of $\mathbf{n}+1$ arguments which bounds the recursion equations determined by γ (as initial function) and χ (as recursion function) locally. Then we can define an arithmetical function φ such that*

(i) $\forall x \vec{y} [\varphi(\emptyset, \vec{y}) \cong_c \gamma(\vec{y}) \text{ and } \varphi(x \cup \{x\}, \vec{y}) \cong_c \chi(x, \varphi(x, \vec{y}), \vec{y})]$
(ii) $\forall x \vec{y} [\varphi(x, \vec{y}) <_c \theta(x, \vec{y})]$
(iii) *The two hypotheses*

   (a) ψ *is arithmetical*
   (b) $\forall x \vec{y} [\psi(\emptyset, \vec{y}) \cong_c \gamma(\vec{y}) \text{ and } \psi(x \cup \{x\}, \vec{y}) \cong_c \chi(x, \psi(x, \vec{y}), \vec{y})]$

   entail the conclusion

   (c) $\forall x \vec{y} [\varphi(x, \vec{y}) \cong_c \psi(x, \vec{y})]$

---

[7] This definition is local.
[8] This definition is Π_1.

Proof The primary task is to define $\lambda x \vec{y} \varphi(x, \vec{y})$ uniformly in the indicated $\mathbf{n+1}$ variables. I shall (in effect) define the value $\varphi(a, \vec{b})$ by a local primitive recursion along the linear ordering ord(a).

Recall that ord(a) (Definition 8.5.3) is the linear ordering whose field is the local cardinal, card(a), of a and whose ordering reflects the sizes of the elements of card(a), so that this ordering is the restriction of $\lambda xy[x \leq_c y]$ to card(a) × card(a).

Now the central difficulty in the argument is this: we need to apply definition by (local) primitive recursion (Theorem 8.4.10) along ord(a) to obtain a *local* function f which satisfies the recursion equations determined by γ and χ on a at the arguments $\vec{b}$. But to apply local primitive recursion we need a *target set*, S, and, instead of the *global* recursion function χ, a *local* recursion function

$$h : \text{card}(a) \times S \to S$$

since $Field(\text{ord}(a)) = \text{card}(a)$.

Of course, the choices of S and h here depend, not only on a, but on the parameters $\vec{b}$ as well. The problem, then, is to define the target set, S, and the local recursion function, h, as (global) functions of the arguments $a, \vec{b}$. Needless to say, the bounding function θ plays a crucial role in these definitions.

The (variable) target set, S, can be given by a global function σ defined by setting

$$\sigma(a, \vec{b}) = card\left(\sum_{x<a} \theta(x, \vec{b})\right)$$

To make my notation more perspicuous let me stipulate that

$$\text{ord}(a) = [S_0, \ldots, S_k]$$

where $S_0 = \emptyset$, $S_{i+1} \cong_c S_i \cup \{S_i\}$, and $S_k \cup \{S_k\} \cong_c a$. Then for every S_i there is, for each $y <_c \theta(S_i, \vec{b})$, exactly one $x \in \sigma(a, \vec{b})$ such that $x \cong_c y$.

The (variable) recursion function, h, can now be given by a global function η defined by stipulating that

$$\eta(a, \vec{b}) : \text{card}(a) \times \sigma(a, \vec{b}) \to \sigma(a, \vec{b})$$

and satisfies

$$\eta(a, \vec{b})'(x, y) = \iota(\{z \in \sigma(a, \vec{b}) : z \cong_c \chi(x, y, \vec{b})\}, \emptyset)$$

9.2 Limited recursion

so that

$$\eta(a,\vec{b})`(x,y) = \begin{cases} \chi(x,y,\vec{b}) & \text{if } (\exists z \in \sigma(a,\vec{b}))[z \cong_c \chi(x,y,\vec{b})] \\ \emptyset & \text{otherwise} \end{cases}$$

By giving $\eta(a,\vec{b})$ the default value $\emptyset$ (which is in $\sigma(a,\vec{b})$) we guarantee that $\eta(a,\vec{b})$: card$(a) \times \sigma(a,\vec{b}) \to \sigma(a,\vec{b})$ when applying local primitive recursion. As we shall see below, this default value never arises in the course of "calculating" with the recursion equations.

By local primitive recursion (Theorem 8.4.10) along the linear ordering ord(a), we can define an auxiliary global function ψ so that $\psi(a,\vec{b})$: card$(a) \to \sigma(a,\vec{b})$ and satisfies the recursion equations

$$\psi(a,\vec{b})`S_0 = \gamma(a,\vec{b})$$
$$\psi(a,\vec{b})`S_{i+1} = \eta(a,\vec{b})`(S_i, \psi(a,\vec{b})`S_i), \text{ for } i \neq k$$

(Note that these equations employ identity, $=$, rather than cardinal equivalence, $\cong_c$.)

Claim For all $x \in \text{card}(a)$

$$\psi(a,\vec{b})`x \cong_c \gamma(\vec{b}) \text{ if } x = \emptyset (= S_0)$$
$$\psi(a,\vec{b})`x \cong_c \chi(x, \psi(a,\vec{b})`S_i, \vec{b}) \text{ if } x = S_{i+1} \text{ and } i \neq k$$
$$\psi(a,\vec{b})`x <_c \theta(x,\vec{b})$$

Pf The proof proceeds by one-point-extension induction with respect to a.

Basis case ($a = \emptyset$) The claim holds vacuously in this case since

$$\text{card}(a) = \sigma(a,\vec{b}) = \psi(a,\vec{b}) = \emptyset.$$

Induction case ($a' = a \cup \{a\}$) In this case we suppose, as hypothesis of induction, that for all $x \in \text{card}(a)$

$$\psi(a,\vec{b})`x \cong_c \gamma(\vec{b}) \text{ if } x = \emptyset (= S_0)$$
$$\psi(a,\vec{b})`x \cong_c \chi(x, \psi(a,\vec{b})`S_i, \vec{b}) \text{ if } x = S_{i+1} \text{ and } i \neq k$$
$$\psi(a,\vec{b})`x <_c \theta(x,\vec{b})$$

and we must establish that

$$\psi(a',\vec{b})`x \cong_c \gamma(\vec{b}) \text{ if } x = \emptyset (= T_0)$$
$$\psi(a',\vec{b})`x \cong_c \chi(x, \psi(a',\vec{b})`T_i, \vec{b}) \text{ if } x = T_{i+1} \text{ and } i \neq k+1$$
$$\psi(a',\vec{b})`x <_c \theta(x,\vec{b})$$

where, in conformity with the earlier notation in which
$$\mathrm{ord}(a) = [S_0, \ldots, S_k]$$
we suppose that
$$\mathrm{ord}(a') = [T_0, \ldots, T_{k+1}]$$
where $T_i \cong_c S_i$ ($i = 1, \ldots, k$) and $T_{k+1} \cong_c a$.

Since the hypothesis of induction tells us that $\psi(a, \vec{b})$ satisfies the arithmetical recursion equations on $\mathrm{card}(a)$, we may conclude (from the assumption of the theorem that θ bounds those equations) that

(a) $\gamma(\vec{b}) <_c \theta(\emptyset, \vec{b})$

and

(b) If $\mathrm{card}(a) \neq \emptyset$, then $\chi(S_k, \psi(a, \vec{b})`S_k, \vec{b}) <_c \theta(a, \vec{b})$

If $a = \emptyset$, then $a' = \{\emptyset\}$ and $\mathrm{ord}(a') = [\emptyset]$. Moreover

$$\begin{aligned}
\sigma(a', \vec{b}) &= \mathrm{card}\left(\sum_{x < \{\emptyset\}} \theta(x, \vec{b})\right) \\
&= \mathrm{card}\left(\bigcup \{\theta(x, \vec{b}) \times \{x\} : x \in \{\emptyset\}\}\right) \\
&\cong_c \theta(\emptyset, \vec{b})
\end{aligned}$$

Since by (a) $\gamma(\vec{b}) <_c \theta(\emptyset, \vec{b})$, $(\exists z \in \sigma(a', \vec{b}))[z \cong_c \gamma(\vec{b})]$ so
$$\psi(a', \vec{b})`\emptyset \cong_c \gamma(\vec{b}) <_c \theta(\emptyset, \vec{b})$$
as required. The other recursion equation holds vacuously.

The case in which $a \neq \emptyset$ must be treated differently. From the size relations between the S_is and T_is, and since θ is arithmetical, we may conclude that
$$\sigma(a', \vec{b}) \cong_c \sigma(a, \vec{b}) + \theta(a, \vec{b})$$
This fact enables us to deduce (by a straightforward induction along $\mathrm{card}(a)$) that, for all $S_i \in \mathrm{card}(a)$,
$$\psi(a, \vec{b})`S_i \cong_c \psi(a', \vec{b})`T_i$$
Now the hypothesis of induction on a tells us that for all $x \in \mathrm{card}(a)$

$\psi(a, \vec{b})`x \cong_c \gamma(\vec{b})$, if $x = \emptyset (= S_0)$
$\psi(a, \vec{b})`x \cong_c \chi(x, \psi(a, \vec{b})`S_i, \vec{b})$ if $x = S_{i+1}$ and $i \neq k$
$\psi(a, \vec{b})`x <_c \theta(x, \vec{b})$

9.2 Limited recursion

From this it follows (since we are assuming θ bounds the recursion equations locally, as a hypothesis of the theorem) that

$$\chi(S_k, \psi(a,\vec{b})`S_k, \vec{b}) <_c \theta(a,\vec{b})$$

and therefore (since both χ and θ are assumed by the hypotheses of the theorem to be arithmetical, and since $\psi(a,\vec{b})$ and $\psi(a',\vec{b})$ agree on card(a))

$$\chi(T_k, \psi(a',\vec{b})`T_k, \vec{b}) <_c \theta(a',\vec{b})$$

It follows that

$$(\exists z \in \sigma(a',\vec{b}))[z \cong_c \chi(T_k, \psi(a',\vec{b})`T_k, \vec{b})]$$

and therefore that

$$\psi(a',\vec{b})`T_{k+1} = \eta(a,\vec{b})`(T_k, \psi(a',\vec{b})`T_k) \cong_c \chi(T_k, \psi(a',\vec{b})`T_k, \vec{b})$$

But since, by hypothesis of induction on a, for all $x \in$ card(a)

$$\psi(a,\vec{b})`x \cong_c \gamma(\vec{b}) \text{ if } x = \emptyset (= S_0)$$
$$\psi(a,\vec{b})`x \cong_c \chi(x, \psi(a,\vec{b})`S_i, \vec{b}) \text{ if } x = S_{i+1} \text{ and } i \neq k$$
$$\psi(a,\vec{b})`x <_c \theta(x,\vec{b})$$

and since for all $S_i \in$ card(a)

$$\psi(a,\vec{b})`S_i \cong_c \psi(a',\vec{b})`T_i$$

we may conclude that for all $x \in$ card(a')

$$\psi(a',\vec{b})`x \cong_c \gamma(\vec{b}) \text{ if } x = \emptyset (= T_0)$$
$$\psi(a',\vec{b})`x \cong_c \chi(x, \psi(a',\vec{b})`S_T, \vec{b}) \text{ if } x = T_{i+1} \text{ and } i \neq k+1$$
$$\psi(a',\vec{b})`x <_c \theta(x,\vec{b}) \text{ if } x \neq k+1$$

Thus $\psi(a',\vec{b})$ satisfies the recursion equations on card(a'), and since θ locally bounds those equations

$$\psi(a',\vec{b})`T_{k+1} \cong_c \chi(T_{k+1}, \psi(a',\vec{b})`T_{k+1}, \vec{b}) <_c \theta(a',\vec{b})$$

as well. This completes the proof of the induction step of the claim and the latter now follows by one-point-extension induction.

We may now define the global function φ required by the theorem by setting

$$\varphi(a,\vec{b}) = \psi(a \cup \{a\}, \vec{b})`Last(\text{ord}(a \cup \{a\}))$$

It is clear that φ so defined is arithmetical. For suppose that $a \cong_c c$ and $\vec{b} \cong_c \vec{d}$ (i.e. $b_1 \cong_c d_1, \ldots, b_n \cong_c d_n$). Then, where

$$\text{ord}(a') = [S_0, \ldots, S_{k+1}] \text{ and ord}(c') = [T_0, \ldots, T_{k+1}]$$

we have (since $S_i \cong_c T_i$ for $i = 1, \ldots, k+1$) that

$$\psi(a', \vec{b})`S_{k+1} \cong_c \psi(c', \vec{d})`T_{k+1}$$

and we may therefore conclude that

$$\varphi(a, \vec{b}) = \psi(a', \vec{b})`S_{k+1} \cong_c \psi(c', \vec{d})`T_{k+1} = \varphi(c, \vec{d})$$

as required.

Part (iii) can be established by a straightforward one-point-extension induction. $\square$

I have already observed that the definition schemata for the functions of class $\mathscr{E}^3$ in the Grzegorczyk hiearchy can be derived as definition schemata for arithmetical functions. In Grzegorczyk's own treatment[9] all the classes $\mathscr{E}^n$ are defined to be the closure of a certain class of initial functions under the schema of limited recursion. Thus if g, and h, lie in $\mathscr{E}^n$, and if f is defined by the recursion equations

$$f(x_1, \ldots, x_m, 0) = g(x_1, \ldots, x_m)$$
$$f(x_1, \ldots, x_m, y+1) = h(x_1, \ldots, x_m, y, f(x_1, \ldots, x_m, y))$$

then f lies in $\mathscr{E}^n$ as well, provided there is a function j lying in $\mathscr{E}^n$ for which the inequality

$$f(x_1, \ldots, x_m, y) < j(x_1, \ldots, x_m, y)$$

holds generally[10].

But if we compare this with the theorem on limited recursion in Euclidean arithmetic (Theorem 9.2.2) we see that the logical role of the limiting function θ in that theorem is quite distinct from that of the function j in the condition given above.

In the theorem it is the function θ that *guarantees* that the recursion is efficacious, i.e. that there is a function satisfying the recursion equations. But in the conventional account of limited recursion just given, the function j simply serves to admit the function f to the appropriate Grzegorczyk hierarchy class by bounding it, *whereas the existence of f is regarded as already guaranteed by the recursion equations on their*

---

[9] In his 1953 paper "Some classes of recursive functions".
[10] H. E. Rose, *Subrecursion: Functions and Hierarchies*, p.33.

own. For the conventional view on these matters is that the recursion equations simply supply instructions for carrying out actual calculations (*in principle*, of course), so there can't be any question of there not being a function defined here[11].

9.3 S-ary decompositions and numerals

We need now to generalise familiar facts of decimal notation to the general theory of *S-ary decompositions* of sets and *S-ary numerals*, where S is a *base* for numeration in the sense of the following Definition 9.3.1. We need this theory for the treatment of simply infinite systems in Chapter 11.

9.3.1 Definition *Let S be a set. Then by definition*

$$S \text{ is a (numeration) base}$$

if, and only if,

  (i) $1 <_c S$
  (ii) $(\forall X \in S)[X \text{ is a set and } X <_c S]$
  (iii) $(\forall XY \in S)[X \neq Y \text{ implies } X \not\cong_c Y]$

So a base for numeration is a set of sets of different sizes, each smaller in size than the base itself. In fact we can show that a base S contains exactly one set of each cardinality less than the cardinality of S itself.

9.3.2 Theorem *Let S be a base for numeration and let T be any set with $T <_c S$. Then there is a unique element $X \in S$ such that $X \cong_c T$.*

Proof Let S be a numeration base, and let $\leq_S$ be the natural linear ordering on S based on the cardinalities of its members. Then since $\text{card}(S) \cong_c S$, $\text{ord}(S) \cong_o \leq_S$, and there is a unique order isomorphism $f : \text{card}(S) \to S$. It now follows, by induction along $\text{ord}(S)$, that $f`x \cong_c x$, for all $x \in \text{card}(S)$. The theorem now follows from the corresponding theorem for $\text{card}(S)$ (Theorem 8.5.4).

□

---

[11] Of course by means of suitable codings, an analogue of Theorem 9.2.2 could be proved in conventional natural number arithmetic, but that does not affect the point I am making here which concerns underlying *foundational* attitudes.

I shall now show how, given a numeration base S, to define *S-ary numerals* analogous to the familiar decimal numerals of simple arithmetic.

Note that there are numeration bases of all possible sizes greater than 1, since given any set, S, with two or more elements, the local cardinal, card(S), of S is a numeration base of the same size as S. Note also that every von Neumann natural number beyond 1 is a numeration base.

The utility of S-ary numerals rests on the possibility of S-ary *decompositions* in the sense of the next definition.

9.3.3 Definition *Let S a base for numeration and T be any set. An S-ary decomposition of T is defined to be a sequence $\langle T_0, \ldots, T_k \rangle$ such that*

(i) *For each $i \in \{0, \ldots, k\}$*
   (a) $T_i \subseteq P(T)$
   (b) $(\forall X \in T_i)[X \cong_c S^{\{0,\ldots,i-1\}}]$
   (c) $T_i <_c S$
(ii) $T_k \neq \emptyset$
(iii) *For all $i, j \in \{0, \ldots, k\}$, $(\forall U \in T_i)(\forall V \in T_j)[U \neq V$ implies $U \cap V = \emptyset]$*
(iv) $T = \bigcup\bigcup\{T_i : i \in \{0, \ldots, k\}\}$

Recall that a sequence $\langle T_0, \ldots, T_k \rangle$ is an ordered pair $(f, [0, \ldots, k])$, where $[0, \ldots, k]$ is a linear ordering and f is a function whose domain is the field, $\{0, \ldots, k\}$, of $[0, \ldots, k]$, and whose value, $f\text{'}i$, at i is T_i, for all $i \in \{0, \ldots, k\}$.

Here, of course, the symbols $0, \ldots, k$ do not denote natural numbers. Whether there are such things in Euclidean set theory, even in the attenuated sense in which they exist in the Cantorian theory, is not yet determined. No, these symbols are intended to denote the quite arbitrary objects that form the field of $[0, \ldots, k]$, and other notation employed in 9.3.3 (e.g. "$\{0, \ldots, i-1\}$") is intended to be understood in a similar manner[12]. An S-ary decomposition of T thus determines a partition of T into ones, S's, S^2's, S^3's, etc., just as a conventional decimal decomposition of T determines a partition of T into ones, tens, hundred, thousands, etc., in the familiar way.

Of course it is well known that every Euclidean finite set admits of a decimal decomposition. Moreover, any two such decompositions of the same set are "equivalent", in the obvious sense. The same applies to S-ary decompositions. We need to formulate all this precisely and prove it.

---

[12] In particular, "$\{0, \ldots, 0-1\}$" stands for the empty set.

9.3 S-ary decompositions and numerals

9.3.4 Definition Let S be a base for numeration and let $\langle U_0, \ldots, U_k \rangle$ and $\langle V_0, \ldots, V_m \rangle$ be S-ary decompositions of the sets U and V, respectively. Then by definition

$$\langle U_0, \ldots, U_k \rangle \cong_{S\text{-ary}} \langle V_0, \ldots, V_m \rangle$$

if, and only if,

(i) $[0, \ldots, k] \cong_o [0, \ldots, m]$
(ii) If $f : \{0, \ldots, k\} \to \{0, \ldots, m\}$ is the unique order isomorphism from the linear ordering $[0, \ldots, k]$ to the linear ordering $[0, \ldots, m]$, then for all $i \in \{0, \ldots, k\}$, $U_i \cong_c V_{f^i}$.

The expression

$$\langle U_0, \ldots, U_k \rangle \cong_{S\text{-ary}} \langle V_0, \ldots, V_m \rangle$$

may be read

$\langle U_0, \ldots, U_k \rangle$ and $\langle V_0, \ldots, V_m \rangle$ are equivalent as S-ary decompositions.

It is obvious that sets having equivalent S-ary decompositions are of the same cardinality. Moreover, if two sets have the same cardinality and one of them has an S-ary decomposition, then the other has an equivalent S-ary decomposition. It remains to be shown that S-ary decompositions of a given set are unique up to equivalence, and that every set has an S-ary decomposition relative to every numeration base S.

9.3.5 Definition Let S be a numeration base. Then an S-ary numeral is a sequence $\langle d_0, \ldots, d_k \rangle$ such that

(i) $(\forall i \in \{0, \ldots, k\})[d_i \in S]$
(ii) $d_k \neq \emptyset$

Note that the order in which the "digits" occur here is the reverse of the usual one, so that if $S = \{0, 1, 2, 3, 4, 5, 6, 7, 8, 9\}$ is a ten which is a numeration base (where $0 = \emptyset$, 1 is a singleton, 2 a pair, etc.) then any S-ary numeral of the form $\langle 0, 4, 6, 9 \rangle$ corresponds to the ordinary decimal numeral 9640. In our notation, as in Arabic but not in English, decimal numerals are read in the natural order: first units, then tens, then hundreds, and so on.

Of course we are not interested in S-ary numerals as particular, individual sequences, but rather as representatives of their equivalence classes under the obvious equivalence relation induced on the species of all these numerals by the global equivalence relation $\cong_o$ on the species of their associated linear orderings.

9.3.6 Definition Let S be a numeration base and let $\langle d_0,\ldots,d_k\rangle_{r_1}$ and $\langle c_0,\ldots,c_m\rangle_{r_2}$ be S-ary numerals, based on the linear orderings $r_1 = [0,\ldots,k]_{r_1}$ and $r_2 = [0,\ldots,m]_{r_2}$, respectively. Then by definition

$$\langle d_0,\ldots,d_k\rangle \cong_{S\text{-ary}} \langle c_0,\ldots,c_m\rangle$$

if, and only if,

(i) $r_1 \cong_o r_2$
(ii) *If* $f : \{0,\ldots,k\}_{r_1} \to \{0,\ldots,m\}_{r_2}$ *is the unique order isomorphism from* $r_1 = [0,\ldots,k]_{r_1}$ *to* $r_2 = [0,\ldots,m]_{r_2}$, *then* $f`k = m$ *and* $d_i = c_{f`i}$, *for all* $i \in \{0,\ldots,k\}_{r_1}$.

By analogy with the previous case, "$\langle d_0,\ldots,d_k\rangle \cong_{S\text{-ary}} \langle c_0,\ldots,c_m\rangle$" may be read "$\langle d_0,\ldots,d_k\rangle$ and $\langle c_0,\ldots,c_m\rangle$ are equivalent as S-ary numerals". We want to use S-ary numerals to stand for numerical *species*: that is the object of the next definition.

9.3.7 Definition Let S be a base for numeration, T be a set, and $\langle d_0,\ldots,d_k\rangle$ be an S-ary numeral. Then by definition

$$T \text{ is a } \langle d_0,\ldots,d_k\rangle$$

if, and only if, there is an S-ary decomposition, $\langle T_0,\ldots,T_k\rangle$, *of T based on* $[0,\ldots,k]$ *such that* $T_i \cong_c d_i$, *for all* $i \in \{0,\ldots,k\}$.

Clearly, if T is a $\langle d_0,\ldots,d_k\rangle$ and $\langle d_0,\ldots,d_k\rangle \cong_{S\text{-ary}} \langle c_0,\ldots,c_m\rangle$, then T is a $\langle c_0,\ldots,c_m\rangle$. Moreover

9.3.8 Theorem Let S be a base for numeration, let $\langle d_0,\ldots,d_n\rangle$ be an S-ary numeral, and let T_0 and T_1 be sets. If both T_0 and T_1 are $\langle d_0,\ldots,d_n\rangle$'s, then $T_0 \cong_c T_1$.

What we need now is a sample set of "size" $\langle d_0,\ldots,d_k\rangle$ for each S-ary numeral $\langle d_0,\ldots,d_k\rangle$. The most familiar and straightforward way to do this is to take the set of smaller numerals as the sample set. Of course there is no such *set*; I am speaking modulo $\cong_{S\text{-ary}}$, so to speak, so I must pick representatives of the equivalence classes of the global relation $\cong_{S\text{-ary}}$ in some suitable way. I shall do this by defining the set, $Num(S,r)$, of S-ary numerals definable using a given linear ordering r. The idea is to take $Num(S,r)$ to be the set of S-ary numerals *defined on initial segments of r*, that is to say, S-ary numerals (f,r_0), where the linear ordering r_0 is an initial segment of r.

9.3 S-ary decompositions and numerals

9.3.9 Definition *Let S be a numeration base and let* $r = [0,\ldots,k]$ *be a linear ordering. Then by definition*

$$Num(S,r) = \bigcup\{\{(g, [0,\ldots,i]) : g \in \{f \in S^{\{0,\ldots,i\}} : f \text{ is an S-ary numeral}\}\} : i \in \{0,\ldots,k\}\}$$

We now need to define the natural ordering on $Num(S,r)$.

9.3.10 Definition *Let S be a base for numeration and let* $r = [0,\ldots,k]$ *be a linear ordering. Then by definition*

$$\leq_{Num(S,r)} = \{(x,y) \in Num(S,r) \times Num(S,r) : \Phi(x,y)\}$$

where $\Phi(a,b)$ *if, and only if,*

(i) $a = b$ or
(ii) $a \neq b$ and $lh(a) <_o lh(b)$ or
(iii) $a \neq b$ and $lh(a) = lh(b)$ and there exists $u \in lh(a)$ such that

1.) $(\forall v \in Field(lh(a)))[u <_{lh(a)} v \text{ implies } (a)_v = (b)_v]$ and
2.) $(a)_u <_c (b)_u$.

Thus if $S = \{0', 1', 2', 3', 4', 5', 6', 7', 8', 9'\}$ is a ten which is a numeration base (so that $0' = \emptyset$, $1'$ is a singleton, $2'$ a pair, etc.), then the S-ary numeral $\langle 9', 2', 1', 4', 5' \rangle$ (i.e. 54129 in conventional notation), defined on the linear ordering $[0,1,2,3,4]$, say, precedes $\langle 7', 1', 3', 4', 5' \rangle$ (i.e. 54317 in conventional notation) also defined on $[0,1,2,3,4]$; this follows by 9.3.10 since both numerals are defined on $[0,1,2,3,4]$ and agree on $v >_r u = 2$, but

$$(\langle 9', 2', 1', 4', 5' \rangle)_u = 1' <_c 3' = (\langle 7', 1', 3', 4', 5' \rangle)_u.$$

It is obvious that if S is a numeration base, $[0,\ldots,k]$ is a linear ordering, a and b both lie in $Num(S,[0,\ldots,k])$, and neither is shorter than the other, then they are both defined over the same initial segment $[0,\ldots,j]$, of $[0,\ldots,k]$. If $a \neq b$ then there must be a last place at which they differ, and that will determine, via clause (iii) of 9.3.10, which of them precedes the other in the ordering relation $\leq_{Num(S,r)}$. This observation proves the following theorem.

9.3.11 Theorem *Let S be a numeration base, and r a linear ordering. Then* $\leq_{Num(S,r)}$ *is a linear ordering whose field is the set,* $Num(S,r)$, *of S-ary numerals defined on initial segments of r.*

We can now define a sample set of size $\langle d_0,\ldots,d_k\rangle$ for every S-ary numeral $\langle d_0,\ldots,d_k\rangle$, namely, the set of all S-ary numerals based on the same linear ordering, $[0,\ldots,k]$, that are strictly smaller than $\langle d_0,\ldots,d_k\rangle$.

9.3.12 Definition Let S be a numeration base and $n = \langle d_0,\ldots,d_k\rangle$ be an S-ary numeral. Then by definition

$$[\ldots n)_S = \{x \in Num(S,[0,\ldots,k]) : x <_{Num(S,[0,\ldots,k])} n\}$$

"$[\ldots n)_S$" should be read "the S-ary numerals up to n". I shall omit the subscript S if it is clear from the context.

Next we need to establish the most important facts about these S-ary numerals, the facts that allow us to use them for "counting out" sets, and for naming their sizes.

9.3.13 Theorem Let S be a numeration base and $n = \langle d_0,\ldots,d_k\rangle$ be a S-ary numeral. Then

(i) $[\ldots \langle d_0,\ldots,d_k\rangle)$ is a $\langle d_0,\ldots,d_k\rangle$

(ii) $[\ldots \langle d_0,\ldots,d_k\rangle) \cong_c \sum\limits_{i \in \{0,\ldots,k\}} d_i \times S^{\{0,\ldots,i-1\}}$

Proof To simplify the argument let us identify all the numerals in $[\ldots \langle d_0,\ldots,d_k\rangle)$ with finite sequences defined over the whole of $[0,\ldots,k]$ by allowing sequences which terminate in a tail of 0's. Under this identification, for each $i \in \{0,\ldots,k\}$ and each $x \in S$ let

$$D_{ix} = \{\langle x_0,\ldots,x_{i-1},x,d_{i+1},\ldots,d_k\rangle : x_0,\ldots,x_{i-1} \in S\}$$

Then

(1) For each $i \in \{0,\ldots,k\}$ and each $x \in S$ with $x <_c d_i$

$$D_{ix} \subseteq [\ldots \langle d_0,\ldots,d_k\rangle)$$

(For $i = k$ and $x = 0$ this depends upon the identification laid down at the beginning of the proof.)

(2) For each $i,j \in \{0,\ldots,k\}$ and each $x,y \in S$ with $x <_c d_i$ and $y <_c d_j$

$$i \neq j \text{ or } x \neq y \; : \text{implies} : \; D_{ix} \cap D_{jy} = \emptyset$$

(3) For each $i \in \{0,\ldots,k\}$ and each $x \in S$, $D_{ix} \cong_c S^{\{0,\ldots,i-1\}}$.

(4) For each $y \in [\ldots \langle d_0,\ldots,d_k\rangle)$ there is an $i \in \{0,\ldots,k\}$ and an $x \in S$ with $x <_c d_i$ such that $y \in D_{ix}$.

Now, for each $i \in \{0,\ldots,k\}$ define

$$D_i = \{D_{ix} : x <_c d_i\}$$

Then

(5) $D_i \cong_c d_i$, for all $i \in \{0,\ldots,k\}$.

From properties (1) – (4) we may immediately conclude that $\langle D_0,\ldots,D_k\rangle$ is an S-ary decomposition of $[\ldots\langle d_0,\ldots,d_k\rangle)$, and property (5) then guarantees that $[\ldots\langle d_0,\ldots,d_k\rangle)$ is a $\langle d_0,\ldots,d_k\rangle$, so that (i) holds.

But (ii) follows immediately, since

$$[\ldots\langle d_0,\ldots,d_k\rangle) = \bigcup\bigcup\{D_{ix} : x <_c d_i\}$$
$$\cong_c \sum_{i \in \{0,\ldots,k\}} \bigcup\{D_{ix} : x <_c d_i\}$$
$$\cong_c \sum_{i \in \{0,\ldots,k\}} d_i \times S^{\{0,\ldots,i-1\}}$$

□

Now we can prove that S-ary numerals suffice to name the sizes of all sets.

9.3.14 Theorem *Let S be a base for numeration and let T be any non-empty set. Then there is an S-ary numeral*

$$\langle d_0,\ldots,d_n\rangle \in Num(S, \operatorname{ord}(T))$$

defined on an initial segment of the local ordinal, $\operatorname{ord}(T)$, of T such that T is a $\langle d_0,\ldots,d_n\rangle$.

Proof If $T <_c S$ then there is an element $X \in S$ such that $T \cong_c X$, and T is an $\langle X\rangle$. Otherwise, let $\operatorname{Ord}(T) = [0,\ldots,k]$. Then

$$Num(S, [0,\ldots,k]) \cong_c S^{\{0,\ldots,k\}}$$

Since $T <_c S^{\{0,\ldots,k\}}$ and $Num(S, [0,\ldots,k])$ is the field of $<_{Num(S,[0,\ldots,k])}$, there must be a numeral $\langle d_0,\ldots,d_n\rangle \in Num(S, [0,\ldots,k])$ such that $[\ldots\langle d_0,\ldots,d_n\rangle) \cong_c T$. Hence by 9.3.13, T is a $\langle d_0,\ldots,d_m\rangle$, and therefore $\langle d_0,\ldots,d_n\rangle$ is the S-ary numeral defined on an initial segment of the local ordinal $\operatorname{ord}(T)$ required by the theorem. □

Of course if T is empty then T is a $\langle 0\rangle$, where $\langle 0\rangle$ is defined on any linear ordering of length one, e.g. on $\operatorname{ord}(\{T\})$. Thus in every case there is an S-ary numeral which "names" the number species to which T belongs.

But we must be careful here. These "numerals" are not numerals in the ordinary sense: in the first place, they are not inscriptions, and in the second, what we ordinarily think of as a numeral corresponds not to a particular S-ary numeral $\langle d_0, \ldots, d_n \rangle$, but to the $\cong_{S\text{-}ary}$ equivalence class $(\langle d_0, \ldots, d_n \rangle)_{\cong_{S\text{-}ary}}$ of that numeral. Both of these facts have important theoretical consequences, as we shall see in the sequel.

10
The Euclidean Theory of Simply Infinite Systems

10.1 Simply infinite systems

I have already dealt with simply infinite systems in the Cantorian theory[1]. The naive idea that the notion of a simply infinite system is intended to capture in the Euclidean case is that of an unlimited sequence

$$0, \sigma(0), \sigma(\sigma(0)), \sigma(\sigma(\sigma(0))), \ldots$$

of distinct terms generated from an arbitrary chosen initial term, 0, by iterated application of an appropriate *global* function, σ, called the *successor function*.

As in the Cantorian case, the primary purpose of the definition is to eliminate the explicit appeal to the notion of the "finite iteration" of a function employed in the naive description. There are, however, special difficulties arising in the Euclidean case owing to the circumstance that the successive terms of such a simply infinite system together constitute, not a set, but a genuinely (or absolutely) infinite species.

One complication that this imposes is that we must characterise such a system through the species of its proper initial segments rather than through the species of its terms[2].

---

[1] In Section 5.1.

[2] In Cantorian set theory if $(N, s, 0)$ is a simply infinite system, then there is a unique predecessor function $p : N - \{0\} \to N$ such that

$$(\forall x \in N - \{0\})[s`p`x = x] \text{ and } (\forall x \in N)[p`s`x = x]$$

But there is no way, in general, to define a global inverse function on the species of values of a global injection. In particular cases this may be possible, but there is no obvious way of going from a global proposition

$$\forall xy[x \neq y \text{ implies } \sigma(x) \neq \sigma(y)]$$

to the definition of a global inverse for σ.

10.1.1 Definition *Let a be any object, σ any first level global function of one argument, and r any linear ordering. Then by definition*

σ generates r from a : iff : $r = [\ ]$ .or. $[r \neq [\ ]$ and
$First(r) = a$ and $(\forall x \in Field(r) - \{Last(r)\})[Next(r, x) = \sigma(x)]$

Thus a non-empty linear ordering r is generated from a by σ if its initial term is a and each of its non-initial terms (if any) is obtained from its immediate predecessor in r by an application of the function σ.

The empty ordering, $[\ ]$, is generated from any object by any global function. Similarly, the one-termed ordering $[a]$ is generated from a by any such function. If $\sigma(a) = a$ these will be the only orderings generated from a by σ. In general, σ will generate a sequence of orderings

$$[\ ], [a], [a, \sigma(a)], [a, \sigma(a), \sigma(\sigma(a))], \ldots$$

which may or may not terminate depending on whether or not σ ever repeats itself. (Recall that the terms of a linear ordering must be distinct.) If the sequence does not terminate then σ is said to *generate a simply infinite system from a*.

Of course all of this is unsatisfactorily vague when expressed in this operationalist fashion. But the following definition makes the notion mathematically precise without recourse to temporal concepts or metaphors.

10.1.2 Definition *Let a be any object and let σ be any first level global function of one argument. Then to say that σ generates a simply infinite system from a is to say that*

$\forall x[x$ is generated from a by σ and $x \neq [\ ]$ .implies.
$\sigma(Last(x)) \notin Field(x)]$

Note that the *definiens* here is Π_1.

Given *any* unary global function σ and *any* object a we may speak of the *species (of linear orderings) generated from a by σ*. The proposition that a linear ordering r lies in the species generated from a by σ is local (and hence decidable); but in general that species is indefinite or infinite. If it is infinite then it forms a simply infinite system in the sense of the definition just given.

I shall on occasion speak directly of a simply infinite system $\mathcal{N}$ without explicitly mentioning its first term or its successor function. It is always to be understood in such circumstances that there is a successor function $\sigma_\mathcal{N}$ and an initial term $a_\mathcal{N}$ that are what actually determine the simply infinite system in question.

10.1 Simply infinite systems

The *elements* of a simply infinite system $\mathcal{N}$ are the linear orderings that *lie in it* (or *belong to it*), i.e. those linear orderings which are generated from its initial term by its successor function. The *terms* of $\mathcal{N}$ are the terms of its constituent linear orderings. Given any element of $\mathcal{N}$ we can define the set of terms that compose it and all of its predecessors in $\mathcal{N}$.

Following Dedekind's example[3] we may regard a simply infinite system as a concrete representation of a natural number sequence[4]. Accordingly, I shall sometimes represent the elements of a simply infinite system $\mathcal{N}$ by the symbols

$$0_{\mathcal{N}}, 1_{\mathcal{N}}, 2_{\mathcal{N}}, \ldots, i_{\mathcal{N}}, j_{\mathcal{N}}, k_{\mathcal{N}}, \ldots, m_{\mathcal{N}}, n_{\mathcal{N}}, \ldots$$

with the numerals in italic, or by

$$0_{\sigma,a}, 1_{\sigma,a}, 2_{\sigma,a}, \ldots, i_{\sigma,a}, j_{\sigma,a}, k_{\sigma,a}, \ldots, m_{\sigma,a}, n_{\sigma,a}, \ldots$$

where $\mathcal{N}$ is generated from a by σ, or simply by

$$0, 1, 2, \ldots, i, j, k, \ldots, m, n, \ldots$$

omitting the subscripts if the context permits. I shall designate the corresponding terms by

$$\boldsymbol{0}_{\mathcal{N}}, \boldsymbol{1}_{\mathcal{N}}, \boldsymbol{2}_{\mathcal{N}}, \ldots, \boldsymbol{i}_{\mathcal{N}}, \boldsymbol{j}_{\mathcal{N}}, \boldsymbol{k}_{\mathcal{N}}, \ldots, \boldsymbol{m}_{\mathcal{N}}, \boldsymbol{n}_{\mathcal{N}}, \ldots$$

with the numerals in bold italic, or by

$$\boldsymbol{0}_{\sigma,a}, \boldsymbol{1}_{\sigma,a}, \boldsymbol{2}_{\sigma,a}, \ldots, \boldsymbol{i}_{\sigma,a}, \boldsymbol{j}_{\sigma,a}, \boldsymbol{k}_{\sigma,a}, \ldots, \boldsymbol{m}_{\sigma,a}, \boldsymbol{n}_{\sigma,a}, \ldots$$

or just by

$$\boldsymbol{0}, \boldsymbol{1}, \boldsymbol{2}, \ldots, \boldsymbol{i}, \boldsymbol{j}, \boldsymbol{k}, \ldots, \boldsymbol{m}, \boldsymbol{n}, \ldots$$

where possible. Thus

$$0 = [\,]$$
$$1 = [\boldsymbol{0}]$$
$$2 = [\boldsymbol{0}, \boldsymbol{1}]$$
$$3 = [\boldsymbol{0}, \boldsymbol{1}, \boldsymbol{2}]$$

In general, if $n \neq 0$

$$n = [\boldsymbol{0}, \boldsymbol{1}, \ldots, \boldsymbol{n-1}]$$

---

[3] *Was sind und was sollen die Zahlen?* §73.
[4] We cannot speak of "the" natural number sequence here.

and, of course

$$n - 1 = Last(n)$$
$$n = \sigma(Last(n))$$
$$n + 1 = \sigma(n)$$

10.1.3 Definition *Let $\mathcal{N}$ be a simply infinite system and let n lie in $\mathcal{N}$. Then by definition*

$$\mathcal{N}_n = \{y \subseteq n : y \text{ lies in } \mathcal{N} \text{ and } y \neq n\}$$

Hence

$$\mathcal{N}_n = \{0, 1, \ldots, n-1\} \ (= \{0_{\mathcal{N}}, 1_{\mathcal{N}}, \ldots, (n-1)_{\mathcal{N}}\})$$

i.e. $\mathcal{N}_n$ is the set of all predecessors of n in $\mathcal{N}$.

Finally, if σ and a are clear from the context and r lies in $\mathcal{N}(\sigma, a)$, I shall often write "$r +_{\mathcal{N}} 1$" or just "$r + 1$" instead of "$r * [\sigma(Last(r))]$".

Each element, n, of a simply infinite system $\mathcal{N}$ is, by definition, a linear ordering, and the natural ordering $\leq_{\mathcal{N}}$ on $\mathcal{N}$ is, in fact, just the restriction of $\leq_o$ to the elements of $\mathcal{N}$.

10.1.4 Definition *Let $\mathcal{N}$ be a simply infinite system and let m and n lie in $\mathcal{N}$. Then by definition*

$$m \leq_{\mathcal{N}} n \text{ .iff. } m \leq_o n$$

Now I can establish the legitimacy of proof by mathematical induction.

10.1.5 Theorem *(The Principle of Mathematical Induction for Simply Infinite Systems) Let $\mathcal{N}$ be a simply infinite system and let Φ be any first level global propositional function of one argument. Then the hypotheses*

(i) $\Phi(0_{\mathcal{N}})$

(ii) $\forall x[x \text{ lies in } \mathcal{N} \text{ .and. } \Phi(x) . : \text{implies} : . \Phi(x +_{\mathcal{N}} 1)]$

together entail the conclusion

iii.) $\forall x[x \text{ lies in } \mathcal{N} \text{ .implies. } \Phi(x)]$

The proof of $\Phi(r)$, for r in $\mathcal{N}$, proceeds by local induction along the linear ordering $r +_{\mathcal{N}} 1$.

We can use induction to show that if m and n both lie in a simply infinite system $\mathcal{N}$ and $Field(m) \cong_c Field(n)$ then $m = n$. Since $\leq_{\mathcal{N}}$ is just $\leq_o$ restricted to members of $\mathcal{N}$, this establishes the following theorem.

10.1 Simply infinite systems

10.1.6 Theorem *Let $\mathcal{N}$ be a simply infinite system. Then*

$$\forall xy [x \text{ and } y \text{ lie in } \mathcal{N} \ . : implies : . \ [x <_\mathcal{N} y \ .or. \ x = y \ .or. \ y <_\mathcal{N} x]]$$

Theorem 10.1.6 thus guarantees that each element n of $\mathcal{N}$ is the unique representative of its order type in $\mathcal{N}$, and I shall often write "$|n|$" for "$Field(n)$", so that $|n|$ represents the corresponding cardinality.

The elementary facts about the ordering $<_\mathcal{N}$ are now easily established, e.g. that

$$m <_\mathcal{N} n + 1 \text{ implies } m \leq_\mathcal{N} n$$

or that

$$m + 1 \leq_\mathcal{N} n \text{ implies } m <_\mathcal{N} n$$

or that

$$m \leq_\mathcal{N} n \leq_\mathcal{N} m + 1 \ .implies. \ [n = m \text{ or } n = m + 1]$$

and so forth[5].

Perhaps some examples are in order. The system, $\mathscr{V}\mathcal{N}$, of *von Neumann natural numbers* is the simply infinite system whose initial term is the empty set and which is generated by the global function $\lambda x(x \cup \{x\})$. It is obvious that $\lambda x(x \cup \{x\})$ generates a simply infinite system from $\emptyset$, for if

$$[\emptyset, \{\emptyset\}, \ldots, k]$$

is generated from $\emptyset$ by $\lambda x(x \cup \{x\})$, then a straightforward induction along $[\emptyset, \ldots, k]$ shows that

$$(\forall x \in \{\emptyset, \{\emptyset\}, \ldots, k\})[x \in x + 1 \text{ and } x \subseteq x + 1]$$

and therefore

$$(\forall x \in \{\emptyset, \{\emptyset\}, \ldots, k\})[x \in k \text{ or } x = k]$$

so it follows that $k \cup \{k\} \notin \{\emptyset, \{\emptyset\}, \ldots, k\}$ by Regularity.

In contrast to the general case, the species of terms of $\mathscr{V}\mathcal{N}$ is locally definable since it has as its defining property the standard definition of von Neumann well-ordering (Definition 4.10.1).

Another familiar example is the system, $\mathscr{Z}$, of *Zermelo natural numbers*. This is the simply infinite system generated from the empty set by the

---

[5] Of course all these results require the hypothesis that $\mathcal{N}$ is a simply infinite system and that m and n both lie in $\mathcal{N}$.

global function $\lambda x(\{x\})$. Again it is obvious that we have a simply infinite system here, for if

$$[\emptyset, \{\emptyset\}, \{\{\emptyset\}\}, \ldots, k]$$

is generated from $\emptyset$ by $\lambda x(\{x\})$, then a straightforward induction along the inverse ordering $[\emptyset, \{\emptyset\}, \{\{\emptyset\}\}, \ldots, k]^{-1}$ shows that

$$(\forall x \in \{\emptyset, \{\emptyset\}, \ldots, k-1\})[x \text{ is a constituent of } k]$$

so that $\{k\} \notin \{\emptyset, \{\emptyset\}, \ldots, k\}$.

As in the case of $\mathscr{V}\mathscr{N}$, the species of terms of $\mathscr{L}$ is locally definable: a non-empty set S is a term of that system if there is a linear ordering composed of constituents of S whose first term is $\emptyset$, each of whose non-initial terms (if any) is the singleton of the immediately preceding term, and whose last term is the sole member of S.

Simply infinite systems can be viewed as natural number systems. The basic use to which natural numbers are put is in counting and indexing the sizes of (Euclidean finite) sets. Questions concerning the sizes of sets naturally give rise to questions concerning the basic arithmetical operations on natural numbers – addition, multiplication, and exponentiation – and to the general theory of these operations – number theory. All of these matters will be addressed in the next section.

10.2 Measures, scales, and elementary arithmetical operations

Let us say that a linear ordering, r, *measures* a set, S, if

$$|r| (= Field(r)) \cong_c S$$

A simply infinite system thus provides us with a species of "canonical" measures for the sizes of sets.

10.2.1 Definition *Let μ be a first level global function of one argument.*

(i) *Let $\mathscr{S}$ be a locally definable species[6] of linear orderings and $\mathscr{T}$ a locally definable species of sets. Then to say that μ is a (global) measure for $\mathscr{T}$ with respect to $\mathscr{S}$ is to say that*

$$\forall x[x \text{ lies in } \mathscr{T} : \text{implies} : \mu(x) \text{ lies in } \mathscr{S} \text{ and } \mu(x) \text{ measures } x]$$

---

[6] A species is *locally definable* if its defining property can be expressed by a local formula, i.e. one that does not contain global quantifiers. Membership in such a species can thus be expressed by a sentence free of global quantifiers.

10.2 Measures, scales, and elementary arithmetical operations

(ii) Let $\mathscr{S}$ and $\mathscr{T}$ be locally definable species of linear orderings. Then to say that μ is a (global) measure for $\mathscr{T}$ with respect to $\mathscr{S}$ is to say that

$$\forall x[x \text{ lies in } \mathscr{T} : \text{implies} : \mu(x) \text{ lies in } \mathscr{S} \text{ and } \mu(x) \text{ measures } |x|]$$

If $\mathscr{S}$ is a simply infinite system and μ is a measure (in either sense (i) or sense (ii)) for $\mathscr{T}$ with respect to $\mathscr{S}$, then we say that $\mathscr{S}$ is a *scale for* $\mathscr{T}$ *with measure* μ. If $\mathscr{T}$ is the species of all sets (or of all linear orderings), then $\mathscr{S}$ is a *universal scale* and μ is a *universal measure with respect to* $\mathscr{S}$.

It should be clear that we have no reason simply to *assume* that, given a simply infinite system $\mathscr{N}$, we can define a measure μ which is a universal measure with respect to $\mathscr{N}$. Indeed, we have no reason to suppose, in the absence of a proof, that, given simply infinite systems $\mathscr{N}$ and $\mathscr{N}_1$, we can define a measure μ which measures either with respect to the other.

Let us now turn to the problem of defining the basic arithmetical operations in a simply infinite system.

10.2.2 Definition Let $\mathscr{N}$ be a simply infinite system, and let φ be a binary first level global function that is arithmetical. Then to say that the binary first level global function, ψ, *defines* φ *in* $\mathscr{N}$ is to say that

$$\forall xy[x, y \text{ lie in } \mathscr{N} : \text{implies} : \psi(x, y) \text{ lies in } \mathscr{N} \text{ and}$$
$$\psi(x, y) \text{ measures } \varphi(|x|, |y|)]$$

This definition can be extended to functions of different numbers of arguments in the obvious way.

In particular, to say that a binary global function φ *defines addition* in $\mathscr{N}$ is to say that

$$\forall xy[x, y \text{ lie in } \mathscr{N} : \text{implies} :$$
$$\varphi(x, y) \text{ lies in } \mathscr{N} \text{ and } \varphi(x, y) \text{ measures } |x| +_c |y|]$$

and to say that a binary global function ψ *defines multiplication* in $\mathscr{N}$ is to say that

$$\forall xy[x, y \text{ lie in } \mathscr{N} : \text{implies} :$$
$$\psi(x, y) \text{ lies in } \mathscr{N} \text{ and } \psi(x, y) \text{ measures } |x| \times_c |y|]$$

and to say that a binary global function χ *defines exponentiation* in $\mathcal{N}$ is to say that

$$\forall xy[x, y \text{ lie in } \mathcal{N} : \text{implies} :$$
$$\chi(x, y) \text{ lies in } \mathcal{N} \text{ and } \chi(x, y) \text{ measures } |x|^{|y|}]$$

Of course we cannot simply assume that, given a simply infinite system $\mathcal{N}$, we can find a binary global function that defines addition, let alone multiplication or exponentiation. We can, however, establish obvious relationships between the definability conditions for these basic arithmetical operations.

10.2.3 Theorem *If multiplication is definable in a simply infinite system then so is addition, and if exponentiation is definable then so is multiplication.*

Proof Let S and T be sets both of which contain at least two elements. If S is at least as large as T then there are a series of injections

$$S +_c T \to S +_c S \to S \times_c \{a,b\} \to S \times_c T \to$$
$$S \times_c S \to S^{\{a,b\}} \to S^T \to S^S$$

where $a, b \in T$.

Thus there are injections $S +_c T \to S \times_c T$ and $S \times_c T \to S^T$. On the other hand, if T is larger than S then we have a series of injections

$$S +_c T \to T +_c T \to \{a,b\} \times_c T \to S \times_c T \text{ (where } a, b \in S)$$

and another series of injections

$$S \times_c T \to T \times_c T \to T^{\{a,b\}} \to T^T \text{ (where } a, b \in T)$$

Suppose now that χ is a binary first level global function that defines exponentiation in the simply infinite system $\mathcal{N}$. Then we can define, for m and n lying in $\mathcal{N}$ with $2 <_\mathcal{N} m, n$,

$\psi(m, n) =$ the unique $k \leq_\mathcal{N} \max_\mathcal{N}\{\chi(m, m), \chi(n, n)\}$ such that
k measures $\text{Field}(m) \times \text{Field}(n)$

$\varphi(m, n) =$ the unique $k \leq_\mathcal{N} \psi(m, n)$ such that k measures
$\text{Field}(m) +_c \text{Field}(n)$

defining $\psi(m, n)$ and $\varphi(m, n)$ for $m, n \leq_\mathcal{N} 2$ in the obvious way, and setting

$$\psi(a, b) = \varphi(a, b) = \emptyset$$

if either a or b lies outside $\mathcal{N}$. Then ψ defines multiplication and φ addition in $\mathcal{N}$. □

10.3 Limited recursion

Let $\mathcal{N}$ be a simply infinite system. The familiar recursion equations for addition, multiplication, and exponentiation can be set out in the following forms appropriate to $\mathcal{N}$:

Addition:

$$\varphi(m, 0) = m$$
$$\varphi(m, n +_{\mathcal{N}} 1) = \varphi(m, n) +_{\mathcal{N}} 1$$

Multiplication:

$$\psi(m, 0) = 0$$
$$\psi(m, n +_{\mathcal{N}} 1) = \varphi(\psi(m, n), m)$$

Exponentiation:

$$\chi(m, 0) = 1$$
$$\chi(m, n +_{\mathcal{N}} 1) = \psi(\chi(m, n), m)$$

Here, of course, the symbols m and n stand for arbitrary elements of $\mathcal{N}$. A straightforward induction argument now establishes.

10.2.4 Theorem *A necessary and sufficient condition that addition, multiplication, or exponentiation be definable on a simply infinite system $\mathcal{N}$ is that the proposed defining function or functions satisfy the appropriate recursion equations over $\mathcal{N}$.*

The induction required here is along the simply infinite system $\mathcal{N}$.

Notice that the basic facts about the arithmetic of arithmetical functions (e.g. the facts contained in Theorem 9.1.6) immediately carry over to the arithmetic of their analogues in simply infinite systems. In particular, the "logarithm" function, log, has an analogue, $\log^{\mathcal{N}}$, in each simply infinite system $\mathcal{N}$ which defines it in $\mathcal{N}$. The facts about log given in Theorem 9.1.8 carry over immediately to $\log^{\mathcal{N}}$ (insofar as they can be interpreted in $\mathcal{N}$).

10.3 Limited recursion

Let us turn now to the general problem of defining functions by recursion along a simply infinite system. The difficulties are similar to those we encountered in Section 9.2, and the solution is essentially the same. But first we need to single out those global functions that are analogous to the *arithmetical* functions of Chapter 9.

10.3.1 Definition *Let $\mathcal{N}$ be a simply infinite system and let φ be a first level global function of one argument. Then to say that φ is defined on $\mathcal{N}$ (or is $\mathcal{N}$-arithmetical) is to say that*

$$\forall x[x \text{ lies in } \mathcal{N} \text{ implies } \varphi(x) \text{ lies in } \mathcal{N}]$$

A similar definition applies to functions of more than one argument.

Now we need to define the notion of a global function's *locally bounding* a system of recursion equations

$$\varphi(0_\mathcal{N}, \vec{b}) = \gamma(\vec{b})$$
$$\varphi(a +_\mathcal{N} 1, \vec{b}) = \chi(a, \varphi(a, \vec{b}), \vec{b})$$

In these equations γ is the *initial function* and χ the *recursion function*. The problem is to determine under what circumstances the equations can be "solved" for an $\mathcal{N}$-arithmetical φ.

The definition of local bounding is analogous to Definition 9.2.1.

10.3.2 Definition *Let $\mathcal{N}$ be a simply infinite system, and let γ, χ, and θ be first level global functions defined on $\mathcal{N}$ and of $\mathbf{n}$, $\mathbf{n+2}$, and $\mathbf{n+1}$ arguments, respectively. Then*

(i) *Let $a, b_1, \ldots, b_n$ all be elements of $\mathcal{N}$. Then f is said to satisfy the recursion equations determined by γ (as initial function) and χ (as recursion function) on a at the arguments $b_1, \ldots, b_n$ ($=\vec{b}$) if, and only if, f is a function defined on $\mathcal{N}_a$ all of whose values lie in $\mathcal{N}$ and*

   (a) $f`0_\mathcal{N} = \gamma(\vec{b})$
   (b) $(\forall x \in \mathcal{N}_a)[x \neq \text{Last}(\mathcal{N}_a) \text{ .implies. } f`(x +_\mathcal{N} 1) = \chi(x, f`x, \vec{b})]$

(ii) *θ is said to bound the recursion equations determined by γ (as initial function) and χ (as recursion function) locally if, and only if, for all $x, y_1, \ldots, y_n$ ($=\vec{y}$) in $\mathcal{N}$, and all z, if z satisfies the recursion equations determined by γ (as initial function) and χ (as recursion function) on x at the arguments $y_1, \ldots, y_n$, then*

   (a) $\gamma(\vec{y}) <_\mathcal{N} \theta(0_\mathcal{N}, \vec{y})$
   (b) $(\forall u \in \mathcal{N}_a)[\chi(u, z`u, \vec{y}) <_\mathcal{N} \theta(u, \vec{y})]$

As in the case of 9.2.1, (i) is local and (ii) is global.

We can now establish a Principle of Definition by Limited Recursion analogous to Theorem 9.2.2.

10.3 Limited recursion

10.3.3 Theorem *(The Principle of Definition by Limited Recursion along a Simply Infinite System)* Let $\mathcal{N}$ be a simply infinite system, let γ and χ be functions of $\mathbf{n}$ and $\mathbf{n}+2$ arguments, respectively, defined on $\mathcal{N}$, and let θ be a function of $\mathbf{n}+1$ arguments defined on $\mathcal{N}$ and increasing in its first argument[7] which locally bounds the recursion equations determined by γ (as initial function) and χ (as recursion function).

Then we can define a global function φ of $\mathbf{n}+1$ arguments on $\mathcal{N}$ such that

(i) $\forall x \vec{y} [x, \vec{y} \text{ lie in } \mathcal{N} : \text{implies} : \varphi(0, \vec{y}) = \gamma(\vec{y}) \text{ and } \varphi(x +_{\mathcal{N}} 1, \vec{y}) = \chi(x, \varphi(x, \vec{y}), \vec{y})]$

(ii) $\forall x \vec{y} [x, \vec{y} \text{ lie in } \mathcal{N} : \text{implies} : \varphi(x, \vec{y}) <_{\mathcal{N}} \theta(x, \vec{y})]$

(iii) *The hypotheses*

ψ *is defined on* $\mathcal{N}$
$\forall x \vec{y} [x, \vec{y} \text{ lie in } \mathcal{N} : \text{implies} : \psi(0, \vec{y}) = \gamma(\vec{y}) \text{ and }$
$\psi(x +_{\mathcal{N}} 1, \vec{y}) = \chi(x, \psi(x, \vec{y}), \vec{y})]$

entail the conclusion

$\forall x \vec{y} [x, \vec{y} \text{ lie in } \mathcal{N} : \text{implies} : \varphi(x, \vec{y}) = \psi(x, \vec{y})]$

Proof Suppose γ, χ, and θ are given as in the theorem, and let $\vec{b}$ all lie in $\mathcal{N}$.

Claim *Let a lie in $\mathcal{N}$. Then there is exactly one function*

$$f_a : \mathcal{N}_a \to \mathcal{N}_{\theta(a, \vec{b})}$$

which satisfies the recursion equations determined by γ (as initial function) and χ (as recursion function) on a at $\vec{b}$[8].

Pf. The proof is by induction on a along $\mathcal{N}$. Such a proof is possible because the condition on a is local. We avoid a global existential quantifier over local functions satisfying the recursion equations at a because the (increasing) bounding function θ allows us to specify the target set $\mathcal{N}_{\theta(a, \vec{b})}$ for such functions.

Basis case $(a = 0)$ In this case $\mathcal{N}_a = [\,] = \emptyset$, so $\emptyset$ is the f_a required since it satisfies the requirements vacuously.

Induction case Suppose, as hypothesis of induction, that there is exactly

---

[7] i.e. $\forall u, v, \vec{y} [u, v \text{ lie in } \mathcal{N} \text{ and } u <_{\mathcal{N}} v : \text{implies} : \theta(u, \vec{y}) <_{\mathcal{N}} \theta(v, \vec{y})]$. This extra condition is inessential, but simplifies the presentation of the proof.

[8] Note that the function f_a referred to here depends not only on a but also on $\vec{b}$; for simplicity of expression I have suppressed the parameters $\vec{b}$.

one function $f_a : \mathcal{N}_a \to N_{\theta(a,\vec{b})}$ such that

$$(\forall u \in N_a)[[u = 0 \text{ implies } f_a{}^\backprime u = \gamma(\vec{b})] \; : and :$$
$$[u \neq 0 \text{ implies } f_a{}^\backprime u = \chi(u-1, f_a{}^\backprime(u-1), \vec{b})]]$$

We must show that there is exactly one function $f_{a+1} : \mathcal{N}_a \to \mathcal{N}_{\theta(a+1,\vec{b})}$ such that

$$(\forall u \in \mathcal{N}_{a+1})[[u = 0 \text{ implies } f_{a+1}{}^\backprime u = \gamma(\vec{b})] \; : and :$$
$$[u \neq 0 \text{ implies } f_{a+1}{}^\backprime u = \chi(u-1, f_{a+1}{}^\backprime(u-1), \vec{b})]]$$

Since θ bounds the recursion equations, we may conclude that

$$\gamma(\vec{b}) <_{\mathcal{N}} \theta(0, \vec{b}) \text{ and } (\forall u \in \mathcal{N}_a)[\chi(u, f{}^\backprime u, \vec{b}) <_{\mathcal{N}} \theta(x, \vec{b})]$$

Now if $a = 0$, $\mathcal{N}_{a+1} = \{0\}$ and we may set $f_{a+1} = \{(0, \gamma(\vec{b}))\}$. Otherwise $\mathcal{N}_{a+1}$ has a as largest element and, since θ bounds the recursion equations, $\chi(a-1, f{}^\backprime(a-1), \vec{y}) <_{\mathcal{N}} \theta(a, \vec{y})$. We may thus set

$$f_{a+1} = f_a \cup \{(a, \chi(a-1, f{}^\backprime(a-1), \vec{b}))\}$$

f_{a+1} so defined clearly maps $\mathcal{N}_a \to \mathcal{N}_{\theta(a+1,\vec{b})}$ and satisfies the recursion equations: at u for which $u <_{\mathcal{N}} a$ by hypothesis of induction, and at a since $(a, \chi(a-1, f{}^\backprime(a-1), \vec{b})) \in f_{a+1}$.

The claim now follows by induction along $\mathcal{N}$.

We may now define the required function φ by setting

$$\varphi(a, \vec{b}) = f_{a+1}{}^\backprime a$$

Part (iii) follows by induction along $\mathcal{N}$. □

As in the corresponding result for arithmetical functions (Theorem 9.2.2), the bounding function θ in 10.3.3 actually serves to guarantee that the recursion equations succeed in defining a function on $\mathcal{N}$.

This completes our basic development of the general theory of simply infinite systems. We must now turn to the study of techniques for defining particular such systems.

10.4 Extending simply infinite systems

We can use the linear orderings that are the elements of a simply infinite system $\mathcal{N}$ to form numerals over a given numeration base S[9].

---

[9] Definiton 9.3.1. This whole section is based on the analysis of S-ary decompositions and numerals carried out in Section 9.3.

10.4 Extending simply infinite systems

10.4.1 Definition *Let $\mathcal{N}$ be a simply infinite system, S a base for numeration, s a sequence, and T a set. Then by definition*

(i) *s is an S-ary decomposition of T over $\mathcal{N}$ if, and only if, s is an S-ary decomposition*[10] *of T and s is defined on a linear ordering which lies in $\mathcal{N}$.*

(ii) *s is an S-ary numeral over $\mathcal{N}$ if, and only if, s is an S-ary numeral*[11] *and s is defined on a linear ordering which lies in $\mathcal{N}$.*

Since no two members of $\mathcal{N}$ are of the same length, each such S-ary numeral over $\mathcal{N}$ has a unique representation in the sense that two S-ary numerals over $\mathcal{N}$ that are S-ary equivalent[12] are identical.

10.4.2 Definition *Let $\mathcal{N}$ be a simply infinite system, let S be a numeration base, let k lie in $\mathcal{N}$, and let s and t be S-ary numerals over $\mathcal{N}$. Then by definition*

(i) $Num(\mathcal{N}, S, k) = Num(S, k)$[13]

(ii) $\leq_{Num(\mathcal{N},S,k)} \; = \; \leq_{Num(S,k)}$[14]

(iii) $s \leq_{\mathcal{N},S} t \; : iff : \; s \leq_{Num(\mathcal{N},S,l)} t$, *where* $l = \max_{\mathcal{N}}\{lh(s), lh(t)\}$

Where the context allows, I shall usually suppress the arguments $\mathcal{N}$ and S, writing, for example, "$Num(k)$" instead of "$Num(\mathcal{N}, S, k)$", "$\leq_{Num(k)}$" instead of "$\leq_{Num(\mathcal{N},S,k)}$", and "$s \leq_S t$" instead of "$s \leq_{\mathcal{N},S} t$" (in this last case the "S" is retained as essential, since it tells us that we are dealing with S-ary numerals).

$Num(k)$ is the set of all S-ary numerals over $\mathcal{N}$ defined on the linear orderings $1, \ldots, k$ (i.e. $1_{\mathcal{N}}, \ldots, k_{\mathcal{N}} = k$), that is to say, those of (non-zero) length k or less; $\leq_{Num(k)}$ is the natural ordering on the S-ary numerals that compose $Num(k)$, given that under the conventions I laid down in Chapter 9 these S-ary numerals are given in the order of *increasing* powers of S instead of in the conventional order in which the powers of S decrease when the numeral is read from left to right; $\leq_S$ is the natural ordering on the species of *all* S-ary numerals over $\mathcal{N}$.

The really elementary facts about these numerals are so well known that it seems otiose actually to spell them out in complete detail with accompanying proofs. But some care must be taken because what arithmetical "calculations" we can perform on S-ary numerals over a simply

---

[10] Definition 9.3.3.
[11] Definition 9.3.5.
[12] Definition 9.3.6.
[13] Definition 9.3.9.
[14] Definition 9.3.10.

infinite system $\mathcal{N}$ depend on the closure properties of $\mathcal{N}$, as we shall soon see.

The successor operation on $\leq_{Num(k)}$, namely, $Next(\leq_{Num(k)})$, maps

$$Num(k) - \{Last(\leq_{Num(k)})\}$$

one to one onto

$$Num(k) - \{First(\leq_{Num(k)})\}$$

so that, for each S-ary numeral x in $Num(k)$, $Next(\leq_{Num(k+1)})`x$ is the S-ary numeral that follows next after x in the natural ordering on S-ary numerals. This permits us to define a successor function for S-ary numerals over $\mathcal{N}$ globally on the species of all such numerals.

10.4.3 Definition *Let $\mathcal{N}$ be a simply infinite system, let S be a numeration base, and let d be an S-ary numeral over $\mathcal{N}$ defined on the linear ordering k lying in $\mathcal{N}$. Then by definition*

$$\sigma_{\mathcal{N}[S]}(d) = Next(\leq_{Num(k+1)})`d$$

10.4.4 Theorem *Let $\mathcal{N}$ be a simply infinite system and S be a numeration base. Then $\sigma_{\mathcal{N}[S]}$ generates a simply infinite system from $\langle\emptyset\rangle_{1_{\mathcal{N}}}$.*[15]

My choice of the notation "$\sigma_{\mathcal{N}[S]}$" here is deliberate. Because of the definition of simply infinite system we have adopted here (Definition 10.1.2), the S-ary numerals on a simply infinite system $\mathcal{N}$ cannot be construed directly as together composing the *elements* of a simply infinite system; but we can define a simply infinite system of which those numerals are the *terms*.

Indeed, given a simply infinite system $\mathcal{N}$ and a numeration base S, let us now define the S-ary expansion, $\mathcal{N}[S]$, of $\mathcal{N}$ to be the species of linear orderings composed of S-ary numerals over $\mathcal{N}$ arranged in their natural order without gaps. This is an important notion so an "official" definition is called for.

10.4.5 Definition *Let $\mathcal{N}$ be a simply infinite system and S be a numeration base. Then $\mathcal{N}[S]$ (the S-ary expansion of $\mathcal{N}$) is the simply infinite system generated from the initial element $\langle\emptyset\rangle_{1_{\mathcal{N}}}$ by the unary global function $\sigma_{\mathcal{N}[S]}$.*

---

[15] Recall (Definition 9.3.5) that the set of "digits" out of which S-ary numerals are formed are the *members* of S and that for every "size" smaller than the size of S itself S has exactly one set of that size as a member. In particular, $\emptyset$ is the S-ary "digit" 0. $1_{\mathcal{N}}$ is, of course, the one-termed linear ordering $[\mathbf{0}_{\mathcal{N}}]$.

10.4 Extending simply infinite systems

Clearly the species of *terms* of the simply infinite system $\mathcal{N}[S]$, that is to say, the species of S-ary numerals over $\mathcal{N}$, is locally definable. This provides us with the option of doing natural number arithmetic directly on these numerals using the familiar calculation algorithms. We can then translate that arithmetic into arithmetic on the simply infinite system $\mathcal{N}[S]$.

It is completely straightforward to define the notion of addition for S-ary numerals over a simply infinite system $\mathcal{N}$. We know that the standard S-ary algorithm for addition always yields a numeral at most one digit longer than the longer of the two numerals to which it is applied. Moreover, if d is an S-ary numeral then $[\ldots d)$ is a d^{16}. These observations lead to the following definition.

10.4.6 Definition *Let $\mathcal{N}$ be a simply infinite system, S a numeration base, a and b S-ary numerals over $\mathcal{N}$, and $k = \max_{\mathcal{N}}(lh(a), lh(b))$. Then by definition*

$$a +_S b = \imath\{x \in Num(k+1) : [\ldots x) \cong_c [\ldots a) +_c [\ldots b)\}$$

A straightforward induction argument suffices to establish the following theorem.

10.4.7 Theorem *Let $\mathcal{N}$ be a simply infinite system, let S be a numeration base, and let a and b be S-ary numerals over $\mathcal{N}$. Then*

$$a +_S \langle 0_S \rangle = a$$
$$a +_S (b +_S \langle 1_S \rangle) = (a +_S b) +_S \langle 1_S \rangle$$

Thus we have addition on S-ary numerals over $\mathcal{N}$. Now we need to use this to define addition on the simply infinite system $\mathcal{N}[S]$.

We may think of the S-ary numerals on $\mathcal{N}$ as *names* of the elements ("numbers") lying in $\mathcal{N}[S]$.

10.4.8 Definition *Let $\mathcal{N}$ be a simply infinite system, and S a numeration base.*

(i) *Let a be an element of $\mathcal{N}[S]$. Then by definition*

$$Name_S(a) = \begin{cases} \langle 0_S \rangle & \text{if } a = 0 \\ \sigma_{\mathcal{N}[S]}(Last(a)) & \text{if } a \neq 0 \end{cases}$$

---

[16] i.e. the set of S-ary numerals from zero up to, but not including, d (Definition 9.3.12) is a set of the "size" named by d (Theorem 9.3.13).

(ii) *Let b an S-ary numeral on $\mathcal{N}$. Then by definition*

$$Value_S(b) = \{(x, y) \in [\ldots b) \times [\ldots b) : x \leq_{Num(lh(b))} y\}$$

The intention behind these definitions should be clear: where a lies in $\mathcal{N}[S]$, $Name_S(a)$ is the S-ary numeral on $\mathcal{N}$ that names the $\mathcal{N}[S]$-number a, namely $Name_S(a) = \langle 0_S \rangle$ if $a = 0_{\mathcal{N}[S]}$, and $Name_S(a) = Last(a) +_S \langle 1_S \rangle$ otherwise. Conversely, where b is an S-ary numeral on $\mathcal{N}$, $Value_S(b)$ is the $\mathcal{N}[S]$-number named by b, namely, the linear ordering consisting of the S-ary numerals on $\mathcal{N}$ which precede b, arranged in their natural order. Thus these global functions are "inverses" of one another (except at "don't care" arguments):

$$Name_S(Value_S(b)) = b \text{ for } S\text{-ary numerals } b \text{ over } \mathcal{N}$$

and

$$Value_S(Name_S(a)) = a \text{ for elements } a \text{ lying in } \mathcal{N}[S]$$

Now we can define addition on $\mathcal{N}[S]$ using these two key global functions as follows:

10.4.9 Definition *Let $\mathcal{N}$ be a simply infinite system, let S be a numeration base, and let a and b lie in $\mathcal{N}[S]$. Then by definition*

$$a +_{\mathcal{N}[S]} b = Value_S(Name_S(a) +_S Name_S(b))$$

As an immediate corollary to Theorem 10.4.7 we have

10.4.10 Theorem *Let $\mathcal{N}$ be a simply infinite system and let S be a numeration base. Then addition is definable in $\mathcal{N}[S]$.*

Proof By Theorem 10.4.7, the binary global function $\lambda xy[x +_{\mathcal{N}[S]} y]$ satisfies the recursion equations for addition in $\mathcal{N}[S]$. It follows, by Theorem 10.2.4, that $\lambda xy[x +_{\mathcal{N}[S]} y]$ defines addition in $\mathcal{N}[S]$. □

The simply infinite system $\mathcal{N}[S]$ is at least as "long" as $\mathcal{N}$, as the following theorem shows.

10.4.11 Theorem *Let $\mathcal{N}$ be a simply infinite system, and let S be a numeration base. Then $\mathcal{N}[S]$ is a scale for $\mathcal{N}$.*

10.4 Extending simply infinite systems

Proof It will suffice to define a global function $\mu_{\mathcal{N}}$ of one argument such that, for k lying in $\mathcal{N}$, $\mu_{\mathcal{N}}(k)$ lies in $\mathcal{N}[S]$ and $|k| = |\mu_{\mathcal{N}}(k)|$. This can be accomplished by setting

$\mu_{\mathcal{N}}(a) = \emptyset$, if a does not lie in $\mathcal{N}$.

$\mu_{\mathcal{N}}(a) = [\,]$ if $a = 0_{\mathcal{N}}$ ($= [\,]$).

$\mu_{\mathcal{N}}(a) = \imath(\{x \in \{(\leq_{Num(a)})_y : y \in Num_{\mathcal{N}}(a)\} : x \cong_o a\})$ if $0_{\mathcal{N}} <_{\mathcal{N}} a$.

That $\mu_{\mathcal{N}}$ so defined is a measure for $\mathcal{N}$ with respect to $\mathcal{N}[S]$ follows from the fact that if a lies in $\mathcal{N}$ and $0_{\mathcal{N}} <_{\mathcal{N}} a$, then $\leq_{Num(a)}$ lies in $\mathcal{N}[S]$ and $a <_o \leq_{Num(a)}$. □

As we shall see, $\mathcal{N}$ cannot, in the general case, be shown to measure $\mathcal{N}[S]$.

Let us now turn our attention to the problem of defining multiplication in the simply infinite system $\mathcal{N}[S]$. The proof of Theorem 10.4.7 suggests that we should first investigate the problem of defining S-ary multiplication, $\times_S$, on S-ary *numerals* over $\mathcal{N}$.

It is a well-known (and easily established) fact about S-ary numerals, for arbitrary numeration bases S, that the length of the product numeral is $\leq_o$ the sum of the lengths of the factor numerals. But since the lengths of our S-ary numerals over $\mathcal{N}$ must be linear orderings lying in $\mathcal{N}$, we need to assume that addition is definable on $\mathcal{N}$. If we make that assumption then we can define multiplication for S-ary numerals over $\mathcal{N}$ as in Definition 10.4.6.

10.4.12 Definition *Let $\mathcal{N}$ be a simply infinite system, S be a numeration base, $\lambda xy(x +_{\mathcal{N}} y)$ define addition on $\mathcal{N}$, a and b be S-ary numerals over $\mathcal{N}$, and $k = lh(a) +_{\mathcal{N}} lh(b)$. Then by definition*

$$a \times_S b = \imath\{x \in Num(k+1) : [\ldots x) \cong_c [\ldots a) \times_c [\ldots b)\}$$

This definition of multiplication for S-ary numerals allows us to establish recursion equations for the species of all S-ary numerals over $\mathcal{N}$.

10.4.13 Theorem *Let $\mathcal{N}$ be a simply infinite system closed under addition and let S be a numeration base. Then*

$$a \times_S \langle 0_S \rangle = \langle 0_S \rangle$$
$$a \times_S (b +_S \langle 1_S \rangle) = (a \times_S b) +_S a$$

As an immediate consequence of this theorem we can establish the following fundamental result.

10.4.14 Theorem *Let $\mathcal{N}$ be a simply infinite system and let S be a numeration base. Then a necessary and sufficient condition that multiplication be definable in $\mathcal{N}[S]$ is that addition be definable in $\mathcal{N}$.*

Proof (*Necessity*) Suppose that multiplication is definable in $\mathcal{N}[S]$, and let k, l lie in $\mathcal{N}$. Let $a_k = \langle 0, 0, \ldots 0, 1 \rangle_{k+1}$ and $a_l = \langle 0, 0, \ldots 0, 1 \rangle_{l+1}$[17]. Then $|Values(a_k)| \cong_c |S|^{|k|}$ and $|Values(a_l)| \cong_c |S|^{|l|}$, and therefore

$$|Values(a_k) \times_{\mathcal{N}[S]} Values(a_l)| \cong_c |S|^{|k|} \times_c |S|^{|l|} \cong_c |S|^{|k|+_c|l|}$$

and it follows that $a_k \times_S a_l = \langle 0, 0, \ldots 0, 1 \rangle_{m+1}$, where m lies in $\mathcal{N}$ and $|m| \cong_c |k| +_c |l|$. Thus given k and l in $\mathcal{N}$, we can define m in $\mathcal{N}$ such that $|m| \cong_c |k| +_c |l|$, i.e. addition is definable in $\mathcal{N}$.

(*Sufficiency*) Suppose that addition is definable in $\mathcal{N}$. Then Definition 10.4.12 allows us to define the global function $\lambda xy(x \times_S y)$ which, by Theorem 10.4.13, satisfies the usual recursion equations for multiplication on the species of S-ary numerals on $\mathcal{N}$. Hence if we define a binary global function $\lambda xy(x \times_{\mathcal{N}[S]} y)$ on elements a and b of $\mathcal{N}[S]$ by setting

$$a \times_{\mathcal{N}[S]} b = Values(Name_S(a) \times_S Name_S(b))$$

then $\lambda xy(x \times_{\mathcal{N}[S]} y)$ can easily be shown to satisfy the relevant recursion for multiplication, and thus defines multiplication in $\mathcal{N}[S]$ by Theorem 10.2.4. □

Thus given any simply infinite system $\mathcal{N}$, Theorem 10.4.10 guarantees that addition is definable in $\mathcal{N}[S]$, for any numeration base S, and Theorem 10.4.14 that multiplication is definable in $(\mathcal{N}[S])[T]$, for any numeration bases S and T. Moreover, by Theorem 10.4.11, $(\mathcal{N}[S])[T]$ measures $\mathcal{N}[S]$, which, in turn, measures $\mathcal{N}$.

We can now establish that if exponentiation is not already definable on a simply infinite system, $\mathcal{N}$, then, provided that S is measured by $\mathcal{N}$, we cannot achieve the definability of exponentiation by taking an S-ary expansion, $\mathcal{N}[S]$, of that system[18].

---

[17] The S-ary numerals a_k and a_l are sequences defined on the linear orderings $k+1$ and $l+1$, respectively.

[18] The proviso that S be measurable by $\mathcal{N}$ is not empty, as I shall show in Chapter 11.

10.4 Extending simply infinite systems 343

10.4.15 Theorem *Let $\mathcal{N}$ be a simply infinite system and let S be a numeration base measured by some element, s, of $\mathcal{N}$. Then the following are equivalent:*

(i) *Exponentiation is definable in $\mathcal{N}[S]$.*
(ii) *$\mathcal{N}$ is a scale for $\mathcal{N}[S]$.*
(iii) *Exponentiation is definable in $\mathcal{N}$.*

Proof ((i) entails (ii)) Suppose that exponentiation is definable in $\mathcal{N}[S]$. Then by Theorem 10.2.3 so are addition and multiplication. We must define a global measure, μ, such that if m lies in $\mathcal{N}[S]$, then $\mu(m)$ lies in $\mathcal{N}$ and $|\mu(m)| \cong_c |m|$. Given m in $\mathcal{N}[S]$, define

$$\mu(m) = 0_\mathcal{N} \text{ if } m = 0_{\mathcal{N}[S]}$$
$$\mu(m) = lh(Last(\exp_{\mathcal{N}[S]}(Values(\langle 0, 1\rangle_{2_\mathcal{N}}), m))) \text{ if } m \neq 0_{\mathcal{N}[S]}$$

Then $|\mu(m)| \cong_c |lh(Last(\exp_{\mathcal{N}[S]}(Values(\langle 0, 1\rangle_{2_\mathcal{N}}), m)))| \cong_c |m|$, since the last term of $\exp_{\mathcal{N}[S]}(Values(\langle 0, 1\rangle_{2_\mathcal{N}}), m)$ will be the S-ary numeral consisting of m consecutive T's, where $T = \max_c(S)$ is the largest element of S.

((ii) entails (iii)) Suppose $\mathcal{N}$ measures $\mathcal{N}[S]$ with measure μ. Since addition is definable in $\mathcal{N}[S]$ (Theorem 10.4.10) it is also definable in $\mathcal{N}$. Hence multiplication is definable in $\mathcal{N}[S]$ (Theorem 10.4.14) and therefore in $\mathcal{N}$. Given p and q in $\mathcal{N}$, let b be the S-ary numeral

$$b = \langle 0_S, \ldots, 0_S, 1_S \rangle_{((p \times_\mathcal{N} q) +_\mathcal{N} 1_\mathcal{N})}$$

Then $Values(b)$ lies in $\mathcal{N}[S]$ and therefore $\mu(Values(b))$ lies in $\mathcal{N}$ with

$$|\mu(Values(b))| \cong_c S^{|p \times_\mathcal{N} q|} \cong_c S^{(|p| \times |q|)}$$

Hence

$$|p|^{|q|} \leq_c S^{(|p| \times |q|)} \cong_c |\mu(b)|$$

It follows that exponentiation is definable in $\mathcal{N}$.

((iii) entails (i)) Suppose that exponentiation is definable in $\mathcal{N}$. Then so are addition and multiplication (by Theorem 10.2.3).

Now let p and q lie in $\mathcal{N}[S]$. It will suffice to define a k in $\mathcal{N}$ such that

$$|p|^{|q|} <_c S^{|k|}$$

Since p and q lie in $\mathcal{N}[S]$, both p and q are linear orderings whose lengths, $lh(p)$ and $lh(q)$, lie in $\mathcal{N}$. If we define

$$a = lh(p) +_{\mathcal{N}} 1_{\mathcal{N}}$$
$$b = lh(q) +_{\mathcal{N}} 1_{\mathcal{N}}$$

then a and b both lie in $\mathcal{N}$ and, moreover

$$|p| <_c S^{|a|}$$
$$|q| <_c S^{|b|}$$

and therefore

$$|p|^{|q|} <_c (S^{|a|})^{(S^{|b|})}$$

But

$$(S^{|a|})^{(S^{|b|})} \cong_c S^{(|a| \times S^{|b|})}$$

Now define

$$k = a \times_{\mathcal{N}} \exp_{\mathcal{N}}(s, b))$$

where s is the element of $\mathcal{N}$ that measures S.[19] Then

$$|k| \cong_c |a| \times S^{|b|}$$

and therefore

$$|p|^{|q|} <_c S^{|k|}$$

as required. □

That (i) is equivalent to (iii) shows us that we cannot define a simply infinite system in which exponentiation is definable using the method of S-ary expansions unless we start with a system in which exponentiation is already defined.

Theorem 10.4.15 corresponds to a familiar fact of practical computation, namely the practical "infeasibility" the operation of exponentiation. We cannot *actually* compute decimal numerals for numbers that we can easily describe exactly using exponential notation: $729^{10^{729}}$ has more than 2.86×10^{729} digits.

Let me try to make the analogy here a little more exact. Let us suppose that $\mathcal{N}$ corresponds, under the analogy, to the species of *stroke numerals* (linear juxtapositions of the stroke symbol "|" – we can think of these as numerals to the base 1) that we could ("in principle") inscribe. Then

---

[19] This is the only place in the proof that the assumption that S is measured by $\mathcal{N}$ is used.

10.4 Extending simply infinite systems   345

$\mathcal{N}[10]$ would correspond to the species of decimal numerals that we could (again "in principle") inscribe. (Take a stroke numeral and replace each stroke by a decimal digit, making sure the first stroke is not replaced by **0**, e.g. replace ||| by **729**. Naively, it is no more difficult to write down a decimal numeral of a given length than it is to write down the stroke numeral of that length.)

Pursuing the analogy, the practical fact that we can easily write down a decimal numeral (e.g. with one thousand digits) which names a particular size, although we cannot write down a stroke numeral which names the same size, this practical fact, I maintain, corresponds, under the analogy, to the proposition that $\mathcal{N}$ is not a scale for $\mathcal{N}[10]$. But from the latter proposition we may infer, using Theorem 10.4.15, that exponentiation is not definable in $\mathcal{N}$, nor in $\mathcal{N}[10]$, nor in $(\mathcal{N}[10])[10]$, and so on. This series of propositions corresponds, under the analogy, to the infeasibility of exponentiation, and, indeed, provides a kind of measure of it.

If we take an S-ary extension of any simply infinite system, we obtain a system in which addition is definable. If we take an S-ary extension of a simply infinite system in which addition is definable, we obtain a system which admits multiplication. What happens if we extend a system in which multiplication is already definable? The previous theorem tells that we *don't* get exponentiation unless we already had it at the beginning. What, then, do we get after such an extension? This question is answered by the following theorem.

10.4.16 Theorem *Let $\mathcal{N}$ be a simply infinite system and let S be a numeration base. Then a necessary and sufficient condition that multiplication be definable in $\mathcal{N}$ is that the arithmetical function*

$$\lambda xy(\mathrm{lexp}_1(S, x, y))$$

be definable in $\mathcal{N}[S]$.

Proof (*Sufficiency*) Suppose that $\lambda xy(\mathrm{lexp}_1 1(S, x, y))$ is definable in $\mathcal{N}[S]$. Then there is a global function ψ_S such that for all x, y in $\mathcal{N}[S]$, $\psi_S(x, y)$ lies in $\mathcal{N}[S]$ and

$$|\mathrm{lexp}_1(S, |x|, |y|)| \cong_c |\psi_S(x, y)|$$

It will clearly suffice to define a global function, ϕ_S, such that for any x and y in $\mathcal{N}$

$$|\phi_S(x, y)| \geq_c |x| \times |y|$$

To that end, define ϕ_S over $\mathcal{N}$ by setting, for each x and y in $\mathcal{N}$

$$\phi_S(x, y) = lh(\psi_S(Values(\langle 0, \ldots, 0, 1 \rangle_x), Values(\langle 0, \ldots, 0, 1 \rangle_y)))$$

Now suppose that x and y are non-zero elements of $\mathcal{N}$. Then (recalling that $\log_S(S^T) = \text{card}(T) +_c 1$)

$$|\phi_S(x, y)| = |lh(\psi_S(Values(\langle 0, \ldots, 0, 1 \rangle_{x+1}), Values(\langle 0, \ldots, 0, 1 \rangle_{y+1})))|$$

$$\cong_c \log_S(|\psi_S(Values(\langle 0, \ldots, 0, 1 \rangle_{x+1}), Values(\langle 0, \ldots, 0, 1 \rangle_{y+1}))|)$$

$$\cong_c \log_S(\text{lexp}_1(S, |Values(\langle 0, \ldots, 0, 1 \rangle_{x+1})|, |Values(\langle 0, \ldots, 0, 1 \rangle_{y+1})|)))$$

$$\cong_c \log_S((S^{|x|})^{\log_S(S^{|y|})})$$

$$\cong_c \log_S((S^{|x|})^{(|y|+_c 1)})$$

$$\geq_c \log_S(S^{(x \times y)})$$

$$\cong_c |x| \times |y| +_c 1$$

Thus using ϕ_S we can define multiplication in $\mathcal{N}$, as required.

(*Necessity*) Suppose multiplication is definable in $\mathcal{N}$. To show that $\lambda xy(\text{lexp}_1(S, x, y))$ is definable in $\mathcal{N}[S]$ it will suffice to define ψ_S on $\mathcal{N}[S]$ such that, if x and y both lie in $\mathcal{N}[S]$,

$$|\psi_S(|x|, |y|)| \geq_c \text{lexp}_1(S, |x|, |y|) \cong_c |x|^{\log_S(|y|)}$$

To that end, define ψ_S by setting

$$\psi_S(x, y) = Values(\langle 0, \ldots, 0, 1 \rangle_{(lh(x) \times_{\mathcal{N}} lh(y))+1})$$

Then

$$|\psi_S(x, y)| \cong_c (S^{\log_S(|x|)})^{\log_S(|y|)} \geq_c |x|^{\log_S(|y|)}$$

as required. □

This theorem has an important corollary. But first we must lay down a definition.

10.4.17 Definition Let $\mathcal{N}$ be a simply infinite system, and let S and T be numeration bases with $S \leq_c T$. Then to say that T is *accessible to* S *over* $\mathcal{N}$ is to say that T is measured by $\mathcal{N}[S]$.

10.4 Extending simply infinite systems

Note that in this definition the *definiens* is Σ_1. Note also that in the absence of a proof that $\mathcal{N}[S]$ is a universal scale we cannot simply *assume* that T is measured by $\mathcal{N}[S]$, for arbitrary S and T, so the definition is necessary.

If it is the case both that T is accessible to S over $\mathcal{N}$ and that S is accessible to T over $\mathcal{N}$, then let us say that S and T *have the same order of magnitude over* $\mathcal{N}$. Clearly, if $S <_c T$, then the proposition that T is accessible to S over $\mathcal{N}$ entails, and is entailed by, the proposition that $\mathcal{N}$ measures $\log_S(T)$.

10.4.18 Corollary *Let $\mathcal{N}$ be a simply infinite system in which multiplication is definable and let S and T be numeration bases having the same order of magnitude over $\mathcal{N}$. Then $\mathcal{N}[S]$ measures $\mathcal{N}[T]$.*

Proof If $S \geq_c T$ then the required conclusion obviously follows, so we may assume that $S <_c T$. It will suffice to define a global function $\phi_{S,T}$ on $\mathcal{N}$ such that, for all k in $\mathcal{N}$, $\phi_{S,T}(k)$ lies in $\mathcal{N}$ and

$$T^{|k|} \leq_c S^{|\phi_{S,T}(k)|}$$

By the theorem just proved (10.4.16), $\lambda xy(\text{lexp}_1(S,x,y))$ is defnable in $\mathcal{N}[S]$ (since $\mathcal{N}$ is closed under multiplication). Hence we can define a global function ψ_S such that, for all x and y in $\mathcal{N}[S]$, $\psi_S(x,y)$ lies in $\mathcal{N}[S]$ and

$$|\psi_S(x,y)| \cong_c |x|^{\log_S(|y|)}$$

Now define, for k lying in $\mathcal{N}$,

$$\phi_{S,T}(k) = lh(\psi_S(\langle 0,\ldots,0,1\rangle_{k+1}, a_{\mathcal{N}[S]})) +_{\mathcal{N}} 1_{\mathcal{N}}$$

where

$$|a_{\mathcal{N}[S]}| \cong_c T \times S$$

Then $\phi_{S,T}(k)$ lies in $\mathcal{N}$ whenever k does. Moreover

$$\phi_{S,T}(k) \cong_c \log_S(|\psi_S(\langle 0,\ldots,0,1\rangle_{k+1}, a_{\mathcal{N}[S]})|) +_c 1$$
$$\cong_c \log_S((S^{|k|})^{\log_S(T \times S)}) +_c 1$$
$$\cong_c \log_S((S^{\log_S(T \times S)})^{|k|}) +_c 1$$
$$\geq_c \log_S(T^{|k|}) +_c 1$$

Hence $T^{|k|} \leq_c S^{|\phi_{S,T}(k)|}$ as required.

□

Let us take stock of what has been established in this last group of theorems (10.4.10 to 10.4.18). If we start with a simply infinite system $\mathcal{N}$ and numeration base S, we can form the system $\mathcal{N}[S]$ whose *terms* are the S-ary numerals defined on the linear orderings which are elements of $\mathcal{N}$. Then we can define addition in $\mathcal{N}[S]$ (10.4.10), and if T is a numeration base, we can form the T-ary expansion, $\mathcal{N}[S][T]$, of $\mathcal{N}[S]$ in which we can define multiplication (10.4.14). Again, if U is a numeration base we can form the U-ary expansion, $\mathcal{N}[S][T][U]$ over $\mathcal{N}[S][T]$, in which $\lambda xy\,(\text{lexp}_1(U, x, y))$ is definable (10.4.16).

In all of these cases the expansion is a scale for the system of which it is an expansion, but the base theory is a scale for the expansion only if it admits exponentiation (10.4.15). In fact, exponentiation is definable in *all* the theories $\mathcal{N}$, $\mathcal{N}[S]$, $\mathcal{N}[S][T]$, and $\mathcal{N}[S][T][U]$ if it is definable in *any* of them (10.4.15).

Once we have reached the second expansion, $\mathcal{N}[S][T]$, then whatever numeration base we choose for a third expansion, we obtain essentially the same simply infinite system that we should obtain by picking another numeration base of same order of magnitude over $\mathcal{N}[S][T]$.

What happens if we continue to carry out such expansions? Theorem 10.4.15 shows that we do not obtain exponentiation unless we had it in the first place. A more precise answer is provided by the next theorem, which generalises Theorem 10.4.16.

10.4.19 Theorem *Let $\mathcal{N}$ be a simply infinite system and let S be a numeration base. Then a necessary and sufficient condition that*

$$\lambda xy(\text{lexp}_k(S, x, y))$$

be definable in $\mathcal{N}$ is that

$$\lambda xy(\text{lexp}_{k+1}(S, x, y))$$

be definable in $\mathcal{N}[S]$.

Proof The argument here is similar to that given in the proof of 10.4.16.

(*Sufficiency*) Suppose that $\lambda xy(\text{lexp}_{k+1}(S, x, y))$ is definable in $\mathcal{N}[S]$. Then there is a global function ψ_S such that for all x, y in $\mathcal{N}[S]$, $\psi_S(x, y)$ lies in $\mathcal{N}[S]$ and

$$|\text{lexp}_{k+1}(S, |x|, |y|)| \cong_c |\psi_S(x, y)|$$

We must now exhibit a global function, ϕ_S, which defines

$$\lambda xy(\text{lexp}_k(S, x, y))$$

10.4 Extending simply infinite systems

on $\mathcal{N}$, so that, for x and y lying in $\mathcal{N}$,

$$|\phi_S(x,y)| \geq_c \text{lexp}_k(S, |x|, |y|)$$

To that end define ϕ_S over $\mathcal{N}$ by setting, for each x and y in $\mathcal{N}$,

$$\phi_S(x,y) = lh(\psi_S(n_{\mathcal{N}[S]}, Values(\langle 0, \ldots, 0, 1 \rangle_{\max(x,y)+1})))$$

Now suppose that x and y lie in $\mathcal{N}$. Then

$$|\phi_S(x,y)| = |lh(\psi_S(Values(\langle 0,1 \rangle)), Values(\langle 0, \ldots, 0, 1 \rangle_{\max(x,y)+1})))|$$

$$\cong_c \log_S(|\psi_S(Values(\langle 0,1 \rangle)), Values(\langle 0, \ldots, 0, 1 \rangle_{\max(x,y)+1}))|)$$

$$\cong_c \log_S(\text{lexp}_{k+1}(S, |Values(\langle 0,1 \rangle)|, |Values(\langle 0, \ldots, 0, 1 \rangle_{\max(x,y)+1})|))$$

$$\cong_c \log_S(S^{\text{lexp}_k(S, \log_S(|Values(\langle 0,\ldots,0,1 \rangle_{\max(x,y)+1})|))})$$

$$\geq_c \log_S(S^{\text{lexp}_k(S, |x|, |y|)})$$

$$\cong_c \text{lexp}_k(S, |x|, |y|) +_c 1$$

as required.

(*Necessity*) Suppose that ϕ_S defines $\lambda xy\, (\text{lexp}_k(S, x, y))$ in $\mathcal{N}$, so that, for x and y lying in $\mathcal{N}$,

$$|\phi_S(x,y)| \cong_c \text{lexp}_k(S, |x|, |y|)$$

To show that $\lambda xy(\text{lexp}_{k+1}(S, x, y))$ is definable in $\mathcal{N}[S]$ it will suffice to define ψ_S on $\mathcal{N}[S]$ such that, if x and y both lie in $\mathcal{N}[S]$,

$$|\psi_S(|x|, |y|)| \geq_c \text{lexp}_{k+1}(S, |x|, |y|) = |x|^{\text{lexp}_k(S, \log_S(|y|), |y|)}$$

To that end, define ψ_S by setting

$$\psi_S(x,y) = Values(\langle 0, \ldots, 0, 1 \rangle_{((lh(x) +_\mathcal{N} 1) \times_\mathcal{N} \phi_S(lh(y), lh(y)))})$$

Then

$$|\psi_S(x,y)| \cong_c (S^{\log_S(|x|+_c 1)})^{\text{lexp}_k(S, \log_S(|y|), \log_S(|y|))}$$

But

$$(S^{\log_S(|x|+_c 1)})^{\text{lexp}_k(S, \log_S(|y|), \log_S(|y|))} \geq_c (|x|)^{\text{lexp}_k(S, \log_S(|y|), \log_S(|y|))}$$

So

$$|\psi_S(|x|, |y|)| \geq_c |x|^{\text{lexp}_k(S, \log_S(|y|), \log_S |y|)} = \text{lexp}_{k+1}(S, |x|, |y|)$$

as required.

$\square$

Successive extensions of this sort thus provide us with simply infinite systems in which stronger and stronger forms of logarithmic exponentiation are definable. This suggests that we have the means for constructing a *hierarchy* of these extensions. I shall take up this topic in the next section.

10.5 The hierarchy of S-ary extensions

Let me begin by introducing some useful notation.

10.5.1 Definition *Let $\mathcal{N}$ be a simply infinite system and let S be a numeration base. Then by definition*

(i) $\mathcal{N}_0[S] = \mathcal{N}$

(ii) *Suppose that $\mathcal{N}_\mathbf{n}[S]$ has been defined. Then*

$$\mathcal{N}_{\mathbf{n}+1}[S] = (\mathcal{N}_\mathbf{n}[S])[S]$$

Like the definition of $\text{lexp}_\mathbf{n}$ in Chapter 9 (Definition 9.1.12) Definition 10.5.1 is by "syntactic recursion", that is to say, the definition gives explicit instructions on how to carry out a succession of *particular* actual definitions. The boldface letter "**n**" is not varying over "natural numbers" but rather over (potential) *inscriptions* of written decimal numerals, and the expression "**n** + **1**" denotes the written decimal numeral which immediately follows the decimal numeral denoted by "**n**".

Let us say that the succession of simply infinite systems

$$\mathcal{N}_0[S], \mathcal{N}_1[S], \ldots, \mathcal{N}_\mathbf{n}[S], \mathcal{N}_{\mathbf{n}+1}[S], \ldots$$

constitutes the *fundamental hierarchy of S-ary extensions based on $\mathcal{N}$*.

On the basis of the axioms for Euclidean set theory laid down so far we can infer that

I. The successor function $\lambda x(x +_c 1)$ is definable in $\mathcal{N}_0[S]$.

II. Addition is definable in $\mathcal{N}_1[S]$.

III. Multiplication is definable in $\mathcal{N}_2[S]$.

IV. The logarithmic exponential $\lambda xy\,(\text{lexp}_\mathbf{n}(S, x, y)$, for $\mathbf{n} \geq \mathbf{1}$, is definable in $\mathcal{N}_{2+\mathbf{n}}[S]$.

10.5 The hierarchy of S-ary extensions

By virtue of Corollary 10.4.18, if S and T are sufficiently small[20] numeration bases, then the fundamental hierarchy of S-ary expansions based on $\mathcal{N}$ is equivalent to the hierarchy of T-ary expansions in the sense that each of $\mathcal{N}_n[S]$ and $\mathcal{N}_n[T]$ measures the other.

How should we regard this fundamental hierarchy? Suppose we start with a simply infinite system $\mathcal{N}$ in which exponentiation is not definable. Notice, to begin with, that it is not immediately clear how to give an exact mathematical formulation to this supposition from a Euclidean finitary standpoint, for issues of global logic are clearly relevant here[21].

Let us, however, defer consideration of this difficulty and assume that a precise meaning can be given to the supposition that exponentiation is not definable in $\mathcal{N}$. Theorem 10.4.15 then tells us that, for any numeration base, S, exponentiation is not definable in the system $\mathcal{N}[S] (= \mathcal{N}_1[S])$ of S-ary numerals over $\mathcal{N}$ either. Moreover, $\mathcal{N}(= \mathcal{N}_0[S])$ is in some sense "shorter" than $\mathcal{N}_1[S]$, for although $\mathcal{N}_1[S]$ measures $\mathcal{N}_0[S]$, $\mathcal{N}_0[S]$ does not measure $\mathcal{N}_1[S]$ (again we must ignore the obvious logical difficulties in giving an exact mathematical formulation of this observation). Repeating the same (so far inexactly grounded) reasoning, exponentiation is not definable in $\mathcal{N}_2[S]$, and $\mathcal{N}_1[S]$ is "shorter" than $\mathcal{N}_2[S]$, for whereas the latter measures the former, the former does not measure the latter.

In general, when we arrive at $\mathcal{N}_{n+1}[S]$ we may conclude that exponentiation is not definable in that system, but that it too is "longer" than its predecessor $\mathcal{N}_n[S]$ in the sense already described. Thus starting with a simply infinite system $\mathcal{N}$ in which exponentiation is not definable, we may successively carry out as many S-ary expansions as we wish – or as we can – and in the course of this procedure we arrive at longer and longer simply infinite systems.

If we assume that not even addition is definable in $\mathcal{N}$, $\mathcal{N}_1[S]$ admits addition, but not multiplication, $\mathcal{N}_2[S]$ admits both addition and multiplication, but not the simplest version of logarithmic exponentiation, namely, $\lambda xy \text{lexp}_1(S, x, y)$. In general, for **n** beyond **2**, $\mathcal{N}_n[S]$ admits addition, multiplication, and logarithmic exponentiation in the forms

$$\lambda xy \text{lexp}_1(S, x, y), \lambda xy \text{lexp}_2(S, x, y), \ldots, \lambda xy \text{lexp}_{n-2}(S, x, y)$$

---

[20] I shall not attempt to spell out here what "sufficiently small" means, but clearly 2 and 10, for example, are sufficiently small in the sense required.

[21] One way to formulate this supposition is as follows

$\neg\neg\exists xy[x, y \text{ lie in } \mathcal{N} \wedge \neg\neg\forall z[z \text{ lies in } \mathcal{N} \text{ .implies. } z <_c (|x|^{|y|})]]$

but that raises the question of the meaning of $\neg$ ("it is impossible that"?).

but not

$$\lambda xy \text{lexp}_{n-1}(S, x, y)$$

which is, however, admitted by $\mathcal{N}_{n+1}[S]$. Exponentiation is not definable in any of these systems, and, in consequence, none of them constitutes a universal scale.

10.6 Simply infinite systems that grow slowly in rank

Let me begin by adding a new type of simply infinite system to our present stock.

10.6.1 Definition *Let U be a set of individuals ($U = \emptyset$ is allowed). Then the cumulative hierarchy, $\mathcal{H}(U)$, over U, is the simply infinite system generated by the global function, $\lambda x(U \cup P(x))$, from the initial term U.*

The linear orderings generated by $\mathcal{H}(U)$ have the form

$$[U, U \cup P(U), \ldots, k, U \cup P(k)]$$

If r lies in $\mathcal{H}(U)$ then a straightforward induction along r proves that

$$(\forall x \in (Field(r) - \{Last(r)\}))[x <_c Last(r)]$$

from which it follows that $P(Last(r)) \notin Field(r)$ and therefore $\mathcal{H}(U)$ is, indeed, a simply infinite system.

The species of *terms* of $\mathcal{H}(U)$ is locally definable since a set S is a term of $\mathcal{H}(U)$ if, and only if, there exists a linear ordering r with $Field(r) \subseteq S$ such that

(i) $First(r) = U$
(ii) $(\forall x \in (Field(r) - \{Last(r)\}))[Next(r, x) = U \cup P(x)]$
(iii) $U \cup P(Last(r)) = S$

Of course, given a *term* of $\mathcal{H}(U)$ we can locally recover the *element* of $\mathcal{H}(U)$ of which it is the last term.

The fact that we can locally define the species of terms of $\mathcal{H}(U)$ is of significance because from the Cantorian standpoint the *terms* (not the elements) of $\mathcal{H}(U)$ are the Euclidean finite levels of the cumulative hierarchy over the set U of individuals, that is, if r is an element of $\mathcal{H}(U)$ then r is of the form

$$r = [V_0(U), \ldots, V_k(U)]$$

10.6 Simply infinite systems that grow slowly in rank

Thus from the Cantorian standpoint every hereditarily Euclidean finite set, S, with $Support(S) \subseteq U$ – that is to say, every member, S, of $V_\omega(U)$ – is a member of some term of $\mathcal{H}(U)$. But in the Euclidean case we do not know how "long" $\mathcal{H}(U)$ is, so we do not know whether every set whose support is a subset of U is a member of some term of $\mathcal{H}(U)$.

In fact, it is easy to see that $\mathcal{H}(U)$ is bounded in length by both $\mathcal{V}\mathcal{N}$ and $\mathcal{L}$. For if r is an element of $\mathcal{H}(U)$ then, for example

$$\{x \in Last(r) : x \text{ is a von Neumann natural number}\}$$

ia a von Neumann natural number that measures $Field(r)$. As we shall see[22] this means that $\mathcal{H}(U)$ is relatively "short".

The simply infinite system $\mathcal{H}(U)$ thus represents an attempt to "construct" the Euclidean cumulative hierarchy of sets over U. There are as many levels in that hierarchy as there are elements in the simply infinite system $\mathcal{H}(U)$. Note that if U_1 and U_2 are both sets of individuals and $U_1 \subseteq U_2$, then given any element, r, of $\mathcal{H}(U_2)$, we can define an element r' of $\mathcal{H}(U_1)$ with $r \cong_0 r'$[23]. It follows that $\mathcal{H}(U_1)$ measures $\mathcal{H}(U_2)$. It is not clear, at this point, whether, or in what circumstances, $\mathcal{H}(U_2)$ measures $\mathcal{H}(U_1)$, and thus whether the hierarchies over different sets, U, of individuals are all of the same length.

In any case, I shall now show how, given a linear ordering l whose field is U, we can use $\mathcal{H}(U)$ to define a simply infinite system $\mathcal{ACH}(U,l)$, *the Ackermann simply infinite system over U determined by l*[24], which corresponds to an attempt to "construct" the Euclidean universe of sets over U directly, set by set.

The idea of the "construction" is that, given an element

$$r = [V_0(U), \ldots, V_k(U)]$$

of $\mathcal{H}(U)$ and a linear ordering, l, of U we can systematically linearly order each of the terms of the given ordering, r, in such a way that each of the successive linear orderings

$$s_0(=l), \ldots, s_k$$

---

[22] See Corollaries 11.2.6 and 11.2.8.

[23] A straightforward induction along r shows that each of its terms contains the corresponding term of $\mathcal{H}(U_1)$ as a subset.

[24] The system $\mathcal{ACH}(\emptyset, [\,])(= \mathcal{ACH}(\emptyset, \emptyset))$ was first defined by S. Popham. He called it "*Slow*" for reasons that I shall shortly explain. I have called it "$\mathcal{ACH}(\emptyset, [\,])$" because it generates pure sets in order of their Ackermann codes. See Ackermann's "Die Widerspruchsfreiheit der allgemeinen Mengenlehre".

of the sets

$$V_0(U)(=U),\ldots,V_k(U)$$

respectively, is an initial segment of all those that follow. The essential trick here is to define s_{i+1} from l and the *lexicographic ordering* on s_i given by the next definition.

10.6.2 Definition *Let r be a linear ordering. Then $lex(r)$ (the lexicographic ordering on $P(Field(r))$ induced by r) is the binary relation on $P(Field(r))$ given by specifying that for all $x, y \subseteq P(Field(r))$ $x \leq_{lex(r)} y$ if, and only if,*

(i) $x = y$

or

(ii) $x \neq y$ *and* $\max_r((x - y) \cup (y - x)) \in y$

If $r = [r_0, \ldots, r_k]$, then each $S \subseteq Field(r)$ can be associated with its *local Ackermann code*, namely, the sequence f_S defined on $[r_0, \ldots, r_k]$ by stipulating that

$$f_S`r_i = \begin{cases} 0 & \text{if } r_i \notin S \\ 1 & \text{if } r_i \in S \end{cases}$$

Each such f_S can be identified with a binary numeral of length r (possibly terminating in zeros[25]) which codes the characteristic function of S as a subset of $Field(r)$ with respect to the ordering of $Field(r)$ given by r. The lexicographic ordering of $P(Field(r))$ induced by r is then given by

$$S \leq_{lex(r)} T \text{ .iff. } f_S \leq_{binary} f_T$$

The next theorem shows how we can weave these lexicographic orderings together when the base orderings are of sets of individuals.

10.6.3 Theorem *Let r_1 and r_2 be linear orderings whose terms are individuals. If r_1 is an initial segment of r_2, then $lex(r_1)$ is an initial segment of $lex(r_2)$.*

Proof Note first that since $P(Field(r_1)) \subseteq P(Field(r_2))$

$$Field(lex(r_1)) \subseteq Field(lex(r_2))$$

Now let $S, T \subseteq Field(r_1)$. Then

$$S \leq_{lex(r_1)} T \text{ iff } S \leq_{lex(r_2)} T$$

---

[25] Recall that our binary numerals are read in reverse order. We may call f_S the *local Ackermann code for S relative to r.*

10.6 Simply infinite systems that grow slowly in rank

since

$$\max_{r_1}((S-T)\cup(T-S)) = \max_{r_2}((S-T)\cup(T-S))$$

Thus $lex(r_1)$ is a subordering of $lex(r_2)$.

But $lex(r_1)$ is actually an initial segment of $lex(r_2)$, for if $S \subseteq Field(r_1)$ and $T - Field(r_1) \neq \emptyset$ then $S <_{lex(r_2)} T$. For in that case

$$\max_{r_2}((S-T)\cup(T-S)) \in T$$

The theorem is thus established.

□

If either r_1 or r_2 contains sets in its field complications arise. Let us, however, consider a simple example that will illustrate how to weave lexicographic orderings together to obtain a linear ordering of the terms of the simply infinite system, $\mathscr{H}(\emptyset)$. Starting with the empty linear ordering we may define

$s_0 = [\,] (= \emptyset)$

$s_1 = lex(s_0) = [\emptyset]$

Then $P(Field(s_1)) = P(\{\emptyset\}) = \{\emptyset, \{\emptyset\}\}$ so we may define

$s_2 = lex(s_1) = [\emptyset, \{\emptyset\}]$

Note that the local Ackermann codes for $\emptyset$ and $\{\emptyset\}$ are 0 and 1, respectively. Continuing in the same way we obtain

$s_3 = lex(r_1) = [\,\emptyset, \{\emptyset\}, \{\{\emptyset\}\}, \{\emptyset, \{\emptyset\}\}\,]$

Note that the terms of s_3 have local Ackermann codes 00, 10, 01, 11, in that order.

$s_4 = [A_{s_3}(0000), A_{s_3}(1000), \ldots, A_{s_3}(1111)]$

where, for example

$$A_{s_3}(1010) = \{\emptyset, \{\{\emptyset\}\}\}$$

In each of these examples the size of s_i is given by an exponential tower of i 2's topped with a 0. Thus s_3 has $2^{2^{2^0}} = 4$ terms, s_4 has 16, s_5 will contain 256 terms, and s_6 2^{256}, etc. Note that if a set T appears in s_i then its local Ackermann code remains fixed in all subsequent s_j's, but supplied with a tail of redundant 0's. Thus each s_i is an initial segment of all that follow it. Note also, that in this particular case of the *pure* sets, the local Ackermann coding corresponds exactly to the conventional Ackermann coding of pure sets in natural number arithnmetic.

Now let us return to the problem of defining the simply infinite system $\mathscr{ACH}(U,l)$. We start with a key definition.

10.6.4 Definition *Let U be a set of individuals, l be a linear ordering of U, and r be an element of $\mathscr{H}(U)$. Then by definition s is an Ackermann sequence for r based on l if, and only if,*

(i) *s is a sequence defined on r.*
(ii) *If $r \neq []$ then*
   (a) $s_{First(r)} = l$
   (b) $\forall x \in (Field(r) - \{Last(r)\})[s_{Next(r,x)} = l * lex(x)]$

Now we must establish the basic properties of Ackermann sequences.

10.6.5 Theorem *Let U be a set of individuals, l be a linear ordering of U, and r be a linear ordering which lies in $\mathscr{H}(U)$. Then*

(i) *If s is an Ackermann sequence for r based on l, and x lies in the field of r, then s_x is a linear ordering of x, and if $x \neq Last(r)$, then s_x is an initial segment of $s_{Next(r,x)}$.*
(ii) *There is exactly one Ackermann sequence for r based on l.*

Proof Let us suppose that $r = [V_0(U), \ldots, V_k(U)]$ and $l = [l_1, \ldots, l_n]$.[26]

(i) We proceed by induction along r. The proposition clearly holds for $V_0(U)$ since $s_{V_0(U)}$ is l and $s_{V_1(U)}$ is $l * lex(l)$. In this case, therefore, the field of $lex(l)$ consists of subsets of U, none of which can appear as elements of U since the latter is composed exclusively of individuals. Hence $lex(s_{V_0(U)})$ linearly orders $P(V_0(U))$ and l linearly orders $V_0(U)$, so $l * lex(V_0(U))$ linearly orders $V_1(U)$.

Let us now consider the case for $V_i(U) \in Field(r)$, where $0 < i < k$. We may assume, as hypothesis of induction, that for $j \leq i$, $s_{V_j(U)}$ linearly orders $V_j(U)$, and that for $j < i$, $s_{V_j(U)}$ is an initial segment of $s_{V_{j+1}(U)}$. We must then prove that $s_{V_{i+1}(U)}$ is a linear ordering of $V_{i+1}(U)$ and that $s_{V_i(U)}$ is an initial segment of $s_{V_{i+1}(U)}$.

The problem here is that some subsets of $V_{i+1}(U)$ will already have been placed in order by $s_{V_i(U)}$ and they will receive *new* ordering relations in $lex(s_{V_{i+1}(U)})$. We must therefore begin by showing that for $S, T \in V_i(U)$

$$S \leq_{s_{V_i(U)}} T \text{ iff } S \leq_{s_{V_{i+1}(U)}} T$$

---

[26] The notation "$V_i(U)$" for terms of $\mathscr{H}(U)$ is merely a mnemonic device intended to suggest their structure and has no deeper or technical significance.

10.6 Simply infinite systems that grow slowly in rank

To that end, suppose that S and T are sets with $S, T \in V_i(U)$. If $S \leq_{s_{V_i(U)}} T$, then since S and T are sets and since we have assumed that $0 < i$, $i = j + 1$ for some j, it follows that $S \leq_{lex(s_{V_j(U)})} T$, i.e.

$$\max{}_{s_{V_j(U)}}((S - T) \cup (T - S)) \in T$$

Therefore, since by hypothesis of induction $s_{V_j(U)}$ is an initial segment of $s_{V_i(U)}$, we may conclude that

$$\max{}_{s_{V_i(U)}}((S - T) \cup (T - S)) \in T$$

so $S \leq_{lex(s_{V_i(U)})} T$ and it follows that $S \leq_{s_{V_{i+1}(U)}} T$, as required.

Conversely, suppose that $S \leq_{s_{V_{i+1}(U)}} T$. Since S and T are sets

$$S \leq_{lex(s_{V_i(U)})} T$$

and therefore

$$\max{}_{s_{V_i(U)}}((S - T) \cup (T - S)) \in T$$

But since members of members of $V_i(U)$ are members of $V_j(U)$

$$(S - T \cup (T - S)) \subseteq V_j(U)$$

and since by hypothesis of induction $s_{V_j(U)}$ is an initial segment of $s_{V_i(U)}$

$$\max{}_{s_{V_j(U)}}((S - T) \cup (T - S)) \in T$$

It follows that $S \leq_{s_{V_j(U)}} T$, as required.

Hence $s_{V_i(U)}$ is a subordering of $s_{V_{i+1}(U)}$. But, in fact, it is an initial segment of $s_{V_{i+1}(U)}$. For the new sets in $V_{i+1}(U) - V_i(U)$ must all contain at least one set which occurs in $V_i(U)$ but does not occur in any *member* of $V_i(U)$, so if $S \in V_i(U)$ and $T \in (V_{i+1}(U) - V_i(U))$ then

$$\max{}_{s_{V_i(U)}}((S - T) \cup (T - S)) \in T$$

and therefore $S \leq_{V_{i+1}(U)} T$, as required.

But from the fact that $s_{V_i(U)}$ is an initial segment of $s_{V_{i+1}(U)}$ it follows from the definition of the latter that it is a linear ordering of $V_{i+1}(U)$ as required. Part (i) now follows by induction along r.

(ii) This will follow if we can define a sequence s on r by a recursion along r of the following form
$s`First(r) = l$
$s`Next(r, x) = l * lex(s`x)$ if $x \neq Last(r)$
Such a recursion is possible if we can find a target set for s which contains all the necessary values for s at arguments in $Field(r)$. But at each argument $V_i(U)$ the value of s at that argument will be an ordering

of $V_i(U)$, and therefore a subset of $V_i(U) \times V_i(U)$ and hence a member of $V_{i+3}(U)$. Since $V_i(U) \subseteq V_j(U)$ for $i < j$ (in the notation used for r), we can take $V_{k+3} (= P(U \cup P(U \cup P(U \cup (V_k)))))$ as the target set for the values of s in the recursion. This gives us an Ackermann sequence and its uniqueness then follows.

$\square$

Given a term, S, of $\mathcal{H}(U)$ we can define the *element*

$$r = [V_0(U)(= U), V_1(U)(= U \cup P(U)), \ldots, V_k(U)(= S)]$$

of $\mathcal{H}(U)$ whose last term is S, since all the terms of r preceeding S in r are members of S. Thus we can define a *canonical linear ordering based on l* for each such term and each linear ordering, l, of U.

10.6.6 Definition *Let U be a set of individuals, l be a linear ordering of U, S a term of $\mathcal{H}(U)$, r the element of $\mathcal{H}(U)$ whose last term is S, and s the Ackermann sequence for r based on l. Then the canonical ordering, $\leq_{S,l}$, on S induced by l is defined by*

$$\leq_{S,l} = s`Last(r)(= s`S)$$

As an immediate consequence of Theorem 10.6.5 we obtain the following corollary.

10.6.7 Corollary *Let U be a set of individuals, l be a linear ordering of U, and let r_1 and r_2 be linear orderings that are elements of $\mathcal{H}(U)$ with $r_1 \leq_{\mathcal{H}(U)} r_2$. Then $\leq_{Last(r_1),l}$ is an initial segment of $\leq_{Last(r_2),l}$.*

We are now in a position to define the simply infinite system $\mathscr{ACH}(U,l)$ described informally above. It will be convenient to begin by defining the species of linear orderings that are elements of $\mathscr{ACH}(U,l)$ before going on to define the successor function $\lambda x(\sigma_{\mathscr{ACH}(U,l)}(x))$ which generates it.

10.6.8 Definition *Let U be a set of individuals and l be a linear ordering of U. Then, by definition, a linear ordering r lies in $\mathscr{ACH}(U,l)$ if, and only if, the following conditions obtain:*

(i) *There is a subordering $\hat{r}$ of r such that $\hat{r}$ lies in $\mathcal{H}(U)$ and*

$$Field(\hat{r}) = \{x \in Field(r) : x \text{ is a term of } \mathcal{H}(U)\}$$

(ii) *If $\hat{r} = []$, then r is an initial segment of l.*
(iii) *If $\hat{r} \neq []$, then r is an initial segment of $\leq_{\sigma_{\mathcal{H}(U)}(S),l}$.*

10.6 Simply infinite systems that grow slowly in rank

Obviously, if S lies in $\mathscr{H}(U)$ then $\leq_{s,l}$ lies in $\mathscr{ACH}(U,l)$. This makes it clear how to define the global successor function for $\mathscr{ACH}(U,l)$.

10.6.9 Definition *The global function* $\lambda xyz(\sigma_{\mathscr{ACH}(x,y)}(z))$ *of three arguments is defined by the following stipulations:*

(i) *If U is a set of individuals, l is a linear ordering of U, and r lies in $\mathscr{ACH}(U,l)$, then*

    (a) *If $\hat{r} = [\,]$ and r is a proper initial segment of l, then*

$$\sigma_{\mathscr{ACH}(U,l)}(r) = Next(l, Last(r))$$

    (b) *If $\hat{r} = [\,]$, and $r = l$, then*

$$\sigma_{\mathscr{ACH}(U,l)}(r) = \emptyset \,(= First(lex(l)))$$

    (c) *If $\hat{r} \neq [\,]$, then*

$$\sigma_{\mathscr{ACH}(U,l)}(r) = Next(\leq_{\sigma_{\mathscr{H}(U)}(Last(\hat{r})),l}, Last(r))$$

where $\hat{r}$ is the subordering of r which lies in $\mathscr{H}(U,l)$ such that

$$Field(\hat{r}) = \{x \in Field(r) : x \text{ is a term of } \mathscr{H}(U)\}$$

and s is the Ackermann sequence for $\hat{r}$ based on l.

(ii) *If U is not a set of individuals, or l is not a linear ordering of U, or r does not lie in $\mathscr{ACH}(U,l)$, then*

$$\sigma_{\mathscr{ACH}(U,l)}(r) = \emptyset$$

As an immediate consequence of the last two definitions we have the following theorem.

10.6.10 Theorem *Let U be a set of individuals and l a linear ordering of U. Then the global function $\lambda x \sigma_{\mathscr{ACH}(U,l)}(x)$ generates the species $\mathscr{ACH}(U,l)$ from $First(l)$ as a simply infinite system.*

It is obvious that $\mathscr{ACH}(U,l)$ measures $\mathscr{H}(U)$, but, as we shall see in Chapter 11, we cannot define a measure for either $\mathscr{VN}$ or $\mathscr{Z}$ with respect to $\mathscr{ACH}(U,l)$, nor, conversely, can we define a measure for the latter with respect to either of the former[27].

$\mathscr{ACH}(U)$ has strong closure properties.

---

[27] It may be that there are sets U of individuals which themselves "measure" $\mathscr{VN}$ and/or $\mathscr{Z}$ in some sense, but we cannot establish this on the basis of the axioms laid down so far.

10.6.11 Theorem *Let U be a set of individuals and l a linear ordering of U. Then the simply infinite system $\mathscr{ACH}(U,l)$ is closed under addition, multiplication, and exponentiation.*

Proof $\mathscr{ACH}(U,l)$ is closed under the basic set-theoretical operations of pair set, power set, union, comprehension, and cartesian product. □

The system $\mathscr{ACH}(U,l)$ is sweeping out the (initial part of) the species of all sets whose support is a subset of the set U of individuals. Any set belonging to that species that we can *actually name* directly in our conceptual notation[28] will be picked up by it. The successive terms of $\mathscr{ACH}(U,l)$ are growing as slowly in rank as it is possible to grow for a simply infinte sequence whose generating function is monotone increasing in rank as a function of the rank of its arguments.

To make this observation more precise I must lay down the definition of the rank of a set that is appropriate for Euclidean set theory.

10.6.12 Definition *Let a be any set or individual, and let r be a linear ordering. Then, by definition, a is of rank r if there is an epsilon chain r_1 from a of maximal length in the epsilon fan, $\epsilon(a)$, of a such that $r +_o 1 \cong_o r_1$.*[29]

Clearly, if a is both of rank r and of rank r', then $r \cong_o r'$, so that, up to order isomorphism, there is only one linear ordering, r, such that a given object a is of rank r. It makes sense, therefore, to speak of one set's being of *lower rank* than another, and accordingly I shall use the notation "$x \leq_{rank} y$", "$x <_{rank} y$", etc.

10.6.13 Theorem *Let U be a set of individuals and l a linear ordering of U. Let the linear ordering r lie in the simply infinite system $\mathscr{ACH}(U,l)$. Then*

$$(\forall x \in Field(r))[x \leq_{rank} Last(r)]$$

We can prove a kind of converse to this theorem.

10.6.14 Theorem *Let U be a set of individuals. Then if T is of rank r, where r lies in $\mathscr{H}(U)$, and $Support(T) \subseteq U$, then $T \subseteq Last(r)$.*

---

[28] e.g. by writing down an appropriate inscription on paper with pen and ink.
[29] See Definition 8.6.1 and Axiom 8.6.2.

10.6 Simply infinite systems that grow slowly in rank

Proof For each linear ordering r' which is an initial segment of r, let $\Phi(r')$ be the proposition

$$(\forall x \subseteq \text{Constituents}(T))[x \text{ is of rank } r' \text{ .implies. } x \subseteq \text{Last}(r')]$$

Since the definition of $\Phi(r)$ is clearly local, we can use induction along r to prove that for all $S \subseteq \text{Constituents}(T)$

$$S \subseteq \text{Last}(r)$$

Since $T \subseteq \text{Constituents}(T)$ the theorem follows.

□

The import of these last two theorems is that the ranks of terms in the system $\mathscr{ACH}(U,l)$ do not increase until *all* terms of a given rank are first generated.

What, then, are the various relationships of "length" among simply infinite systems of the forms $\mathscr{H}(U)$ and $\mathscr{ACH}(U,l)$ when U and l are allowed to vary? The results that follow give partial answers to this question.

10.6.15 Theorem *If U is a set of individuals and $U_1 \subseteq U$, then $\mathscr{H}(U_1)$ is a scale for $\mathscr{H}(U)$.*

Proof Let the linear ordering r lie in $\mathscr{H}(U)$ and let the function f be defined on the field of r by

$$f`x = \{y \in x : \text{Support}(y) \subseteq U_1\}$$

Then a straightforward induction along r establishes that, for every x in the field of r

$$f`x \text{ is a term of } \mathscr{H}(U_1)$$

and, if $x \neq \text{Last}(r)$,

$$f`\text{Next}(r,x) = \sigma_{\mathscr{H}(U_1)}(f`x)$$

It follows that if

$$r' = [f`\text{First}(r), \ldots, f`\text{Last}(r)]$$

then r' lies in $\mathscr{H}(U_1)$ and $r' \cong_o r$.

□

Notice that the "thinner" hierarchy, $\mathscr{H}(U_1)$, measures the "thicker" one, $\mathscr{H}(U)$, here.

362   *The Euclidean Theory of Simply Infinite Systems*

In the case of simply infinite systems of the form $\mathscr{ACH}(U,l)$, based on the same set U of individuals, the length of the system is independent of the linear ordering l of U, as the following theorem shows,

10.6.16 Theorem *Let U be a set of individuals, and let l_1 and l_2 both be linear orderings of U. Then $\mathscr{ACH}(U,l_1)$ is a scale for $\mathscr{ACH}(U,l_2)$.*

Proof Let r be an element of $\mathscr{ACH}(U,l_2)$. If r is an initial segment of l_2 then there is clealy an initial segment r' of l_1 of the same length; moreover, r' lies in $\mathscr{ACH}(U,l_2)$. Suppose, then, that r is not an initial segment of l_2 and let S be the first term of $\mathscr{H}(U)$ that is not a term of r. Then $Field(r) \subseteq S$ and if r' is the element of $\mathscr{ACH}(U,l_1)$ whose field is S, some initial segment of r' measures r.

$\square$

The next obvious question to ask is whether the length of either $\mathscr{H}(U)$ or $\mathscr{ACH}(U,l)$ depends only on the cardinality of U. I shall take up this question in the next section.

10.7 Further axioms

Up to now I have laid down only three axioms beyond the basic axioms of Chapter 4 as characterising the Euclidean standpoint in set theory: the Axiom of Euclidean Finiteness[30], the Epsilon Fan Axiom[31], and the Euclidean Axiom of Foundation[32]. But the issues raised in the last two sections demand that we must consider the possibility that further axioms may be required. For questions about the existence of measures and about the closure properties of simply infinite systems under various arithmetical operations obviously depend upon what global functions can be defined, and therefore, ultimately, upon what further finiteness principles we are forced, or inclined, to adopt.

The first question we must address concerns the simply infinte systems of the form $\mathscr{H}(U)$ which generate the cumulative hierarchy over sets, U, of individuals. It is natural to suppose that the "length" of the simply infinite system $\mathscr{H}(U)$ depends only on the cardinality of its base set U, and not upon the individual natures of the elements of U: after all, from the standpoint of set theory, the *individuals* that compose U have

---

[30] Axiom 8.3.1.
[31] Axiom 8.6.2.
[32] Axiom 8.6.6

no "internal structure". All that is known about them is that they are distinct and collectively comprise a collection having a definite size.

Surely the length of $\mathcal{H}(U)$ depends only on the cardinality of U, so that if $U \cong_c U'$ then $\mathcal{H}(U)$ is a scale for $\mathcal{H}(U')$. Perhaps this is provable from the axioms already laid down for Euclidean set theory, but if so its proof has escaped me. The obvious induction argument fails because the induction property is Σ_1. The system $\mathcal{H}(U \cup U')$ contains initial segments of both $\mathcal{H}(U)$ and $\mathcal{H}(U')$ which are of the same length, but since we don't know the length of $\mathcal{H}(U \cup U')$ we cannot show that those segments coincide with the original systems.

If we regard individuals as cognate to the *units* in a Greek *arithmos*, what we have run up against here is a variant of the problem of *pure units* discussed in Section 2.6. In default of a proof, we must lay down an axiom which says, in effect, that individuals are interchangable.

Accordingly let us introduce a new first level global function constant, υ, of two arguments, and lay down the following axiom

10.7.1 Axiom *(Axiom of Uniformity) Let S be a set and f be a local, 1–1 function whose domain is Support(S) and whose range is composed entirely of individuals. Then there is a bijective local function*

$$g : Constituents(S) \to Constituents(\upsilon(S, f))$$

such that

(i) $(\forall x \in Support(S))[g`x = f`x]$
(ii) $(\forall x \in (Constituents(x) - Support(S)))[g`x = g``x]$
(iii) $\upsilon(S, f) = g``S$

Thus the set $\upsilon(S, f)$ is "constructed" in exactly the same manner as S, except that each individual, x, in the support of S is replaced by the "cognate" individual, $f`x$, determined by the (1 − 1) local function f.

It follows as a consequence of this axiom that if U and U_1 are sets of individuals and $U \cong_c U_1$, then each of $\mathcal{H}(U)$ and $\mathcal{H}(U_1)$ is a scale for the other and, similarly, by Theorem 10.6.16, each of $\mathscr{ACH}(U, l)$ and $\mathscr{ACH}(U_1, l_1)$ (where l and l_1 are linear orderings of U and U_1, respectively) is a scale for the other.

We now know that the length of the hierarchy system $\mathcal{H}(U)$ depends only on the cardinality of U; moreover, if $U <_c U_1$, then $\mathcal{H}(U_1)$ is a scale for $\mathcal{H}(U)$, so, in particular, $\mathcal{H}(\emptyset)$ is a scale for all of them. But how far out does each $\mathcal{H}(U)$ extend? In particular, does it pick up all sets whose support is a subset of U? I have two further candidate axioms to consider, the Weak and Strong Hierarchy Principles, both of which provide answers to these questions.

The Weak Hierarchy Principle is the asumption that, given any set, S, then the species of all sets T whose support is a subset of the support of S and whose rank is smaller than, or equal to, that of S is finite and therefore forms a set, the *cumulative hierarchy over Support(S) up to S*, which we denote by

$$R(S)$$

On this assumption, the correspondence

$$S \mapsto R(S)$$

defines a first level global function which can be characterised by the following axiom:

10.7.2 Axiom *(The Weak Hierarchy Principle)*

$$\forall xy [x \in R(y) \text{ .iff. } Support(\{x\}) \subseteq Support(\{y\}) \text{ and } x <_{rank} y]$$

As in the Cantorian case we can show that for every set, S, the von Neumann natural numbers lying in the set $R(S)$ constitute a von Neumann natural number of the same rank as S. It is thus natural, in the presence of the Weak Hierarchy Principle, to lay down the following definition:

10.7.3 Definition *Let s be any set. Then by definition*

$$rank(S) = \{x \in R(S) : x \text{ is a von Neumann natural number}\}$$

It is not difficult to show that both S and $R(S)$ are of rank $\subseteq_{rank(s)}$, as, indeed, is $rank(S)$ itself.

The Weak Hierarchy Principle is consistent with the von Neumann simply infinite system's not being closed under addition in the following sense: even with the addition of the global function $\lambda x R(x)$ to our list of initially given global functions, it is not possible to define a binary global function which can be proved, from the previously adopted axioms, augmented by the Weak Hierarchy Principle, to define addition in the von Neumann simply infinite system[33].

It is a consequence of the Weak Hierarchy Principle that every set is included in some term of the hierarchy system based on its support, for

---

[33] This is a *Cantorian* meta-theorem. See Corollaries 11.1.2 and 11.2.6 in the next chapter.

10.7 Further axioms

$R(S)$ picks out the first term of the hierarchy system $\mathscr{H}(Support(S))$ of which S is a subset. It follows that any Ackermann system based on the set U of individuals is a scale for the species of all sets whose support is a subset of U. In fact, we can establish a stronger theorem.

10.7.4 Theorem *Let S be any set. Then S occurs as a term in any Ackermann system of the form $\mathscr{ACH}(U,l)$ where U is a set of individuals with $Support(S) \subseteq U$ and l linearly orders U.*

Equally significant is the the fact that from the Weak Hierarchy Principle it follows that all hierarchy systems have the same length.

10.7.5 Theorem *Assume the Weak Hierarchy Principle as hypothesis, and let U_1 and U_2 be sets of individuals with $U_1 \subseteq U_2$. Then $\mathscr{H}(U_2)$ is a scale for $\mathscr{H}(U_1)$.*

Proof Let S be a term of $\mathscr{H}(U_1)$, so that $R(S) = S$. It follows that $Support(S \cup U_2) = U_2$ and $rank(S \cup U_2) = rank(S)$. Hence

$$[U_1, \ldots, S] \cong_o [U_2, \ldots, R(S \cup U_2)]$$

$\square$

This theorem, together with the Axiom of Uniformity and Theorem 10.6.14, immediately yields

10.7.6 Corollary *Let U_1 and U_2 be sets of individuals. Then each of $\mathscr{H}(U_1)$ and $\mathscr{H}(U_2)$ is a scale for the other.*

Again as in the Cantorian case there is a stronger version of the hierarchy principle available. Formally we introduce a new global function, $\lambda xy V_y(x)$, by means of the following axiom.

10.7.7 Axiom *(The Strong Hierarchy Principle) Let S be a set and r a linear ordering. Then by definition*

$$\forall x[x \in V_r(S) \text{ .iff. } Support(\{x\}) \subseteq Support(\{S\}) \text{ and } x \text{ is of rank } < r]$$

Since $rank(V_r(S)) \cong_o r$ for any linear ordering r, it is clear that the Strong Hierarchy Principle entails that the von Neumann simply infinite system forms a universal scale. As in the Cantorian case the Strong Hierarchy Principle is equivalent, in the presence of the Weak Hierarchy Principle, to the assumption that every linear ordering is order isomorphic to a von Neumann ordinal.

These two axioms dealing with the global structure of the universe of Euclidean set theory must be *taken into consideration* as possible new axioms because of the analogy with the Cantorian universe. But should we *adopt* either of them? *That* is an altogether different question.

Both versions of the Hierarchy Principle recommend themselves for consideration as new axioms mainly by virtue of the role their cognates play in characterising the global structure of the universe of sets in the Cantorian theory. But even in the Cantorian case we must have reservations about the inclusion of these axioms[34], reservations based on the operationalist cast of the naive intuitions supporting the legitimacy of these principles. Such reservations must be even more acute in Euclidean set theory.

Let us consider the case for these Hierarchy Principles in some detail. To begin with, notice that we can establish the following theorem without appealing to either of the Hierarchy Principles (or, indeed, to the Uniformity Axiom).

10.7.8 Theorem *Let S be a set and r a linear ordering such that S is of rank r. Then there is a function*

$$f : Field(r) \to P(Constituents(S))$$

such that

(i) $\bigcup (f``Field(r)) = Constituents(S)$
(ii) *For all x in the field of r and all constituents, y, of S*

$$y \in f`x \text{ .iff. } (\forall u \in y)(\exists v <_r x)[u \in f`v]$$

The function $f : Field(r) \to P(Constituents(S))$ whose existence is asserted in Theorem 10.7.8 seems to "construct" or "generate" the set of constituents of S from the bottom, each constituent being "generated" as soon all of *its* constituents have been "generated".

Of course talk of "generating" the set of constituents of S is strictly metaphorical, and, in any case, the set of constituents of S has, in effect, already been been posited directly (in the Epsilon Fan Axiom[35]) and plays an essential role in defining the "constructing" function f by providing a target set of values for the recursion along r that defines f. So the "construction" takes place only *after* the set supposedly constructed is

---

[34] See the discussion in Section 5.6.
[35] Axiom 8.6.2.

10.7 Further axioms

already in place, so to speak. We should therefore think of f as *analysing* rather than as *constructing* the set of constituents of S.

But the intuition underlying the Weak Hierarchy Principle really does involve us in taking the metaphor of "construction" seriously in the case just considered. The argument must run something like this: given that we can construct *some* sets of rank r (e.g. S itself) by first "constructing" their constituents – the members of their transitive closures – in the manner given in Theorem 10.7.8, may we not then conclude that *all* the other sets in $R(S)$ – the cumulative hierarchy up to S – can be "generated" by cognate "constructions" starting from the same set of individuals (i.e. $Support(S)$)?

Of course in *some* sense we can "generate" the cumulative hierarchy in question: it is "generated" by the cumulative hierarchy system $\mathscr{H}(Support(S))$. In fact, if l linearly orders $Support(S)$, then the system $\mathscr{ACH}(Support(S), l)$ "generates" all the sets which occur as elements of terms in the hierarchy system $\mathscr{H}(Support(S))$.

But how do we know that $\mathscr{ACH}(Support(S), l)$ reaches out as far as S, or equivalently, that $\mathscr{H}(Support(S))$ reaches out sufficiently far to "generate" a term containing *all* sets of rank $\leq r$?

In any case, as I have already pointed out, Theorem 10.7.8 doesn't really rest on any intuition of the possibilities of "generation" at all. On the contrary, it is a straightforward consequence of the Epsilon Fan Axiom, and what makes it possible to "generate" the set of constituents of a set "step by step" is that it is already laid out as finite plurality prior to its supposed "generation": this gives us a target set for the recursive definition of its supposed "generator" – the local function f.

A recursion to "generate" the whole cumulative hierarchy up to S would clearly have to be an illegitimate *global* recursion, in default of a target set – in this case, $R(S)$ itself. We are confronting here another form of the operationalist fallacy which simply takes unrestricted global recursion as a *first principle*. If we adopt the Weak Hierarchy Principle as a further axiom we must do so other grounds than these.

Of course this appeal to supposed intuitions of "generation" is much more obvious in the case of the Strong Hierarchy Principle. Indeed, in that case we can get the full effect of the Principle by adding to its weak version the assumption that the von Neumann simply infinite system forms a scale, and thus we come full circle. If we adopt *that* assumption, there doesn't seem to be any principled reason not simply to adopt the assumption that every simply infinite system measures every other, and that all of them are universal scales.

But we have now reached a point at which arguments of this kind over the plausibility (or otherwise) of proposed axioms seem futile – futile *not* because such arguments carry no weight, but rather because they cannot be logically compelling in the way that mathematical arguments, in which the underlying first principles are accepted by all disputants, are compelling.

There is, however, one line of argument that is not vulnerable to this sort of impasse, and it is this: *the weaker theory here is much more interesting mathematically than its stronger rivals.* Euclidean set theory without even the Weak Hierarchy Principle has a much richer structure than the theory that includes it, and consequently much richer possibilities for extension by further axioms.

I am not making a logically trivial point here (weaker theories are compatible with more extensions than stronger ones). On the contrary, I have in mind the fact that the weaker theory in this particular case gives us the option of positing linear orderings – Euclidean finite linear orderings – in which we can embed a whole simply infinite system as a proper (semi-set) initial segment[36]. This, in turn, raises the possibility of powerful new axioms of an essentially "local" character, axioms that might take us far beyond even the strongest Cantorian set theories that have yet been considered. The weaker theory here may, in fact, be the stronger, in the sense of containing more possibilities for natural extension. Indeed, I am convinced that it *is* the stronger in this sense.

---

[36] Just how this can be done in a foundationally coherent way is a difficult problem which involves deep issues in the global logic of set theory. I shall have more to say on this matter in the following chapters and in Appendix 1.

11
Euclidean Set Theory from the Cantorian Standpoint

11.1 Methodology

In this chapter I intend to investigate Euclidean set theory from a resolutely Cantorian standpoint. My intention is to discover what is and what is not provable in that theory from a Cantorian point of view. In this way I can make use of well-known results of logic and model theory.

When we are exploring the consequences of a novel point of view we need all the information we can glean, and we ought not to allow foundational or philosophical scruples to prevent our coming to grips with what is complicated and difficult enough in any case. Besides, what we can establish using Cantorian principles provides a kind of outer boundary for our theory, and gives us an indication of what we can hope to establish using more severe and restrictive principles.

From the Cantorian standpoint, the Euclidean universe consists of the individuals together with those sets which occur in level ω of the cumulative hierarchy. Thus given a set, U, of individuals (which may, but need not, be Euclidean finite), the set

$$\hat{V}_\omega(U) = \bigcup \{V_\omega(S) : S \text{ is Euclidean finite .and. } S \subseteq U\}$$

constitutes, from the Cantorian standpoint, the *Euclidean universe based on the individuals in U*, and provides us with a *standard Cantorian model*, $\hat{V}_\omega(U)$, of that universe, which assigns to the terms and formulas of Euclidean set theory their obvious intended meanings[1].

In particular, from the Cantorian standpoint the Euclidean universe of pure sets is V_ω, which gives rise to the *standard Cantorian model*, V_ω,

---

[1] Although my exposition of the Euclidean theory in Part Four has not been formal, I have given, in Section 3.4, a definition of the terms and formulas of a suitable formal language for the theory, and, in Appendix 1, a description of a suitable system of formal proofs for it.

of the Euclidean theory of pure sets. A straightforward induction over the construction of terms in Euclidean set theory suffices to establish the following theorem.

11.1.1 Theorem *Let U be a set of individuals and t be any term of Euclidean set theory whose free variables are from among the list $a_1, \ldots, a_n$ of distinct free variables. Then there is a von Neumann natural number k such that for all elements $p_1, \ldots, p_n \in V_\omega$*

$$\hat{V}_\omega(U) \vDash [t \text{ is of rank } \subseteq_b] \; [p_1/a_1, \ldots, p_n/a_n, (k+m)/b]$$

where m is the von Neumann natural number which is the maximum of the ranks of the sets $p_1, \ldots, p_n$.

As an immediate corollary we obtain

11.1.2 Corollary *There is no term t of Euclidean set theory for which*

$$\hat{V}_\omega(U) \vDash \lambda xy(t) \text{ defines addition in } \mathscr{V}\mathscr{N}$$

Thus we cannot define addition in the von Neumann simply infinite system with the means made available to us by Euclidean set theory. The same, of course, holds for the Zermelo system, $\mathscr{Z}$, and the system, $\mathscr{H}(\emptyset)$, which generates the cumulative hierarchy based on the empty set. In fact, Theorem 11.1.1 and its corollary continue to hold if we add the Weak Hierarchy Principle to our axioms, so addition is not still not definable in $\mathscr{V}\mathscr{N}, \mathscr{Z}$, or $\mathscr{H}(\emptyset)$, even in the stronger theory.

But we need to look at *nonstandard* models if we are to find answers to the other basic questions about the definability of arithmetical functions and measures that naturally arise, and this confronts us with a practical problem. For Euclidean set theory, as I have presented it, is not couched in a conventional first order language, and has a large set of primitive symbols that must be given interpretations in models of the language of the theory. This will complicate the mathematical treatment of its model theory.

Accordingly, I shall direct my attention, not to Euclidean set theory itself, but to a theory of Euclidean finite sets formulated in a conventional first order (classical) language. Of course, anything provable in the formal version of Euclidean set theory presented in Appendix 1 will be provable in this theory, and therefore anything *not* provable in the latter will not be provable in Euclidean set theory either.

11.1 Methodology

The theory of Euclidean finite sets that I shall work with is the set-theoretical analogue of the theory $I\Sigma_0^{exp}(exp)^2$. Indeed, these theories are equivalent, being mutually interpretable, but I shall not prove that here[3].

The formal set theory is the theory ZF_0 based on the (classical) first order language with identity whose primitive symbols (in addition to the identity sign "=") are the binary relation symbol "∈" and the unary function symbol "P" (for "power set"). The axioms of ZF_0 are the standard Zermelo–Fraenkel axioms of *Extensionality, Regularity, Empty Set, Pair Set, Union, Power Set* (formulated with the function symbol "P"), and *Comprehension* (restricted to bounded quantifier formulas).

I shall not make provision for treating individuals in ZF_0, so the theory is a theory of pure sets built up from the empty set. This will simplify the treatment technically. It is possible to modify the arguments to take individuals into account, but I shall not do that here.

I have not included Replacement among the axioms of ZF_0 since a straightforward induction on the structure of bounded quantifier formulas shows that each bounded quantifier instance of Replacement is already provable in ZF_0.

It can also be shown that the basic global functions of Euclidean set theory can be added conservatively to ZF_0 and can be incorporated conservatively into instances of Comprehension. This is completely straightforward for the pairing and union functions. For the second level functions of Comprehension and Replacement, which in Euclidean set theory are incorporated into the theory using (in effect) term-forming operators, we can incorporate them into ZF_0 by assigning to each *particular instance* of a comprehension or replacement term a *particular* new function symbol of appropriate degree.

Finally, function variables in Euclidean set theory can be interpreted as *syntactic* (meta-) variables when we pass over to ZF_0, so that, for example, a binary function constant of Euclidean set theory can be identified with a lambda term, $\lambda xy(t[x/a, y/b])$, where t is a term in some definitional extension of ZF_0 which has exactly two free variables, a and b. Such lambda terms belong only to the meta-theory.

We may summarise all these observations by saying that ZF_0 *interprets* Euclidean set theory and that for every formal theorem of that theory,

---

[2] See Hajek and Pudlak, *Metamathematics of First-order Arithmetic*, p. 272.
[3] The interpretation of $I\Sigma_0^{exp}(exp)$ in Euclidean set theory uses the results on arithmetical functions in Chapter 9. The interpretation in the other direction goes via the Ackermann coding.

its interpretation is provable in ZF_0[4]. I shall use "ZF_0" to mean either ZF_0 proper or one of its definitional extensions.

The *standard model* of ZF_0, is, of course the structure $(V_\omega, \in_{V_\omega}, P)$, where P is the power set function. Clearly every element of V_ω is named by infinitely many terms of ZF_0, if the latter is extended to include function symbols for pairing and union and a constant for the empty set. In fact, we can always designate an element of V_ω in that extension of ZF_0 by means of a *standard term* built up by listing its members, their members, their members, in turn, ... in the obvious way. Hence we can assume that all nonstandard models of ZF_0 can be taken to be end extensions of $(V_\omega, \in_{V_\omega}, P)$, and can speak unambiguously of the *standard part* of such a model.

Now we need to take a detailed look at some nonstandard models of ZF_0.

11.2 Cumulation models

Let $M = (M, e_M, f_M)$ and $M_1 = (M_1, e_{M_1}, f_{M_1})$ be $\mathscr{L}(ZF_0)$-structures. Then we say that M_1 is a *transitive submodel* of M if M_1 is a substructure of M and

$$(\forall x \in M_1)(\forall y \in M)[M \vDash [y \in x] \text{ implies } y \in M_1]$$

Then we can prove the following theorem.

11.2.1 Theorem *Let $M \vDash ZF_0$ and $M_1 \vDash ZF_0$, where M_1 is a transitive submodel of M, let $\Phi(\vec{x})$ be a Δ_0 formula, and let $\vec{a} \in M_1$. Then*

$$M \vDash \Phi(\vec{a}) \text{ iff } M_1 \vDash \Phi(\vec{a})$$

The proof of this theorem is not entirely routine, but is an essentially straightforward induction on the construction of formulas.

Now we must define the key notion of a *cumulation model*[5].

11.2.2 Definition *Let M be a nonstandard model of ZF_0 and let $b \in M$ be a nonstandard element of M. Then the cumulation submodel of M generated*

---

[4] In fact, the converse also holds, for the conservative extension of ZF_0 obtained by adding the primitive vocabulary of Euclidean set theory is itself conservative over Euclidean set theory, though I shall not prove this here.

[5] The definition of cumulation model, and the following results about such models, are due to S. Popham (*Some Studies in Finitary Set Theory*, Ph.D. Thesis, University of Bristol, 1984).

11.2 Cumulation models

by b is the submodel of M whose universe, $C(M,b)$, is given by

$$C(M,b) = \bigcup_{n \in N} \{x \in M : M \vDash [x \in \underbrace{P(\ldots(P(Constituents(b)))\ldots)}_{n}]\}$$

where N is the standard natural numbers.

That $C(M,b)$ so defined is closed under P^M (the power set in M), and hence is the universe of a submodel of M, must be shown. In fact we can establish more.

11.2.3 Theorem *Let M be a nonstandard model of ZF_0 and let $b \in M$ be a nonstandard element of M. Then $C(M,b)$ is closed under P^M (the power set in M), is a transitive submodel of M, and is a model of ZF_0.*

Proof $C(M,b)$ is closed under P^M, for if

$$M \vDash a \in P^n(Constituents(b))$$

then, since M is a model of ZF_0,

$$M \vDash Trans(P^n(Constituents(b)))$$

so

$$M \vDash a \subseteq P^n(Constituents(b))$$

and therefore

$$M \vDash P(a) \subseteq P^{n+1}(Constituents(b))$$

Thus

$$M \vDash P(a) \in P^{n+2}(Constituents(b))$$

and $P^M(a) \in C(M,b)$ as required.

$C(M,b)$ is thus a submodel of M. Moreover, $C(M,b)$ is transitive, for if $a \in C(M,b)$ then for some n

$$M \vDash a \in P^n(Constituents(b))$$

and, since by the argument just given

$$M \vDash [c \in a \in P^n(Constituents(b)) \supset c \in P^n(Constituents(b))]$$

it follows that

$$M \vDash [c \in a] \text{ :implies: } c \in C(M,b)$$

as required.

To show that $C(M,b) \vDash ZF_0$ we note that extensionality follows from the transitivity of $C(M,b)$. Moreover, closure under pairing (i), union (ii), and transitive closure ($\lambda x Constituents(x)$ – (iii)) are easily established:

(i) If
$$M \vDash x \in P^n(Constituents(b)) \text{ and } M \vDash y \in P^m(Constituents(b))$$
then
$$M \vDash \{x,y\} \in P^{max(n,m)+1}(Constituents(b))$$

(ii) If
$$M \vDash x \in P^n(Constituents(b))$$
then
$$M \vDash \bigcup(x) \in P^n(Constituents(b))$$

(iii) If
$$M \vDash x \in P^n(Constituents(b))$$
then
$$M \vDash Constituents(x) \in P^n(Constituents(b))$$

From the closure of $C(M,b)$ under $\lambda x Constituents(x)$ and the transitivity of $C(M,b)$ the Axiom of Regularity follows.

To show that $C(M,b) \vDash \Delta_0$-Comprehension, we must first establish

(*) For all Δ_0-formulas $\Phi(\vec{x})$ of $\mathscr{L}(ZF_0)$ and all $\vec{a} \in C(M,b)$:
$$M \vDash \Phi(\vec{a}) .\mathit{iff}. C(M,b) \vDash \Phi(\vec{a})$$

This can be proved by induction on the construction of $\Phi(\vec{x})$. The basis for the induction follows from the fact that $C(M,b)$ is a submodel of M, the induction steps corresponding to the propositional connectives go through immediately by the hypothesis of induction, and the induction steps for bounded quantification follow from the induction hypotheses and the transitivity of $C(M,b)$.

Now to prove that $C(M,b)$ satisfies Δ_0-Comprehension, let $\Phi(\vec{x},y)$ be a Δ_0 formula, and let $\vec{a} \in C(M,b)$. Since $M \vDash ZF_0$ we have, for $x \in C(M,b)$,

$$M \vDash \exists z \in P(x)[\forall w \in z[\Phi(\vec{a},w) \wedge w \in x] \wedge \forall w \in x[\Phi(\vec{a},w) \supset w \in z]]$$

Hence by (∗)

$$C(M,b) \vDash \exists z \in P(x)[\forall w \in z[\Phi(\vec{a},w) \wedge w \in x] \wedge \forall w \in x[\Phi(\vec{a},w) \supset w \in z]]$$

and

$$C(M,b) \vDash \Delta_0\text{-Comprehension}$$

as required. The theorem now follows. □

11.2.4 Corollary *If M is a nonstandard model of ZF_0 and b is a nonstandard element of M, then for all Δ_0-formulas $\Phi(\vec{x})$ of $\mathscr{L}(ZF_0)$ and all $\vec{a} \in C(M,b)$*

$$M \vDash \Phi(\vec{a}) \text{ .iff. } C(M,b) \vDash \Phi(\vec{a})$$

Proof By Theorems 11.2.1 and 11.2.3. □

11.2.5 Theorem *Let M be a nonstandard model of ZF_0 and let b be a nonstandard von Neumann (or, alternatively, Zermelo) natural number of M. Then there is no element, c, of $C(M,b)$ such that*

$$\text{For all standard } y, \ C(M,b) \vDash y \in c$$

Proof Let b be a nonstandard von Neumann natural number of M, and suppose, by way of contradiction, that such a c exists in $C(M,b)$. Then $M \vDash b = \text{Constituents}(b)$ so there is a natural number n_0 such that

$$C(M,b) \vDash c \in P^{n_0}(b)$$

But

$$M \vDash (\forall x \in b)[VN(x)]$$

(where "$VN(x)$" means "x is a von Neumann natural number") and therefore

$$C(M,b) \vDash (\forall x \in b)[VN(x)]$$

so

$$C(M,b) \vDash \neg\{\{\emptyset\}\} \in b$$

Since $\{\{\emptyset\}\}$ is standard, so is $P^{n_0}(\{\{\emptyset\}\})$, and our assumption on c guarantees that

$$C(M,b) \vDash P^{n_0}(\{\{\emptyset\}\}) \in c$$

If x is transitive, so is $P(x)$. Therefore, since $C(M,b) \vDash Trans(b)$, we may conclude that $C(M,b) \vDash Trans(P^{n_0}(b))$ and hence that

$$C(M,b) \vDash P^{n_0}(\{\{\emptyset\}\}) \in P^{n_0}(b)$$

Now if $P(x) \in P(y)$ and y is transitive, then $x \in y$. For if $P(x) \in P(y)$, then $P(x) \subseteq y$, so $x \in \{x\} \in P(x) \subseteq y$, and if y is transitive $x \in y$. This little argument shows that

$$ZF_0 \vdash \forall xy[Trans(y) \wedge P(x) \in P(y) . \supset . x \in y]$$

Therefore, since, by Theorem 11.2.3, $C(M,b)$ is a model of ZF_0, we have

$$C(M,b) \vDash P^{n_0}(\{\{\emptyset\}\}) \in P^{n_0}(b)$$
$$C(M,b) \vDash P^{n_0-1}(\{\{\emptyset\}\}) \in P^{n_0-1}(b)$$
$$\vdots$$
$$C(M,b) \vDash P(\{\{\emptyset\}\}) \in P(b)$$
$$C(M,b) \vDash \{\{\emptyset\}\} \in b$$

Since the last of these is impossible, we have arrived at the desired contradiction, so no such c exists, as claimed.

If we take b to be a nonstandard Zermelo natural number we obtain a similar contradiction with

$$\{\emptyset \cup \{\emptyset\}\} \in Constituents(b)$$

$\square$

As an immediate consequence of this theorem we obtain the following corollary.

11.2.6 Corollary *There is a nonstandard model of ZF_0 containing a nonstandard von Neumann (or, alternatively, Zermelo) natural number in which the cumulative hiearchy function R is defined only for standard elements.*

Proof Let M be a nonstandard model of ZF_0, b a nonstandard von Neumann natural number of M, and consider the cumulation model $C(M,b)$. If x is nonstandard in $C(M,b)$, then

$$y <_{rank} x$$

for all standard y. Hence $R(x)$ cannot exist in $C(M,b)$ by Theorem 11.2.5.

$\square$

This corollary shows that the Weak Hierarchy Principle is not derivable in Euclidean set theory. Moreover, we can now see that we cannot

11.2 Cumulation models

establish that the cumulative hierarchy simply infinite system, $\mathscr{H}(\emptyset)$, is a scale for either $\mathscr{V}\mathscr{N}$ or $\mathscr{Z}$, though each of the latter is easily seen to be a scale for it.

Let us now turn to the problem of determining how the von Neumann natural numbers and the Zermelo natural numbers are related to one another.

11.2.7 Theorem *Let M be any nonstandard model of ZF_0, and let $b, c \in M$ be nonstandard elements, where b is a von Neumann natural number and c a Zermelo natural number. Then $C(M, b) \cap C(M, c)$ is the standard part of M.*

Proof Let $d = Constituents(c)$ and suppose that $a \in M$ is given such that

$$a \in C(M, b) \cap C(M, d)$$

Since b and d are both transitive in M, if i, j are standard natural numbers with $i \leq j$

$$M \vDash P^i(b) \subseteq P^j(b)$$

and

$$M \vDash P^i(d) \subseteq P^j(d)$$

Hence for suitable choice of n

$$a \in P^n(b) \cap P^n(d)$$

Now we apply induction to show that, for all natural numbers n,

$$M \vDash \forall y [y \in P^n(b) \cap P^n(d) \supset y \in P^{n+2}(\emptyset)]$$

Basis $n = 0$. In this case

$$M \vDash \forall x [x \in b \cap d \supset [x = \emptyset \vee x = \{\emptyset\}]]$$

Induction Suppose as hypothesis of induction that

$$M \vDash \forall y [y \in P^n(b) \cap P^n(d) \supset y \in P^{n+2}(\emptyset)]$$

Then if

$$M \vDash a \in P^{n+1}(b) \cap P^{n+1}(d)$$

it follows that

$$M \vDash \forall y \in a \, [y \in P^n(b) \cap P^n(d)]$$

Hence, by hypothesis of induction

$$M \vDash \forall y \in d\,[y \in P^{n+2}(\emptyset)]$$

and, therefore

$$M \vDash a \in P^{n+3}(\emptyset)$$

as required.

This completes the induction so

$$M \vDash \forall y [y \in P^n(b) \cap P^n(d) \supset y \in P^{n+2}(\emptyset)]$$

and the theorem follows.

□

As an immediate consequence of this theorem we obtain

11.2.8 Corollary *There is a nonstandard model of ZF_0 in which the Zermelo natural numbers are standard, and the von Neumann natural numbers are nonstandard. Similarly, there is a model in which the von Neumann natural numbers are standard and the Zermelo natural numbers are nonstandard.*

Proof In the notation of the preceding theorem, $C(M,b)$ provides a model in which the Zermelo natural numbers are standard and the von Neumann nonstandard. In $C(M,c)$ the circumstances are reversed.

□

This corollary shows that in Euclidean set theory neither the von Neumann nor the Zermelo simply infinite system can be shown to be a scale for the other: they are, as it were, *incomparable* in length. This observation can strengthened, as the next theorem shows.

11.2.9 Theorem *Let M be a nonstandard model of ZF_0 in which $b, c \in M$ are nonstandard elements, where b is a von Neumann natural number and c is a Zermelo natural number. Then for any terms s, t of Euclidean set theory[6] and Δ_0-formula, $\Phi_{s,t}(x)$, such that*

$$M \vDash [\lambda x(t(x)) \text{ generates a simply infinite system from } s]$$

$$M \vDash \forall x [\Phi_{s,t}(x). \equiv . \ x \text{ is generated by } \lambda x(t(x)) \text{ from } s]$$

---

[6] Every term of Euclidean set theory can, as I have observed above, be introduced conservatively in a suitable definitional extension of ZF_0. This means any proposition expressible in the thus extended language can be proved equivalent in the extended theory to a formula in the unextended theory in which the given term does not occur.

11.2 Cumulation models

the simply infinite system generated by $\lambda x(t(x))$ from s is either standard in $C(M,b)$ or standard in $C(M,c)$, i.e. either

$$\forall x \in C(M,b)\,[C(M,b) \vDash \Phi_{s,t}(x) : implies : x \text{ is standard}]$$

or

$$\forall x \in C(M,c)\,[C(M,c) \vDash \Phi_{s,t}(x) : implies : x \text{ is standard}]$$

Proof Let s, t, $\Phi_{s,t}(x)$, b, and c be given as indicated, and suppose, by way of contradiction, that for some nonstandard $a_1 \in C(m,b)$ and nonstandard $a_2 \in C(m,c)$

$$C(M,b) \vDash \Phi_{s,t}(a_1)$$

and

$$C(M,c) \vDash \Phi_{s,t}(a_2)$$

It follows (Theorems 11.2.1 and 11.2.3) that

$$M \vDash \Phi_{s,t}(a_1) \wedge \Phi_{s,t}(a_2)$$

But by the hypothesis of the theorem

$$M \vDash [\lambda x(t(x)) \text{ generates a simply infinite system from } s]$$

and

$$M \vDash \forall x[\Phi_{s,t}(x). \equiv . \, x \text{ is generated by } \lambda x(t(x)) \text{ from } s]$$

Hence

$$M \vDash [a_1 \subseteq x_2 \vee a_2 \subseteq x_1]$$

Suppose that

$$M \vDash a_1 \subseteq a_2$$

Then for some natural number n

$$M \vDash [a_1 \in P(a_2) \in P^n(c)]$$

Thus, by transitivity,

$$M \vdash a_1 \in P^n(c)$$

Hence $a_1 \in C(M,b) \cap C(M,c)$ and a_1 is a nonstandard element of M common to $C(M,b)$ and $C(M,c)$. This contradicts Theorem 11.2.7. The assumption that $M \vDash a_2 \subseteq a_1$ leads to a similar contradiction. The theorem now follows.

□

As an immediate consequence of this theorem we obtain two important corollaries.

11.2.10 Corollary *Let s and t be terms of Euclidean set theory such that*

$$ZF_0 \vdash [\lambda x(t(x)) \text{ generates a simply infinite system from } s]$$

Then

(i) $ZF_0 \nvdash [\lambda x(t(x)) \text{ defines a scale for } \mathscr{VN}]$
(ii) $ZF_0 \nvdash [\lambda x(t(x)) \text{ defines a scale for } \mathscr{Z}]$

Proof Take M, $C(M,b)$, and $C(M,c)$ as in the previous theorem. If $\lambda x(t(x))$ provably generates a simply infinite system from s in ZF_0 then, by the theorem, all the elements of the simply infinite system generated by $\lambda x(t(x))$ from s in $C(M,b)$ are standard in $C(M,b)$ (so $\lambda x(t(x))$ does not measure $\mathscr{VN}$ in $C(M,b)$), or all the elements it generates from s in $C(M,c)$ are standard in $C(M,c)$ (so it does not measure $\mathscr{Z}$ in $C(M,c)$). Since both these structures model ZF_0, the latter cannot prove both (i) and (ii).

□

Thus we cannot establish the existence of a simply infinite system $\mathscr{N}$ that constitutes an "upper bound" for *both* the von Neumann simply infinite system, $\mathscr{VN}$, *and* the Zermelo simply infinite system, $\mathscr{Z}$.

A fortiori we have, finally,

11.2.11 Corollary *Let s and t be terms of Euclidean set theory such that*

$$ZF_0 \vdash [\lambda x(t(x)) \text{ generates a simply infinite system from } s]$$

Then

$$ZF_0 \nvdash [\lambda x(t(x)) \text{ defines a universal scale}]$$

12
Envoi

12.1 Euclid or Cantor?

The choice between the Cantorian and the Euclidean versions of set theory turns on the meaning of *finite*: does the cardinality of a set increase when we add new elements to it?

We all know the consequences of adopting Cantor's version of finiteness: we get mathematics as it has been practised for the past one hundred years. To opt for the traditional Euclidean answer would seem to be, at least on the face of it, to give up all of the marvellous developments of twentieth century mathematics – not something to be done lightly. Moreover, as I have argued in Part 3, Cantor's radical innovation, though it appears at first to be unjustified, leads not just to a perfectly coherent account of the current practice of mathematics, but to a convincing *justification* of that practice.

For the axioms of set theory, when they are interpreted as principles characterising the notion of a *finite* plurality or multitude, all, with the single exception of Cantor's Axiom, possess that character of *self-evidence* that has traditionally been required of axioms. So if you accept that single novel idea, if, like Cantor, you take it as *obvious* that the species of natural numbers is definite in size[1], then you get the whole of modern mathematics, and, moreover, the whole of modern mathematics *resting on self-evident truths about clear and logically simple basic concepts.*

---

[1] "The absolutely infinite sequence of numbers [i.e. transfinite ordinal numbers] thus seems to me to be an appropriate symbol of the absolute; in contrast the infinity of the first number class [i.e. the natural numbers] which has hitherto sufficed seems to me to dwindle to nothingness by comparison, because I consider it to be a graspable idea (though not a representation)." (*Grundlagen einer allgemeinen Mannigfaltigkeitslehre*, endnote [2])

But is Cantor *right* about the natural numbers? *That* is the central question.

12.2 Euclidean simply infinite systems

The Axiom of Euclidean Finiteness simply asserts that every set is larger than any of its proper subsets. There is *nothing* in this conception to suggest that a set is a plurality that can be "built up" by successive addition of its elements. *Insofar as* that is true, and *in whatever sense* it is true, it must be *proved* from the basic principles of the Euclidean theory.

In Euclidean set theory the naive notion of "the" natural numbers as "what you get when you start from zero and carry on for ever successively adding one" must be replaced by the mathematically exact notion of a simply infinite system. Given a definite starting point, a, and a definite "successor operation" in the form of a suitable global function, σ, of one argument, then it is determined, *in a purely logical manner* from the basic principles of Euclidean set theory, what the successive "natural numbers" in the number system determined by a and σ are.

We are thus not generating abstract things called "natural numbers", indeed, we are not *generating* anything at all, in any literal sense: the definition is "static" and purely logical, and there are no "processes" or "constructions" (mental or otherwise) involved.

In this respect, the Euclidean approach to the natural numbers via the theory of simply infinite systems is of a piece with the Cantorian approach: both represent a further step in a general project, which began with the ϵ–δ definition of "limit", the object of which is to purge mathematics of any *logical* dependence on notions of temporal process. It is thus, in that regard at least, clearly in the mainstream of mathematical development since the 1830s[2].

All of these considerations would seem less pressing were it not for the fact that, as we have seen, the Euclidean theory of simply infinite systems, unlike its Cantorian counterpart, forces us to acknowledge the existence of natural number systems of different lengths.

What are we to make of this? How could *any* simply infinite system, let alone *all* of them, fail to constitute a scale with which to measure the species of all (Euclidean) sets? How are we to come to grips, imaginatively

---

[2] It is curious that old-fashioned operationalism, which has been thoroughly discredited in analysis, should remain entrenched in the foundations of natural number arithmetic, even though a modern theory has been available since Dedekind published his monograph *Was sind und was sollen die Zahlen?* in the mid 1880s.

12.2 Euclidean simply infinite systems

and intuitively, with an endless succession of natural number systems, each longer than its predecessor?

When we define a simply infinite system we are defining a species of "canonical" exemplars of the cardinal equivalence classes of Euclidean finite sets. In such circumstances we may perhaps imagine that we are attempting to define, for each cardinality, a single linear ordering whose field has that cardinality.

In fact it is easy to see that these ideas are equivalent. For suppose that the global function χ assigns a "canonical representative" of each cardinality to each set, so that

$$\forall x[\chi(x) \cong_c x]$$

and

$$\forall xy[x \cong_c y \text{ .implies. } \chi(x) = \chi(y)]$$

Then $\lambda x \chi(x \cup \{x\})$ generates a simply infinite system $\mathcal{N}_\chi$ from the empty set. Clearly $\mathcal{N}_\chi$ is a universal scale with χ as measure: given a set S, $\{\chi(x) : x \in card(S)\}$, ordered by increasing size, lies in $\mathcal{N}_\chi$ and measures S.

But unless we can *actually define* a global function χ which sends an arbitrary set to the canonical representative of its cardinality class, or, equivalently, a simply infinite sytem and a measure μ under which it forms a universal scale, we cannot regard our aim as realised. Perhaps such an aim *cannot* be realised, and perhaps that very fact is a measure of the gulf that divides the finite from the infinite in Euclidean finitism. I am convinced that this is, indeed, the case.

Our reservations, however, are hard to shake off: how *could* the von Neumann natural number system, for example, fail to form a scale, let alone fail to be closed under exponentiation, or even addition? Let us attempt to get at the source of these reservations.

Take an arbitrary (Euclidean finite) linear ordering

$$r = [0_r, 1_r, \ldots, k_r]$$

Let us imagine ourselves as starting from the element $0_r \in Field(r)$ and carrying out the operation

$$x \mapsto Next(r)`x$$

as often as we can. Clearly, if we arrive at k_r we can go no further, as k_r does not lie in the domain of $Next(r)$. But surely we *shall* arrive at

k_r, for all we need do is iterate our operation as many time as there are elements preceding k_r in r.

Similarly, starting from the empty set and carrying out the operation

$$x \mapsto x \cup \{x\}$$

"for ever" or "endlessly", we shall, surely, eventually arrive at, and pass beyond, an element of the von Neumann simply infinite system that measures $Field(r)$: again we need only as many iterations as there are elements in $Field(r)$.

This argument seems powerful and convincing. But if we force ourselves to say what the argument is with mathematical precision, we can see what assumptions *really* underlie it. In fact, the argument to establish that the von Neumann natural numbers can be used to count out *all* the elements of a linear ordering

$$r = [0_r, 1_r, \ldots, k_r]$$

rest on a fallacy that we have encountered before.

To show that every initial segment $[0_r, 1_r, \ldots, j_r]$ of r can be counted by an element of $\mathscr{VN}$ we need an induction with respect to a Σ_1 predicate, namely, the predicate

$$\exists x[x \text{ lies in } \mathscr{VN}.and.\, x \text{ measures } [0_r, 1_r, \ldots, j_r]]$$

and therein lies the difficulty.

For induction with respect to Σ_1 properties cannot be proved, because we cannot use such a property to define a sub*set* of the field of a linear ordering, consisting precisely of the elements of the field which have the property. This is because the Comprehension Principle applies only to properties that are *definite* in Zermelo's sense, and Σ_1 properties do not have this character[3].

Of course if we establish the premises of an induction, we can establish that as many of the initial segments of r as we can count out have the property in question. But that still leaves us with the question whether, or in what sense, we can count out *all* of them – the very question at issue.

The only alternative to this fallacious inductive "proof" is simply to *take it as axiomatic*, that is to say, to take it as simply as *given* that any linear ordering can be measured by some element of the von Neumann system. But what is it that "gives" us this? Of course we all have the

---

[3] This is *Brouwer's Principle* again.

12.2 Euclidean simply infinite systems

conviction that we can go on "constructing" von Neumann ordinals "for ever". This conviction is even justified, in a sense. But what does "for ever" mean? *That* is the very point at issue!

We are all caught up in these careless operationalist habits of thought[4], and they stand in the way of our arriving at an adequate view of these matters.

The "ultra-intuitionists" and "strict finitists", for example, have firmly grasped the stick by the wrong end. It is pointless to attempt to lay down a mathematical account of what we can actually do in the way of counting out or of carrying out calculations of heroic length: what we can actually do is utterly irrelevant to the problem that confronts us, for it is a *logical* and *set-theoretical* problem, not a *pracical* one.

The beginning of wisdom here is to realise that we do not, indeed we cannot, literally *construct* the *elements* of a simply infinite system: the most we can do is to construct *names* for (some of) them. At the most literal level this means *actually* writing them down.

When we go beyond this, when we begin to speak of what we *could* write down *in principle* we enter the unreal world of operationalist fantasy in which we live for ever and have an unlimited capacity for carrying out repetitious tasks without succumbing to boredom or inattention, or, alternatively, in which we have an unrestricted capacity to construct perfect artifacts, and posit computers of a size that dwarfs the cosmos itself and which run "for ever" without error and without requiring human attention.

But I repeat: the very point at issue is what "for ever" means? Is there only *one* way of "going on for ever"? In short, does every simply infinite system measure every other?

All simply infinite systems share the same *local* structure (they satisfy the *first order* axioms in Dedekind's axiomatic definition of simply infinite system), but does this mean that they all have the same *global* structure?

In Cantorian set theory, where such systems are definite in size – *finite* – by definition, we can *prove* that they all have the same structure. But in Euclidean set theory we cannot even *formulate* Dedekind's second order induction axiom, and that is what fixes the global structure of simply infinite systems in the Cantorian theory. You can "go on for ever", so

---

[4] I certainly do not exclude myself from this censure. It required a Dedekind to clear his mind of these fantasies and to carry out a proper mathematical analysis of the notion of iteration.

to speak, in *any* simply infinite system, so the use of this operationalist metaphor has no bearing on the question at issue.

But even if all this be granted, the question remains: how should we *think* about these things, I mean informally? What props to our imagination and intuition will help us to grasp what is going on here?

We must somehow purge ourselves of the simple conception that the elements of a simply infinite system are just "lying there" in an infinite, linear array which we can represent in our imaginations as something resembling an actual line. On this picture, the longer of two simply infinite systems is the one whose representing line is the longer, so to speak.

But such a "geometrical" representation is seriously misleading, and, in fact, it surreptitiously imposes a *Cantorian* conception of the matter on our imaginations. The comparison of the lengths of simply infinite systems $\mathcal{N}_1$ and $\mathcal{N}_2$ is not a matter of laying them out in extension, so to speak, so that we can "see" how far out they stretch. Indeed, talking about "length" here may be seriously misleading, for it may simply reinforce our tendency to think of these issues in terms of inappropriate geometrical metaphors.

What determines the "relative lengths" of the simply infinite systems $\mathcal{N}_1$ and $\mathcal{N}_2$ is the *logical* and *set-theoretical* relations between the *particular* global successor functions, σ_1 and σ_2, that generate them.

To say, for example, that the simply infinite system $\mathcal{N}$ is "shorter than" the *S*-ary extension $\mathcal{N}[S]$ is, at least, to say that we do not have the resources to define the relevant measure μ. And if we do not have those resources what are the grounds on which we can simply *posit* them?

By the same token, to say that $\mathcal{N}$ is "closed" under addition, for example, is to say that we have a set-theoretical "construction" in the form of a binary global function, ψ, which, when applied to elements m and n lying in $\mathcal{N}$, yields an element $\psi(m,n)$, which also lies in $\mathcal{N}$ and measures $|m| +_c |n|$. But if we cannot exhibit such a ψ, then what right have we to say that such a function nevertheless exists?

After all, *no* global function exists in the ordinary, robust sense, for talk about global functions is a mere *façon de parler* which simplifies talk about our linguistic conventions for designating sets, and those conventions, in turn, reflect our basic conception of what it is to be finite.

Having thought about these matters at great length, and over a number of years, I have now reached the point that it seems *obvious* to me that what I have called the "operationalist fallacy" is just that: a *fallacy*[5] The assumption that simply by by giving the conventional operationalist *description* of the natural numbers, we have *thereby* characterised them uniquely will simply not stand up under serious scrutiny. Indeed, once you see this as an *assumption*, which you can dispense with and still make sense of arithmetic, then, I am convinced, you are bound to see that it is unwarranted.

12.3 Speculations and unresolved problems

Is Cantor right about the natural numbers? It is surely obvious that he is not *just wrong*: his version of set theory is too fruitful and too *plausible* to be simply dismissed.

My own view is that that *we do not yet know enough about how mathematics can be developed in Euclidean set theory to make an informed choice between that theory and Cantorian orthodoxy*. But I also believe that the prospects of developing mathematics in that way are promising, and indeed, exciting.

My investions here suggest how arithmetic can be developed without a unique natural number sequence. How to incorporate the rest of mathematics within this unfamiliar framework remains an open problem, though I do have some tentative ideas and suggestions that I shall present shortly.

Mathematical investigation of the possibility of rejecting the idea of a unique natural number system goes back to the pioneering work of Yessenin-Volpin in the 1960s and to the seminal 1971 paper "Existence and feasibility in arithmetic" by Rohit Parikh. More recently, Petr Vopenka[6] and Edward Nelson[7], with their students and collaborators, have been pursuing a similar course.

Neither Vopenka nor Nelson approaches the natural number problem from the foundational standpoint I have adopted here. Indeed, both are formalists of a sort (in Nelson's case, a self-described formalist), but their formalism is not, like Bourbaki's, a mere device for avoiding a serious confrontation with foundational issues. On the contrary, it functions more

---

[5] The fallacy also arises in the theory of the formal syntax of languages.

[6] His *Mathematics in the Alternative Set Theory* is of considerable interest here. The reader will have noticed that I have made extensive use of his concept of *semi-set*.

[7] His *Predicative Arithmetic* contains much of considerable foundational and technical interest, and his beautiful little book *Radically Elementary Probability Theory* points the way to the development of analysis without Cantorian assumptions.

like a working hypothesis that allows them to get to grips, each in his own fashion, with the very difficult business of developing mathematics without presupposing a unique natural number sequence.

If the case for Euclidean foundations is to be made, there are many questions that have to be addressed. Some of them I have listed below, but I don't claim that my list is exhaustive, or, indeed, that I haven't overlooked something of central importance. My questions fall under six principal headings.

I. *Questions to do with the global logic of set theory*

Questions under this heading are relevant to both the Cantorian and the Euclidean versions of set theory. Some are technical, some philosophical, and some a combination of both. At the top of the list here we must place

(I-1) *How are the global quantifiers to be interpreted?*

The most pressing and immediate problem here is the meaning to be given to the global existential quantifier. The problem arises out of the Axiom of Local Definiteness which asserts that every non-empty set is inhabited (in Conceptual Notation: $S \neq \emptyset \vdash \exists x[x \in S]$). Surely this must be accepted as implicit in what is meant by "set".

But in both the Cantorian and the Euclidean versions of the theory this axiom precludes a strong reading of $\exists x A[x/a]$ in accordance with which a *proof* of that statement must be, or include, or entail the existence of, a proof of some statement of the form $A[t/a]$ for some explicitly given term t.

The term t is *naming* an instance here. Maybe it would suffice to name a set which *contains* an instance (by virtue of the Principle of Choice, for example).

Equally pressing, from a practical point of view, is

(I-2) *How are Π_1 and Σ_1 propositions to be negated?*

Again this question is common to both the Cantorian and Euclidean theories. A central problem is with interpretations of the logical operators $\neg\neg\exists$ and $\neg\neg\forall$ which naturally arise when we consider closure properties of simply infinite systems.

The general question here is

(I-3) *How are the global propositional connectives to be interpreted?*

This question has not come up in our discussion so far, but I have described a formal theory employing such connectives in Appendix 1[8]. These questions concerning the meanings of the logical constants – quantifiers and connectives – have been much discussed in the context of intuitionistic arithmetic and analysis, and have proved very difficult[9].

II. Questions to do with arithmetic

The most pressing question here is

(II-1) *How do we make the distinction between large and small sets?*

A natural way to express the proposition that a simply infinite system $\mathcal{N}$ is not a scale is as follows:

$$\neg\neg\exists x \forall y [\, y \text{ lies in } \mathcal{N} \text{ .implies. } |y| <_c x\,]$$

But where do we go from there? Here are two possibilities.

(1) Since our global theory admits a Gentzen interpretation of classical logic[10], we can simply treat the whole theory in the classical subtheory[11], but that perhaps makes the problem of interpreting the meanings of the propositions all the more difficult.

(2) Add a new constant, c, together with a (Π_1 or free variable) formula asserting that c is larger than the field of any element of $\mathcal{N}$. This is cleaty not conservative over our global theory, but may be over the quantifier free part.

(II-2) *What is the connection between the arithmetic of arithmetical functions and relations of Chapter 9 and that of simply infinite systems?*

Of especial interest here are Π_1 propositions. In particular

(II-3) *Given a simply infinite system $\mathcal{N}$, what Π_1 propositions over $\mathcal{N}$ hold in $\mathcal{N}$ while their cognate Π_1 arithmetical propositions over the universe of sets fail?*

We can ask an analogous question about Π_1 propositions in $\mathcal{N}$ and their cognates in $\mathcal{N}[S]$.

To progress on these questions we may need to address a more technical problem.

---

[8] Section A1.3.
[9] See Dummett's discussion in the last chapter of his *Elements of Intuitionism*.
[10] See Theorem A1.3.4 in Appendix 1.
[11] Using the fact that $\neg\neg\exists \supset \neg\forall\neg$ holds in intuitionistic logic.

(II-4) *Can we formulate axiomatic natural number arithmetics that capture the theory of particular simply infinite systems $\mathcal{N}$?*

In one sense this seems straightforward – bounded induction will hold in all of these systems for example. A more stringent measure of success would be to formulate for particular systems $\mathcal{N}$, a system of natural number arithmetic for $\mathcal{N}$, incorporating suitable function constants (and axioms characterising them) so that one could show that in the system $\mathcal{N}$ for which it was designed every $\mathcal{N}$-arithmetical function has a "canonical" definition in the notation of the natural number theory for $\mathcal{N}$.

If this can be successfully accomplished it may bear on the next the next question.

(II-5) *What is the relationship between the basic hierarchy of subtheories[12] $I\Sigma_0 + \Omega_n$ of $I\Sigma_0 + exp$ and the basic hierarchy $\mathscr{V}\mathcal{N}_0[2]$, $\mathscr{V}\mathcal{N}_1[2]$, ... of 2-ary extensions of $\mathscr{V}\mathcal{N}$?*

This question can be generalised to simply infinite systems other than $\mathscr{V}\mathcal{N}$ and numeration bases other than 2.

(II-6) *Can we develop a theory of S-ary extensions for "large" numeration bases S?*

In particular, what is the relation between $\mathcal{N}[S]$ and $\mathcal{N}[T]$ when T is not accessible to to S over $\mathscr{V}\mathcal{N}$ (i.e. $\mathcal{N}[S]$ does not measure T – Definition 10.4.17)?

III. Questions to do with analysis and the foundations of the Calculus

Since there can be no Weierstrassian continuum in Euclidean set theory, analysis will necessarily be "nonstandard".

(III-1) *How should real numbers be defined?*

The problem here is irrational numbers. I am convinced that the way to explain the distinction between the reals and the rationals is in terms of a distinction between large and small numbers (linear orderings).

Suppose that

$$r = [0, 1, \ldots, k]_r$$

is a linear ordering, which we may think of as composing a finite initial

---

[12] See Hajek and Pudlak, *Metamathematics of First-order Arithmetic* p. 272.

12.3 Speculations and unresolved problems 391

system of "natural numbers". Then we can form the ordering

$$Q_r = [-k, \ldots, -1/k, 0, 1/k, \ldots, k]$$

of positive and negative fractions $\pm m/n$ (in lowest terms) for $m, n \in$ Field(r), arranged in order of magnitude.

Now if r is large then it has (semi-set) initial segments that "behave" like natural number systems. In particular we can take initial segments $N <_r N^* <_r k$ so that N^* is an end extension of N. This gives us (semi-set) suborderings Q and Q^* corresponding to N and N^*, both of which "behave" like rational number systems with respective "natural numbers" N and N^*, respectively.

Q^* is a non-archmidean extension of Q and if we identify the "small" elements of Q^* with those elements bounded both above and below by elements of Q and the "infinitesimals" of Q^* with reciprocals of "large" elements, then we have the materials out of which to create "real numbers". The "rationals" are the small elements of Q^* that differ from elements of Q by infinitesimals. The remaining small elements of Q^* are "irrationals".

All this is very vague, but it is not difficult to see how to interpret it from a Cantorian standpoint using nonstandard methods. In fact it is *not* as straightforward as it might at first appear to work out the details in a manner that is natural in Euclidean set theory.

In any event, even in the Cantorian case the following question arises.

(III-2) *What class of functions should we take as fundamental?*

The answer here may be *analytic functions*. This is suggested by the fact that polynomials of large degree over the outer field Q^* can give rise to Taylor series over the field of small numbers (not Q but that part of Q^* that lies within the boundaries of Q). Thus the natural setting for our most basic theory may be the complex numbers.

There are formidable technical problems to solve before we can produce a natural and usable theory. But one must be encouraged by the example of Edward Nelson's *Radically Elementary Probability Theory*, which may be the first treatise which approaches analysis from a standpoint compatible with Euclidean set theory[13].

Finally, in this connection we must ask

(III-3) *How does this approach to analysis relate to Lawvere's smooth infinitesimal analysis?*

---

[13] Jan Mycielski's "Analysis without actual infinity" may also prove useful here.

Lawvere's theory, which is geometrical rather than arithmetical in origin, nevertheless seems to have a natural affinity with the arithmetical approach suggested here.

IV. *Questions to do with the foundations of geometry*

The obvious difficulties surrounding the notion of "infinite divisibility" suggest that even if we can, in some sense, recover the " continuous" spaces of traditional geometry by using analytical methods, it is natural to suppose that the underlying geometry is *discrete* in the sense that at the ultimate level space is composed of *minimal parts* having an "immediate contiguity" or *nearest neighbour* relation that determines the "cohesion" of the space. This raises the immediate question

(IV-1) *Can we recover orthodox Euclidean geometry from a discrete minimal parts geometry in a natural way?*

We assume that the nearest neighbour graph contains a large number of nodes so that that we can use appropriate versions of nonstandard methods to ensure that "continuity" emerges when regions of the graph composed of sufficiently large numbers of minimal parts are considered.

The essential problem here is to define a Euclidean distance function in a manner that relates it naturally to the intrinsic metric on the nearest neighbour graph obtained by simply counting the number of steps, from nodes to nearest neighbour nodes, required to pass from a given node to a distant one.

But this general approach should not be confined to reconstructing traditional Euclidean geometry, for it holds out the possibility a general concept of discrete geometry.

(IV-2) *What conditions must be placed on a large graph for it to constitute the nearest neighbour graph of a geometry?*

In particular, what kinds of uniformity and symmetry conditions must be placed on such a graph?

(IV-3) *How can we classify the geometries determined by nearest neighbour graphs?*

The aim must be to exploit the notion of largeness so that continuity "emerges" naturally from the discrete notion of distance intrinsic to the nearest neighbour graph.

(IV-4) *How can we introduce analytical methods in a natural way so that*

12.3 Speculations and unresolved problems

these discrete geometries have appropriate "continuity" and "smoothness" properties?

Again, it seems natural to compare this sort of approach with Lawvere's.

(IV-5) *How does "minimal parts" geometry relate to Lawvere's synthetic differential geomety?*

Classical differential geometry, based as it is on the Cantorian continuum, arose in the context of a physics in which the infinite divisibility of space was simply taken as given. But developments in modern physics perhaps suggest that this assumption may no longer be warranted.

Here, I believe, is the main source of inspiration and guidance for the development of geometry along the lines briefly outlined here. In the past mathematics and physics were so closely related that was difficult to say where the one left off and the other began, but with the revolution in rigour that took place in the middle of the nineteenth century, the two subjects grew apart.

Perhaps we may look forward to a time when the interests of mathematicians and physicists again converge. Such a development would be beneficial for both subjects.

V. Questions to do with the formulation of classical mathematical logic

The first question here is clearly

(V-1) *How should the Euclidean theory of syntax and formal proof be developed?*

The conventional approach to syntax in which terms and formulas are regarded as Euclidean finite sequences of primitive symbols can obviously be carried over to Eucldean set theory. If we allow *arbitrary* sequences (defined on *arbitrary* linear orderings) we obtain what we might call *absolute* syntax. The complications arise when we put restrictions on the the linear orderings on which these sequences are defined.

An obvious restriction would be to restrict those linear orderings to elements of some simply infinite system $\mathcal{N}$. Unless $\mathcal{N}$ is closed under addition we are clearly stymied. Substitution of terms for variables is clearly going to require that $\mathcal{N}$ be closed under multiplication.

It may be possible to get "hybrid" versions of restricted formal syntax in which, say, different simply infinite systems are used to measure the lengths of terms and the lengths of formulas.

There is a large and interesting literature on the problem of *coding* syntax and formal proof theory in weak systems of arithmetic which is relevant to these questions. The proofs of the central "finitary" metatheorems of first order logic, in particular, Cut Elimination, Herbrand's Theorem, and the Hilbert–Ackermann Consistency Theorem, in their full arithmetical versions all lie just beyond $I\Sigma_0 + exp$, but are provable in $I\Sigma_1$. This raises the question

(V-2) *What versions of the Cut Elimination Theorem, Herbrand's Theorem, and the Hilbert–Ackermann Consistency Theorem can be established?*

Finally, there is the problem of semantics.

(V-3) *Can we formulate a notion of a model of a first order language which will allow us to establish a completeness theorem in some form?*

Since we don't have a unique notion of syntax for formal languages this question splits into many particular questions. That might seem at first to complicate any attempt to find answers, but, in fact, it holds out the possibility that if we pick the right restrictions on the notions of *formal language, formal proof,* and *model* we may be able to answer this question in the affirmative for an interesting class of cases. Perhaps Mycielski's Local Finiteness Theorem may prove of use here.

VI. Questions to do with the theory of computability

The central question here is

(VI-1) *What is the meaning and status of the Church–Turing thesis in Euclidean set theory?*

Given that we have non-equivalent natural number systems there is obvious scope to specialise the mathematical account of theoretical computing machines, for example by requiring the specification of a Turing machine, or the number of steps it can take, or the length of tape it can use, or all of these things, to be measurable by given simply infinite systems.

Obviously what is computable, or what is computable under various restrictions, will vary in accordance with the choice of the sytem used to measure these various parameters. But as in the case of formal syntax, there is an *absolute* formulation of all these notions in terms of arbitrary linear orderings, or arbitrary construction trees.

A particular case here is

12.3 Speculations and unresolved problems

(VI-2) *What is the correct formulation of the P = NP problem in Euclidean set theory?*

Is it possible to obtain "solutions" to the problem by placing restrictions on the simply infinite systems used to measure its various "counting" parameters?

This raise a general issue concerning methodology: is it not possible that our use of the unique natural number system of conventional Cantorian mathematics distorts our thinking about the connection between the practical and the theoretical aspects of this problem and similar ones? Are the practical natural numbers of practical computation closed under the formation of decimal or binary numerals? Are they closed under multiplication? Or even addition? And what does "closure" mean in such a practical context? Perhaps Euclidean set theory, with its wealth of different natural number systems, is a better theoretical setting from which to consider practical problems in computation.

Appendix 1
Conceptual Notation

A1.1 Setting up a conceptual notation

I shall now set up a conceptual notation analogous to Frege's *Begriffsschrift*. To simplify my presentation, I shall conform to the customary conventions of exposition that mathematical logicians have come to employ in such matters, speaking of symbols, terms, formulas, etc., as if they were actually written down. In particular I shall speak of expressions as *strings* of primitive signs, and strings of expressions separated by commas as *lists* of expressions; the expressions that comprise such a list I shall call its *items* or *terms*[1].

The most convenient formulation for our conceptual notation is as a sequent calculus in the style of Gentzen. I have already laid down[2] the definitions of *term* and *formula*, which comprise the *well-formed expressions* of our conceptual notation.

Amongst formulas we must distinguish between *global* formulas, which contain global (unbounded) quantifiers, and *local* formulas, which do not. In the basic conceptual notation I shall require that a global quantifier of either type can occur only in the prefixes of prenex formulas, all other quantifiers in which (if any) are of the same sort. All formulas in the basic conceptual notation are thus either quantifier free, Π_1, or Σ_1. This notation is sufficient to express the propositions I have employed in the body of the text, but there is theoretical interest in considering a wider class of formulas. I shall return to this in the next section.

---

[1] It is well-known how to define a mathematically precise "model" of this conceptual notation in the Cantorian theory. Formal syntax in the Euclidean theory is a much more complex matter which I shall not pursue here.

[2] Section 3.4.

A1.1 Setting up a conceptual notation

A *sequent* is defined, in the usual way, to be an expression of the form

$$\Gamma \vdash \Delta \quad (\text{read "}\Gamma \text{ entails } \Delta \text{"})$$

where Γ, the *antecedent*, and Δ, the *consequent* or *succedent*, are (possibly empty) lists whose items are formulas. But we must impose the additional condition that the list Δ can contain more than one formula only if each of the formulas in Δ is free of global quantifiers. If a formula in Δ contains an occurrence of a global quantifier, then Δ must consist of that formula alone.

The basic conceptual notation contains, in addition to binding letters (bound variables ranging over objects), free variables of three fundamental sorts: individual variables, (ordinary) global function variables (of all degrees), and global propositional function variables (again of all degrees).

The syntactic notion of substituting an individual term **t** for a free individual variable **a** in an expression **E** to obtain the expression **E**[**t**/**a**] has already been explained in Section 3.4. There is an analogous syntactic notion of substituting a function term, of a given type and degree, for a free function variable of that type and degree. To describe this we need to make use of the lambda expressions described in Section 3.4.

Recall that if **E** is a well-formed *local* expression in conceptual notation (i.e. an individual term or a local formula), $a_1, \ldots, a_n$ are distinct individual free variables (each of which may, but need not, occur in **E**), and if $x_1, \ldots, x_n$ are distinct binding letters (bound variables) such that for each $i = 1, \ldots, n$, x_i is free for a_i in **E**, we can form the lambda expression

$$\lambda x_1 \ldots x_n \mathbf{E}[x_1/a_n, \ldots, x_n/a_1]$$

which we can write in a shorter form using "vector" notation as follows

$$\lambda \vec{x} \mathbf{E}[\vec{x}/\vec{a}]$$

where $\vec{x}$ and $\vec{a}$ stand for the lists $x_1, \ldots, x_n$ and $a_1, \ldots, a_n$, respectively.

Now let **A** be a formula (local or global) of conceptual notation, let σ be a first level global function parameter of degree **n** which may, but need not, occur in **A**, and let $\lambda \vec{x} t[\vec{x}/\vec{a}]$ be a λ-expression *free for* σ *in* $\mathbf{A}$[3], and representing a particular **n**-ary global function[4]. Then by the expression

$$\mathbf{A}[\lambda \vec{x} t[\vec{x}/\vec{a}]/\sigma]$$

---

[3] What this means I shall explain shortly.
[4] Thus $\vec{x}$ is a list $x_1, \ldots, x_n$ of distinct binding letters, and $\vec{a}$ is list $a_1, \ldots, a_n$ of the same length composed of distinct free individual variables.

I mean the formula obtained from **A** by replacing each occurrence of a *semi-term*[5] of the form $\sigma(\vec{s})$ (where $\vec{s}$ is the list $s_1, \ldots, s_n$ of semi-terms) by the corresponding semi-term

$$t[\vec{s}/\vec{a}]$$

To say that $\lambda\vec{x}t[\vec{x}/\vec{a}]$ is *free for* σ *in* **A** is to say that for each occurrence of a semi-term $\sigma(s_1, \ldots, s_n)$ in **A**, each of the semi-terms $s_1, \ldots, s_n$ is free for the corresponding free variable $a_1, \ldots, a_n$ in **t**, that is to say, no occurrence of a_i in **t** falls within the scope of a binding occurrence of any binding letter that occurs unbound in the semi-term s_i, for $i = 1, \ldots, n$.

Substitution for free *propositional* function variables

$$A[\lambda\vec{x}B[\vec{x}/\vec{a}]/\Phi]$$

can be handled in an analogous manner, although here we must require that *the formula* **B** *must be local, so that it contains no global quantifiers*.

These definitions are easily extended to lists, and thereby to sequents. Let me emphasise, however, that *these λ-expressions do not belong to the conceptual notation itself* but rather to the informal "*meta*"-notation we employ when discussing the conceptual notation properly so called.

A1.2 Axioms, definitions, and rules of inference

I shall base the purely logical part of Basic Set Theory on a modified subsystem of Gentzen's sequent calculus *LK*, as presented in Takeuti's classic textbook *Proof Theory*.

The changes necessary in the inference rules essentially come to this, namely, that the formulas and sequents in an instance of a rule must conform to the definitions of *formula* and *sequent* laid down in Section A1.1. Thus the the auxiliary formulas in a propositional inference[6] must be free of global quantifiers, and in each of the rules ($\forall$: right) and ($\exists$: right)[7] the auxiliary formula must stand alone as the only formula in the succedent of the premise. Moreover in ($\forall$: right) the auxillary formula must be either local or Π_1 and in ($\exists$: right) either local or Σ_1.

This subsystem of *LK* – let us call it LK_0 – suffices for the underlying general logic for Basic Set Theory. The Cut Elimination Theorem and

---

[5] A *semi-term* (*-formula*) is an expression obtained from an ordinary term (respectively, formula) by replacing one or more occurrences of free individual variables by occurrences of binding letters (bound variables).

[6] *Proof Theory*, Chapter 1, §2.

[7] ibid.

A1.2 Axioms, definitions, and rules of inference

its consequences[8], including Gentzen's Midsequent Theorem[9], carry over to LK_0. We now need to supplement this general logic with axioms of identity and axioms specific to the subject matter of set theory.

With the axioms of identity we can continue to follow Takeuti[10] by extending LK_0 to a system LK_{0e} which stands to LK_0 as Takeuti's LK_e stands to LK. This can be accomplished by taking the equality axioms of LK_e as initial segments[11]. Here we must include instances of these axioms corresponding to free variables for global functions and global propositional functions, which count as "function constants" and "predicate constants", repectively, in Takeuti's terminology.

Now I can lay down the properly set-theoretical axioms of Basic Set Theory. Here, again, I can follow Takeuti's treatment[12], and call a sequent, $\Gamma \vdash \Delta$, a *theorem of Basic Set Theory* if there is a (finite) list, Γ_0, of axioms of Basic Set Theory such that

$$\Gamma_0, \Gamma \vdash \Delta$$

is provable in LK_{0e}.

I. The Axiom Schema of Comprehension

(i) $\forall \vec{u} \forall x \, Set(\{z \in x : A[z/a, \vec{u}/\vec{b}]\})$

(ii) $\forall \vec{u} \forall x \forall y \, [y \in \{z \in x : A[z/a, \vec{u}/\vec{b}]\} \; :iff: \; [y \in x \; and \; A[y/a, \vec{u}/\vec{b}]\,]\,]$

where A is a *local* formula with free variables from among the distinct free variables $a, b_1, \ldots, b_n \; (= a, \vec{b})$ and $y, u_1, \ldots, u_n \; (= a, \vec{b})$ are binding letters free for $= a, \vec{b}$ in A, respectively.

Now I can lay down the definitions of the local quantifiers "$(\forall x \in S)$" and "$(\exists x \in S)$".

II. The Definitions of the Local Quantifiers

(i) $(\forall x \in s)A[x/a] \; :iff: \; s = \{x \in s : A[x/a]\}$

(ii) $(\exists x \in s)A[x/a] \; :iff: \; not \, (\forall x \in s) not \, A[x/a]$

---

[8] *Proof Theory*, Chapter 1, §6.
[9] We must drop the connective "*or*" for this proof to go through. See *Proof Theory*, Chapter 1, §6, Theorem 6.4.
[10] *Proof Theory*, Chapter 1, §7.
[11] ibid., Chapter 1, §7, Definition 7.1.
[12] ibid., Chapter 1, §4.

where **A** is a *local* formula, **s** is any term, and **x** is a binding letter free for the variable **a** in the formula **A**.

Here I shall depart from the conventions followed by Takeuti and treat these definitions formally as *axiom schemata*. This means that the definitions of *term*, *local formula*, and *formula* must be adjusted to accommodate the new notation, and the inference schemata of LK_{0e} must be interpreted as applying to the extended language. The new formulas beginning with local quantifiers are, of course, counted as local formulas.

Now I can proceed to lay down the remaining axioms and definitions of Basic Set Theory.

III. The Axiom of the Empty Set

$\forall \mathbf{x}[\mathbf{x} \notin \emptyset]$

IV. The Axiom of Sethood

$\forall \mathbf{x}[Set(\mathbf{x}) \;.iff.\; [\mathbf{x} = \emptyset \;or\; (\exists \mathbf{y} \in \mathbf{x})[\mathbf{y} \in \mathbf{x}]]]$

V. The Definition of Inclusion

$\mathbf{s} \subseteq \mathbf{t} \;.iff.\; (\forall \mathbf{x} \in \mathbf{s})[\mathbf{x} \in \mathbf{t}]$

Again, I shall treat this definition formally as an axiom schema, and add "⊆" to our language as a new binary relation constant, adjusting the interpretation of the inference schemata as required.

VI. The Axiom of Extensionality

$\forall \mathbf{x} \forall \mathbf{y} \forall \mathbf{z}[\mathbf{x} \subseteq \mathbf{y} \;and\; \mathbf{z} \in \mathbf{x} \;:implies:\; \mathbf{z} \in \mathbf{y}]$

VII. The Axiom of Singleton Selection

(i) $\forall \mathbf{u} \forall \mathbf{v}[(\exists \mathbf{x} \in \mathbf{u})(\forall \mathbf{y} \in \mathbf{u})[\mathbf{y} = \mathbf{x}] \;.implies.\; \iota(\mathbf{u}, \mathbf{v}) \in \mathbf{u}]$
(ii) $\forall \mathbf{u} \forall \mathbf{v}[not(\exists \mathbf{x} \in \mathbf{u})(\forall \mathbf{y} \in \mathbf{u})[\mathbf{y} = \mathbf{x}] \;.implies.\; \iota(\mathbf{u}, \mathbf{v}) = \mathbf{v}]$

VIII. The Axiom of Pair Set

$\forall \mathbf{x} \forall \mathbf{y} \forall \mathbf{z}[\mathbf{z} \in \{\mathbf{x}, \mathbf{y}\} \;.iff.\; \mathbf{z} = \mathbf{x} \;or\; \mathbf{z} = \mathbf{y}]$

Axiom VIII allows us to lay down the following important definition.

IX. The Definition of the Singleton Set

$\{\mathbf{s}\} = \{\mathbf{s}, \mathbf{s}\}$

X. The Axiom Schema of Replacement

A1.2 Axioms, definitions, and rules of inference

(i) $\forall x \, Set(\{t[y/a] : y \in x\})$

(ii) $\forall \vec{u} \forall x \forall z[z \in \{t[y/a, \vec{u}/\vec{b}] : y \in x\} \, .iff.$
$(\exists y \in x)[z = t[y/a, \vec{u}/\vec{b}]]]$

where **t** is a term with free variables from among the distinct free variables $a, b_1, \ldots, b_n \, (= a, \vec{b})$ and $y, u_1, \ldots, u_n \, (= a, \vec{b})$ are binding letters free for $= a, \vec{b}$ in **t**, respectively.

XI. The Axiom of Union

(i) $\forall x \, Set(\bigcup(x))$

(ii) $\forall x \forall y [y \in \bigcup(x) \, :iff: (\exists z \in x)[y \in z]]$

XII. The Definitions of the Boolean Operations

(i) $s \cup t = \bigcup(\{s, t\})$

(ii) $s \cap t = \{z \in s : z \in t\}$

(iii) $s - t = \{z \in s : z \notin t\}$

XIII. The Axiom of Power Set

$\forall x \forall y \, [y \in P(x) \, .iff. \, Set(y) \, and \, y \subseteq x]$

XIV. The Definitions of Ordered Pair and Cartesian Product

(i) $(s, t) = \{\{s\}, \{s, t\}\}$

(ii) $s \times t = \bigcup(\{\{(u, v) : u \in s\} : v \in t\})$

With the definitions of *ordered pair* and *Cartesian product* in place, I could now lay down definitions of the basic notions concerning local relations and functions, but I shall leave that to the reader. Further definitions will, of course, be required in order to state some of the axioms for both the Cantorian and the Euclidean extensions of Basic Set Theory, but that basic theory is now essentially in place.

The system presented here is adequate to carry out the informal arguments presented in the main body of this book. But there is an important difference between the inferences sanctioned in LK_0 and the informal arguments employed earlier. It concerns the logical status and function of the higher level variables (of both types and all degrees).

From the standpoint of the formal theory just presented these variables are treated as what Takeuti calls "constants"[13], but in the informal

---

[13] *Proof Theory*, Chapter 1, §1.

arguments in the main body of the text they are regarded as genuine variables expressing generality.

However, it is possible to use them *as if* they were genuine variables without altering the system as it has been presented so far. Thus consider the following two inference figures:

XV. The Rule of Substitution for (Ordinary) Global Function Variables

$$\frac{\Gamma \vdash \Delta}{\Gamma[\lambda \check{x} t[\check{x}/\check{a}]/\sigma] \vdash \Delta[\lambda \check{x} t[\check{x}/\check{a}]/\sigma]}$$

provided that $\lambda \check{x}(t[\check{x}/\check{a}])$ *is free for* σ *in* Γ *and* Δ.

We may think of ordinary substitution of terms for free individual variables as a special case of this rule if we regard the latter as global function variables of degree zero.

XVI. The Rule of Substitution for Global Propositional Function Variables

$$\frac{\Gamma \vdash \Delta}{\Gamma[\lambda \check{x} B[\check{x}/\check{a}]/\Phi] \vdash \Delta[\lambda \check{x} B[\check{x}/\check{a}]/\Phi]}$$

provided that $\lambda \check{x}(B[\check{x}/\check{a}])$ *is free for* Φ *in* Γ *and* Δ.

Both of these inference figures can be derived in LK_{0e} in the sense of the following theorem.

A1.2.1 Theorem *If the sequent* $\Gamma \vdash \Delta$ *is derivable in* LK_{0e}, *then so are the sequents*

$$\Gamma[\lambda \check{x} t[\check{x}/\check{a}]/\sigma] \vdash \Delta[\lambda \check{x} t[\check{x}/\check{a}]/\sigma]$$

and

$$\Gamma[\lambda \check{x} B[\check{x}/\check{a}]/\Phi] \vdash \Delta[\lambda \check{x} B[\check{x}/\check{a}]/\Phi]$$

provided that $\lambda \check{x} t[\check{x}/\check{a}]$ *is free for* σ *in all the formulas of* Γ *and* Δ *(in the first case), and* $\lambda \check{x} B[\check{x}/\check{a}]$ *is free for* Φ *in all the formulas of* Γ *and* Δ *(in the second).*

Proof Given an LK_{0e} proof, $\Pi(\sigma)$, of $\Gamma(\sigma) \vdash \Delta(\sigma)$ ($= \Gamma \vdash \Delta$), we may assume that the proof is *regular* in Takeuti's sense[14], that is to say, that

---

[14] *Proof Theory*, Chapter 1, §2, Definition 2.9 and Lemma 2.10.

A1.2 Axioms, definitions, and rules of inference

all eigenvariables in $\Pi(\sigma)$ are distinct and each occurs only at or above the premise of the inference of which it is the eigenvariable. We may further assume that none of those eigenvariables occur in the term **t**. A straightforward induction argument can now be used to establish that the configuration $\Pi(\lambda \vec{x} \mathbf{t}[\vec{x}/\vec{a}])$, obtained in the obvious way from $\Pi(\sigma)$, is, in fact, an LK_{0e} proof of the sequent $\Gamma[\lambda \vec{x} \mathbf{t}[\vec{x}/\vec{a}]/\sigma] \vdash \Delta[\lambda \vec{x} \mathbf{t}[\vec{x}/\vec{a}]/\sigma]$. The argument in the other case is exactly the same.

□

As an immediate corollary we obtain

A1.2.2 Corollary *Let LK_{0e}^{+} denote the system obtained from LK_{0e} by adding the two Rules of Substitution for higher level variables. Then LK_{0e}^{+} is conservative over LK_{0e}, that is to say, exactly the same sequents are provable in the two systems.*

Proof Given any proof in LK_{0e}^{+} we can remove an instance of one of the new inference figures above which no other such inference occurs. □

In the system LK_{0e}^{+} all the instances of the equality axioms[15] can be derived from two particular axioms:

The Axiom of Reflexivity

$\forall \mathbf{x}[\,\mathbf{x} = \mathbf{x}\,]$

The Axiom of Substitutivity

$\forall \mathbf{x} \forall \mathbf{y}[\,\mathbf{x} = \mathbf{y}\ :implies:\ [\Phi(\mathbf{x})\ .implies.\ \Phi(\mathbf{y})]\,]$

This last axiom deals with only one substitution of equals for equals in global propositional functions. But the corresponding propositions for two or more such substitutions are consequences of the one substitution case. For if we take Φ to be $\lambda x(\Psi(x,b))$ we may conclude that

$$a = c\ implies\ [\Psi(a,b)\ iff\ \Psi(c,b)]$$

Taking Φ to be $\lambda y(\Psi(c,y))$ we may conclude that

$$b = d\ implies\ [\Psi(c,b)\ iff\ \Psi(c,d)]$$

and from these two propositions we may conclude that

$$a = c\ and\ b = d\ .implies.\ [\Psi(a,b)\ iff\ \Psi(c,d)]$$

---

[15] *Proof Theory*, Chapter 1, §7, Definitions 7.1 and 7.3.

This argument is easily formalised in LK_{0e}^+, and will allow us to establish any instance of part (ii) of the Schema of Substitution of Identicals. We can also deduce all instances of part (i) of that schema by an equally straightforward argument.

In LK_{0e}^+ we can also reduce the Comprehension and Replacement Schemata to single axioms.

I*. The Axiom of Comprehension

(i) $\forall x \, Set(\{z \in x : \Phi(z)\})$
(ii) $\forall x \forall y \, [\, y \in \{z \in x : \Phi(z)\} \; :iff: \; [\, y \in x \text{ and } \Phi(y) \,] \,]$

X*. The Axiom of Replacement

(i) $\forall x \, Set(\{\sigma(y) : y \in x\})$
(ii) $\forall x \forall z [\, z \in \{\sigma(y) : y \in x\} \; .iff. \; (\exists y \in x)[z = \sigma(y)] \,]$

The upshot of these observations is that we could formulate Basic Set Theory in LK_0^+ taking the single axioms just stated, namely, Reflexivity, Substitutivity, Comprehension, and Replacement in place of the corresponding axiom schemata[16]. In this version of the theory the higher level variables function genuinely as variables, not only in the semantics of the theory (in expressing generality with respect to global functions), but also proof-theoretically, in the carrying out of arguments.

In this respect, perhaps, this reformulation is superior to the original treatment. But the latter has the advantage of being formulated in orthodox proof theory.

We need to lay down one further axiom schema to assert that non-empty sets are *inhabited*.

XVII. Axiom Schema of Local Definiteness

$s \neq \emptyset \vdash \exists x \, [x \in s]$

The instances of this schema are to be taken as initial sequents.

There is still one further logical matter that requires clarification: the technique of *global definition*. These definitions cannot be treated in the conventional manner because of the restrictions on the use of global

---

[16] Of course, strictly speaking, these "single" axioms have been formulated as axiom schemata: in the Axiom of Comprehension, for example, the symbol Φ is a *metavariable* ranging over first level global propositional variables, properly so called. But unlike the earlier schemata, all instances of these schemata are mere alphabetic variants of one another, and they all express, when asserted, the same (general) proposition.

quantifiers in Basic Set Theory. In every case the problem is that the *definiens* must employ global quantifiers. Most typically, these definitions define higher level properties of global functions.

Perhaps an example will serve better here than a detailed formal account given in complete generality. Let us consider the definition of *arithmetical function* in Definition 9.1.1.

The Definition of Arithmetical Function

(i) $\forall xy[x \cong_c y \text{ .implies. } \sigma(x) \cong_c \sigma(y)] \vdash Arith_z(\sigma(z))$
(ii) $Arith_z(\sigma(z)) \vdash \forall xy[x \cong_c y \text{ .implies. } \sigma(x) \cong_c \sigma(y)]$

Note that *Arith* is a variable binding operator. Clearly, similar definitions can be laid down to cover global functions of two or more arguments[17].

Formally, we take both of the sequents (i) and (ii) as *initial sequents*. As with ordinary, local definitions, we must add the newly defined symbol to our formal vocabulary and re-interpret our inference figures and meta-theorems accordingly.

The system LK_{0e} constitutes the logical part of our formalised version of set theory. Basic Set Theory, the theory treated in Part Two, is determined by the axioms and definitions given in III–XVII given above.

Cantorian Set Theory (Part Three) is obtained from Basic Set Theory by adding the appropriate new axioms. *Cantor's Axiom* (Axiom 5.2.2), the *Axiom of Choice* (Axiom 5.3.2), the *Cantorian Axiom of Foundation* (Axiom 5.4.2), and the *Axiom of Extensional Analysis* (5.4.4) constitute the most fundamental of these. Indeed, even among these axioms only the first two are in wide use in mathematics outside set theory itself.

The remaining Cantorian axioms constitute a series of increasingly stronger principles, most of which are remote from (but by no means irrelevant to!) the ordinary practice of mathematics. Thus we have the *Weak Hierarchy Principle* (Axiom 5.5.1), the *Strong Hierarchy Principle* (Axiom 5.5.5), the *Principle of Simple Closure* (Axiom 5.6.3), and the *Principle of Strong Closure* (Axiom 5.6.8). The latter can be used to establish the efficacy of definition by transfinite recursion along the transfinite ordinals.

With *Mahlo's Principle* (Axiom 5.7.2) we pass beyond that part of Cantorian Set Theory which is implicit in the Zermelo–Fraenkel axioms, and indeed, as I explained in Section 5.7, beyond what can be justified as acknowledging the existence of "closure" points for global functions.

---

[17] This definition works only in LK_{0e}^+. In LK_{0e} we need the corresponding definition schema.

A weaker axiom than Mahlo's, and one which is widely accepted among set theorists, is *Zermelo's Principle* (Axiom 7.1.5). In fact, as we saw in Chapter 7, Zermelo's Principle underlies the whole theory of axioms of strong infinity, and, practically speaking, represents the starting point of that theory.

Euclidean Set Theory (Part Four) rests on four fundamental axioms beyond Basic Set Theory: the *Axiom of Euclidean Finiteness* (Axiom 8.3.1), the *Epsilon Fan Axiom* (Axiom 8.6.2), the *Euclidean Axiom of Foundation* (Axiom 8.6.6), and the *Axiom of Uniformity* (Axiom 10.7.1).

Of course in a formal exposition of all these three theories (in whichever variants), it is necessary to include formal versions of the key definitions, without which the axioms cannot be given in their present forms.

A1.3 Global propositional connectives

In this section I shall show how to extend the *local* theory of Section A1.2 to a *global* theory in a modified version, LJ_e^*, of Takeuti's LJ_e[18]. I shall retain the vocabulary and rules of LK_{0e} (and its variants), adding a new set of *global propositional connectives* in passing to LJ_e^*.

Thus the primitive vocabulary of LJ_e^* is obtained from that of LK_0 by taking as additional new primitive symbols a set of *constructive* or *intuitionistic* or *global* propositional connectives. The global quantifiers remain as before and, of course, the local quantifiers are introduced by a definition schema (Definition II).

Let us take as our new propositional connectives

$$\wedge \text{ (and)}, \vee \text{ (or)}, \supset \text{ (implies)}, \neg \text{ (not)}$$

Terms and formulas are defined in the usual way, except that we now have two ways of forming propositional combinations.

Formulas containing the new global propositional connectives are to be counted as global formulas, and so cannot be combined using local propositional connectives, nor incorporated in comprehension terms.

The logical inference rules are those of LJ_e (including the Cut rule) and

---

[18] The subscript "e" indicates the presence of the equality axioms, as in Takeuti; the superscript "*" is intended to reflect the fact that the rules of LK_0 are also built into the logic.

LK_0 (including Schema XVII of Local Definiteness), augmented by a new set of initial sequents which assert that local formulas are *decidable*[19].

XVIII. Axiom Schema of Decidability for Local Formulas

$\vdash A \vee \neg A$

where A is a local formula.

The set-theoretical axioms remain as before.

No doubt this doubling up of propositional connectives adds an undesirable prolixity to the theory. But the new connectives have a quite different character from the old – they are, as traditional logicians might have said, mere *homophones* – and therefore exactly analogous propositions built up from atomic propositions using the the different sorts of propositional connectives have different meanings.

In fact the analogy extends beyond propositional connectives to the use of *bounded quantifiers*

$$\forall x [x \in s \supset A[x/a]]$$

and

$$\exists x [x \in s \wedge A[x/a]]$$

in place of the *local quantifiers*

$$(\forall x \in s) A[x/a]$$

and

$$(\exists x \in s) A[x/a]$$

respectively.

Let us call LJ_e^* formulas built up from atomic formulas using global propositional connectives and bounded quantifiers *bounded formulas*[20]. Clearly every bounded formula, A, has a cognate local formula, A^{loc}, and every local formula, A, a cognate bounded formula, A^{bnd}.

The axiom schemata of Local Definiteness (XVII) and Local Decidability (XVIII) allow us to deduce that, in *bounded* formulas, bounded quantifiers and global propositional connectives behave like their local cognates.

---

[19] This use of "decidable" is merely technical. It reflects our decision to regard the truth value of local propositions as logically determinate, and carries with it no implication about what we can *prove*.

[20] Note that complex local formulas may occur in bounded formulas *inside terms* but not as direct subformulas.

A1.3.1 Theorem *Let A be a bounded formula and A^{loc} its local counterpart. Then*

$$\vdash A \equiv A^{loc}$$

is provable in LJ_e^*.

Of course the fact that A and A^{loc} are provably equivalent does *not* imply that they are identical in meaning. On the contrary, the local connective "*implies*", whose meaning is completely captured by its truth table, has an entirely different meaning from the cognate global connective "$\supset$"[21].

With the help of Theorem A1.3.1 we can establish the following key theorem.

A1.3.2 Theorem *Let the formulas in the list Γ, A all be bounded. Then $\Gamma \vdash A$ is provable in LJ_e^* if, and only if, $\Gamma^{loc} \vdash A^{loc}$ is provable in LK_{0e}.*

It is a consequence of Theorems A1.3.1 and A1.3.2 that our theories, as formulated with LJ_e^* as their underlying logic, are *conservative extensions* of the same theories with LK_{0e} as underlying logic.

These results are the key to reformulating our global theory, whose underlying logic is LJ_e^*, in a more orthodox fashion that will enable us to make use of standard reults in model theory and proof theory. To begin with, the set-theoretical axioms proper can be reformulated using *bounded* formulas rather than *local* ones. We can also look on our definitions in the conventional way, either as abbreviations or as giving rise to (conservative) *definitional extensions* of our base theory.

In such a reformulation, the Axiom Schemata of Comprehension and Replacement need to be formulated with some care. The most straightforward way to handle this is take each *particular* instance of either of these axioms as characterising a *particular* new function constant.

Thus, for example, if A is a bounded formula with free variables from among the distinct free variables $a_1, \ldots, a_n$, then we can introduce a new global function constant $\phi_{\lambda \vec{z} A[\vec{z}/\vec{a}]}$ of $n+1$ variables (where $\vec{a} = a_1, \ldots, a_n$) together with the axiom

$$\forall \vec{z} \forall xy [x \in \phi_{\lambda \vec{z} A[\vec{z}/\vec{a}]}(y, \vec{z}) :\equiv: [x \in y \wedge A[\vec{z}/\vec{a}]]]$$

Having made these changes in the set-theoretical axioms we obtain a

---

[21] Indeed, there is a serious problem concerning the meanings of the global connectives, especially "$\vee$", "$\neg$", and "$\supset$". For a thorough and illuminating discussion of these difficulties see Chapter 7 of Michael Dummett's *Elements of Intuitionism*.

A1.3 Global propositional connectives

theory which can be formalised in the standard sequent calculus LJ_e, as the following theorem shows.

A1.3.3 Theorem *Let Γ, A be a list of formulas each of which is either a local formula or a Π_1 or Σ_1 formula whose matrix is a local formula, and let Γ', A' be the list that results from Γ, A when, in each constituent formula, all occurrences of comprenhension and replacement terms are replaced by function terms (in the manner just indicated), and all local quantifiers are replaced by their bounded correlates. Then $\Gamma \vdash A$ is derivable in LJ_e^* if, and only if, $\Gamma' \vdash A'$ is derivable in LJ_e.*

This theorem allows us to reformulate our theory in an orthodox sequent calculus, so that we can carry out meta-mathematical investigations using standard results. In particular we can carry out an analogue of Gentzen's analysis of the relation between classical Peano Arithmetic and the corresponding intuitionistic theory, Heyting Arithmetic.

We assign to each formula, A, its *Gentzen translation*, A^G, as follows:

  (i) If A is atomic, then A^G is A itself.
  (ii) If A is $A_1 \wedge A_2$, then A^G is $(A_1)^G \wedge (A_2)^G$.
  (iii) If A is $A_1 \vee A_2$, then A^G is $\neg[\neg(A_1)^G \wedge \neg(A_2)^G]$.
  (iv) If A is $A_1 \supset A_2$, then A^G is $(A_1)^G \supset (A_2)^G$.
  (v) If A is $\neg A_1$, then A^G is $\neg(A_1)^G$.
  (vi) If A is $\forall x A_1[x/a]$, then A^G is $\forall x (A_1)^G[x/a]$.
  (vii) If A is $\exists x A_1[x/a]$, then A^G is $\neg\forall x \neg(A_1)^G[x/a]$.

All our local versions of set theory (formalised, in the manner sketched above, with bounded quantifiers in place of local ones, and with comprehension and replacement terms replaced by suitable new function constants) can be *classically* globalised using the classical sequent calculus LK_e instead of LJ_e. We can then establish the following theorem[22].

A1.3.4 Theorem *Let T be any of the local versions of set theory under consideration. Then*

  (i) *For any formula A, $\vdash A \equiv A^G$ is provable from T in LK_e.*

  (ii) *For any list Γ, A of formulas, if $\Gamma \vdash A$ is provable from T in LK_e, then $(\Gamma)^G \vdash A^G$ is provable from T in LJ_e.*

---

[22] See Kleene's *Introduction to Metamathematics*, Chapter XV, §81, Theorem 60.

From this theorem, together with Theorem A1.3.3, we can deduce that the classical globalisation of any of these local theories, T, is also conservative over T.

We now have the logical machinery to formulate some of the key global notions in the Euclidean version of set theory. For example, suppose that the simply infinite systems $\mathcal{N}_1$ and $\mathcal{N}_2$ are generated from a_1 and a_2 by the global functions σ_1 and σ_2, respectively. Then a natural way to express the global proposition

$$\mathcal{N}_1 \text{ is not a scale for } \mathcal{N}_2$$

immediately suggests itself, namely

$$\neg\neg\exists x[[x \text{ is generated from } a_2 \text{ by } \sigma_2] : \wedge :$$
$$\forall y[[y \text{ is generated from } a_1 \text{ by } \sigma_1] \supset Field(y) <_c Field(x)]]$$

Surely we cannot simply assert that *there exists* an element r lying in $\mathcal{N}_2$ that is not measured by any element of $\mathcal{N}_1$, for surely no such element could actually be *named*. But we can, it seems to me, assert that the proposition that the existence of such an element is absurd is itself absurd[23].

This suggests the appropriate way to formulate the assumption that no simply infinite system forms a universal scale. This leads to the following formulation.

A1.3.5 Axiom *(Axiom of Absolute Infinity)*

$$\neg\neg\exists x \forall y[\sigma \text{ generates } y \text{ from } a \supset Field(y) <_c x]$$

I have called this the "Axiom of Absolute Infinity" because I want it to imply that the universe is so vast, and contains sets of so enormous a size, that no simply infinite system can possibly count out all of them. Of course the question whether this axiom is properly formulated must await a convincing account of the meanings of the global quantifiers and global negation.

---

[23] Perhaps a better reading of "$\neg\neg$" would be "it is impossible that it is impossible that".

Appendix 2
The Rank of a Set

We need to introduce the notion of a set (or individual) S's *being of rank r*, where r is a well-ordering, before actually defining the function $\lambda x\,(rank(x))$ outright. But first we need to define the notion of a function f's *constructing S along r*. Roughly, this means that f is an r-sequence whose successive terms represent the successive "layers" of $Constituents(S)$, when the elements of the latter are "layered" in order of their "construction" from the support of S. More precisely we lay down the following definition.

A2.1 Definition *Let S be any object, let r be a well-ordering, and let f be a local function such that $f : Field(r) \to P(Constituents(S))$. Then by definition f constructs S along r if, and only if,*

(i) *For all x in the field of r*

$$f`x = \{y \in Constituents(S) : (\forall u \in y)(\exists v <_r x)[u \in f`v]\}$$

(ii) $Constituents(S) = Support(S) \cup \bigcup f``Field(r)$

(iii) *For each proper initial segment s of r*

$$S - (Support(S) \cup \bigcup f``Field(s)) \neq \emptyset$$

Thus if f constructs S along r then, in general,

(a) $f`First(r)$ is the support of S.
(b) If x lies in $Field(r)$ and is not the last element of r, $f`(x+1)_r$ is obtained from $f`x$ by adding in all members of $Constituents(S)$ whose members lie in $f`x$.
(c) If l lies in $Field(r)$ but has no immediate predecessor, then $f`l = \bigcup \{f`x : x <_r l\}$.

Now we can lay down the main definition.

A2.2 Definition *Let S be any set or individual and r a well-ordering. Then, by definition, S is of rank r if, and only if, there is a function* $f : Field(r) \to P(Constituents(S))$ *which constructs S along r.*

Thus if S is of rank r then r measures the number of "steps" required to "construct" all the constituents of S. As I shall now show, this depends only on the "length" (order type) of r (i.e. on its equivalence class modulo $\cong_o$).

A2.3 Theorem
  (i) *If S is not a set or S is the empty set, then S is of rank $\subseteq_\emptyset$*[1].
  (ii) *If S is of rank r, then $P(S)$ is of rank $r +_o 1$.*
  (iii) *If f constructs S along r, and $a, b \in Field(r)$ with $a <_r b$, then $f`a$ is a proper subset of $f`b$.*
  (iv) *If f constructs S along r, and T is a constituent of S, then T is of rank s, where s is the shortest initial segment of r such that $T \subseteq \bigcup f``Field(s)$.*
  (v) *If S is both of rank r and of rank s, then $r \cong_o s$.*

Proof (i) This follows from the fact that if S has no members then S has no constituents, so the empty function constructs S along the empty well-ordering.

(ii) Note first that $Constituents(P(S)) = P(S) \cup Constituents(S)$. Now suppose that f constructs S along r. Since $r +_o 1$ is obtained from r by adding $Field(r)$ as last element, we may define g on $Field(r +_o 1)$ by setting

$$g = f \cup \{(Field(r), P(S) \cup Constituents(S))\}$$

Since $S \in P(S) - Constituents(S)$, condition (iii) of Definition A2.1 is satisfied. Moreover S and all its subsets have their members among $Constituents(S)$, so g constructs $P(S)$ along $r +_o 1$.

(iii) From the definition of ranking function it is obvious that $f`a \subseteq f`b$. But the inclusion must be proper by part (iii) of Definition A2.1.

(iv) Define $g : Field(s) \to P(Constituents(T))$ by setting

$$g`x = f`x \cap Constituents(T)$$

for all $x \in Field(s)$. Then g is a ranking function for T on s.

---

[1] $\subseteq_\emptyset$ is the empty well-ordering whose field is $\emptyset$.

(v) Suppose, by way of contradiction, that $r \not\approx_o s$. Then $r <_o s$ or $s <_o r$ and we may assume, without loss of generality, that $r <_o s$. Since S is of rank r and of rank s, there are functions f and g constructing S on r and s, respectively. Since $r <_o s$ there is a full order morphism $h : Field(r) \to Field(s)$ which embeds r in s as a proper initial segment. A straightforward induction along r verifies that for all $x \in Field(r)$, $f'x = g'h'x$. Since $r <_o s$, $h"Field(r)$ is (the field of) a proper initial segment of s. But $g"h"Field(r) = f"Field(r)$, and $Constituents(S) = \bigcup f"Field(r)$ (Definition A2.1(ii).) and therefore $S - \bigcup g"h"Field(r) = \emptyset$, which violates (iii) of Definition A2.1. Hence g is not a ranking function for S on s, and we have arrived at the required contradiction. □

Part (v) of A2.3 tells us that if S is of rank r it is only the order type of r that is relevant. But does every set *have* a rank? The next theorem answers this question.

A2.4 Theorem *Let S be a non-empty set. Then there is a well-ordering, whose field consists of subsets of the set of constituents of S, along which a function constructing S can be defined.*

Proof The idea here is to use the "stages" in the "construction" of the set of constituents of S under the well-ordering induced by inclusion.

Given S define a well-ordering, r, to be *critical* if

(i) $Field(r) \subseteq P(Constituents(S))$
(ii) $\leq_r = \subseteq_{Field(r)}$
(iii) For all x in the field of r

$$x = \{y \in Constituents(S) : (\forall u \in y)(\exists v <_r x)[u \in v]\}$$

Note that the totality of critical well-orderings is a set.

Claim 1 *If r and s are both critical well-orderings with $r <_o s$, then r is an initial segment of s.*

Pf. Since $r <_o s$ there is a full order morphism $h : Field(r) \to Field(s)$ which embeds r as an initial segment of s. A straightforward induction along r suffices to establish that, for all x in the field of r, $h'x = x$.

Claim 2 *If T is a set of critical well-orderings then $\bigcup T$ is a critical well-ordering.*

Pf. $\bigcup T$ is an ordering and $Field(\bigcup T) = \bigcup \{Field(x) : x \in T\}$. That it is a critical well-ordering follows as a consequence of Claim 1. For example, let U be a non-empty subset of the field of $\bigcup T$. Then there is a critical well-ordering r lying in T such that $U \cap Field(r) \neq \emptyset$. The least element of $U \cap Field(r)$ in r is the least element of U in $Field(\bigcup T)$, by virtue of Claim 1. A similar argument using Claim 1 shows that $\bigcup T$ is critical.

Now set $r = \bigcup \{x : x$ is a critical well-ordering$\}$, and let f be the identity function on $Field(r)$. I claim that f constructs S along r. Claim 2 guarantees that r is a critical well-ordering and, clearly, $f : Field(r) \to P(Constituents(S))$. Moreover, f clearly satisfies clause (i) of Definition A2.1. Suppose, by way of contradiction, that f does not construct S along r. Then either $Constituents(S) \neq \bigcup Field(r)$ (Case 1) or for some proper initial segment s of r, $S - \bigcup Field(s) = \emptyset$ (Case 2).

Case 1. In this case let

$$T = \{y \in Constituents(S) : (\forall u \in y)(\exists v \in Field(r))[u \in v]\}$$

Since $\bigcup Field(s) \subseteq Constituents(S)$ and $Constituents(S) \neq \bigcup Field(r)$, $T \notin Field(r)$. Hence we may add T as a last element to r to obtain a new critical well-ordering s properly extending r. Since this is impossible, $Constituents(S) = f"Field(r)$.

Case 2. In this case s is, by Claim 1, an initial segment of some critical well-ordering t, and this means that all terms in t beyond s are identical, which is impossible.

Thus f constructs S along r and the theorem follows. □

Since all sets can be ranked and that ranking is unique up to order isomorphism, the following definitions of "S and T are of the same rank" ("$S \cong_{rank} T$") and "S is smaller than or equal to T in rank" ("$S \leq_{rank} T$") suggest themselves.

A2.5 Definition *Let S and T be objects. Then by definition*
  (i) *$S \cong_{rank} T$ iff every well-ordering of a subset of $Constituents(S)$ that ranks S also ranks T*
  (ii) *$S \leq_{rank} T$ iff every well-ordering of a subset of $Constituents(T)$ that ranks T has an initial segment that ranks S.*

I shall also use the notation "$S <_{rank} T$" with the obvious meaning.

It is easy to see that $\cong_{rank}$ is a global equivalence relation and $\leq_{rank}$ is a global linear ordering of its equivalence classes.

Bibliography

Ackermann, P.

"Die Widerspruchsfreiheit der allgemeinen Mengenlehre", *Mathematische Annalen*, 1937

Aristotle

De Caelo, Books I and II, translated with commentary by Stuart Leggatt as *On the Heavens I and II*, Aris and Phillips, Warminster, 1995

Metaphysics, Books M and N, translated with introduction and notes by Julia Annas, *Clarendon Aristotle Series*, Oxford University Press, Oxford, 1976

Metaphysics, translated by J.A. Smith and W.D. Ross, Clarendon Press, Oxford, 1908

Physics, translated by Robin Waterfield with introduction and notes by David Bostok, *World's Classics*, Oxford University Press, Oxford, 1996

Barrow, I.

The Usefulness of Mathematical Learning explained and demonstrated: Being Mathematical Lectures read in the Publick Schools at the University of Cambridge, translated from the Latin by John Kirkby, Stephen Austen, London, 1734

Bell, J.L.

A Primer of Infinitesimal Analysis, Cambridge University Press, Cambridge, 1998.

Bernays, P.

"A system of axiomatic set theory", *Journal of Symbolic Logic*, Part I, (**2**), 1937, Part II, (**7**), 1942

"Zur Frage der Unendlichkeitsschema in der axiomatischen Mengenlehre", in Bar–Hillel, Poznanski, Rabin, and Robinson, *Essays on the Foundations of Mathematics*, Magnes Press, Jerusalem, 1961

Axiomatic Set Theory, North Holland, Amsterdam, 1963

Bourbaki, N.

Elements of Mathematics, Theory of Sets, Addison–Wesley, London, 1968

Burnyeat, Miles

"Plato on why mathematics is good for the soul", in *Essays on the History of Philosophy*, edited by Timothy Smiley, Oxford University Press (Forthcoming)

Cantor,G.

Gesammelte Abhandlungen mathematischen und philosophischen Inhalts, edited by E. Zermelo, Georg Olms Verlagsbuchhandlung , Hildesheim, 1966

Cohen, P.J.

"The independence of the Continuum Hypothesis", *Proceedings of the National Academy of the Sciences U.S.A.*, Part I (**50**) 1963, Part II (**51**) 1964

Dedekind, R.

Was sind und was sollen die Zahlen?, Vieweg, Braunschweig, 1893. Translated by Wooster W. Berman as *The Nature and Meaning of Numbers*, Dover, New York, 1963

Stetigkeit und irrationale Zahlen, Vieweg, Braunschweig, 1872. Translated by Wooster W. Berman as *Continuity and Irrational Numbers*, Dover, New York, 1963

Dummett, M.

Elements of Intuitionism, Oxford Logic Guides 2, Oxford University Press, Oxford, 1977

Frege: Philosophy of Mathematics, Duckworth, London, 1991

Eilenberg, S. and Mac Lane, S.

"General theory of natural equivalences", *Transactions of the American Mathematical Society*, **58**, 1945

Euclid

Elements, translated in three volumes and with introduction and commentary by Sir Thomas Heath, Dover, New York, 1956

Ewald, William B.

From Kant to Hilbert: A Source Book in the Foundations of Mathematics, (2 volumes), Oxford University Press, Oxford, 1996

Field, H.

Science Without Numbers, Oxford University Press, Oxford, 1980

Frege, G.

Begriffsschrift, eine der arithmetischen nachgebildete Formalsprache des reinen Denkens, Nebert, 1879. Translated by Stephan Bauer–Mengelberg in van Heijenoort, *From Frege to Gödel* (q.v.)

Die Grundlagen der Arithmetik, eine logische–mathematische Untersuchung über den Begriff der Zahl, Koebner, Berlin, 1884. Translated by J.L. Austin as *The Foundations of Arithmetic: a logico–mathematical enquiry into the concept of number* (2nd edition), Basil Blackwell, Oxford, 1953

Grundgesetze der Arithmetik, begriffsschriftlich abgeleitet (2 volumes), H. Pohle, Jena, 1893 (Vol. I) and 1903 (Vol. II). Reprinted by Georg Olms, Hildesheim, 1962

Über Sinn und Bedeutung, *Zeitschrift für Philosophie und philosophische*

Kritik, (new series) **100**, translated as "On sense and reference" in Geach and Black (q.v.)

"Was ist eine function?" in *Ludwig Boltzmann gewidmet zum sechzigten Geburtstage 20 Februar 1904*, Ambrosius Barth, Leipzig, 1904, translated as "What is a function?" in Geach and Black (q.v.)

Friedman, Harvey M.

"Higher set theory and methematical practice", *Annals of Mathematical Logic* **2**, 1971

Geach, P.T. and Black, M.

Translations from the Philosophical Writings of Gottlob Frege, Basil Blackwell, Oxford, 1970

Gentzen, Gerhard

Collected Papers of Gerhard Gentzen (ed. M.E. Szabo), North Holland, Amsterdam, 1969

Gödel, K.

"Die Vollständigkeit der Axiome des logischen Functionenkalküls", Monatshefte für Mathematik und Physik, **37**, 1930, translated into English by Sthephan Bauer-Mengelberg in van Heijenoort *From Frege to Gödel* (q.v.)

"Remarks before the Princeton bicentennial conference on problems in mathematics", in M. Davis (ed.) *The Undecidable*, Raven Press, New York, 1965

"What is Cantor's continuum problem?", in Benecerraf, P. and Putnam, H. (eds.) *Philosophy of Mathematics: selected readings* 2nd edition, Cambridge University Press, Cambridge, 1983

"Consistency proof for the Generalized Continuum Hypothesis", *Proceedings of the National Academy of Sciences of the U.S.A* **24** (1939)

Grzegorczyk, A.

"Some classes of recursive functions", *Rozprawy Matematiczne* IV, Warsaw, 1953

Hajek, Petr and Pudlak, Pavel

Metamathematics of First-order Arithmetic, in the series *Perspectives in Mathematical Logic*, Springer-Verlag, Berlin, 1993

Hallett, Michael

Cantorian Set Theory and Limitation of Size, Oxford Logic Guides 10, Clarendon Press, Oxford, 1984

Halmos, Paul R.

Naive Set Theory, D. van Nostrand, Princeton, 1968

Homolka, V.

A System of Finite Set Theory Equivalent to Elementary Arithmetic, Ph.D. Thesis, University of Bristol, 1983

van Heijenoort, Jean

From Frege to Gödel: a Source Book in Mathematical Logic, 1879–1931, Harvard University Press, Boston, 1967

Husserl, Edmund

Formal and Transcendental Logic, translated by Dorian Cairns, Martinus Nijhoff, The Hague, 1969

The Crisis of European Sciences and Transcendental Phenomenology: An Introduction to Phenomenological Philosophy, translated by David Carr, Northwestern University Press, Evanston, 1970

Kant, Immanuel

Critique of Pure Reason, translated by Paul Guyer and Allen W. Wood, *The Cambridge Edition of the Works of Immanuel Kant*, Cambridge University Press, Cambridge, 1997

Prolegomena to Any Future Metaphysics that Will be Able to Present Itself as a Science, translated by Peter G. Lucas, Manchester University Press, Manchester 1966

Kleene, S.C.

Introduction to Metamathematics, D. van Nostrand, Princeton, 1962

Klein, Jacob

Greek Mathematical Thought and the Origin of Algebra, Dover, New York, 1992

Kluge, E–H.W.

On the Foundations of Geometry and Formal Theories of Arithmetic, Yale University Press, New Haven, 1971

Kreisel, G.

"Informal rigour and completeness proofs", in *Philosophy of Mathematics* (ed. J. Hintikka), Oxford University Press, 1970

Lawvere, F.W.

"Categorical dynamics", in *Topos Theoretic Methods in Geometry*, Aarhus Math. Inst. Var. Publ., Series 30

Lewis, David

Parts of Classes, Basil Blackwell, Oxford, 1991

Mac Lane, S.

Categories for the Working Mathematician, Springer–Verlag, New York, 1971

Martin, Donald A.

"Borel determinacy", *Annals of Mathematics*, **102**, 1975

Mayberry, J.P.

"On the consistency problem for set theory: an essay in the Cantorian foundations of mathematics", *British Journal for the Philosophy of Science*, **28**, 1977

"Global quantification in Zermelo-Fraenkel set theory", *Journal of Symbolic Logic*, **50**, 1985

"What is required of a foundation for mathematics?", *Philosophia Mathematica* (3) Vol. 2, 1994

Moerdijk, I. and Reyes, G.E.

Models for Smooth Infinitesimal Analysis, Springer–Verlag, New York, 1991

Bibliography

Mycielski, Jan
: "Analysis without actual infinity", *Journal of Symbolic Logic*, **46**, 1981

: "Locally finite theories", *Journal of Symbolic Logic*, **51**, 1986

Nelson, Edward
: *Predicative Arithmetic*, Number 32 in Mathematical Notes, Princeton University Press, Princeton, 1986

: *Radically Elementary Probability Theory*, Annals of Mathematics Studies 117, Princeton University Press, 1987

Newton, Isaac
: *Universal Arithmetic: or a Treatise of Arithmetical Composition and Resolution*, London, 1728. Reprinted in *The Mathematical Works of Isaac Newton*, volume 2 (ed. Derek T. Whiteside), Johnson Reprint Corporation, New York, 1966

Ockham (See William of Ockham)

Parikh, Rohit
: "Existence and feasibility in arithmetic", *Journal of Symbolic Logic* **36**, 1971

Plato
: *Republic*, translated with introduction and notes by F.M. Cornford, Oxford University Press, Oxford, 1966

Popham, S.
: *Some Results in Finitary Set Theory*, Ph.D. Thesis, University of Bristol, 1984

Pritchard, Paul
: *Plato's Philosophy of Mathematics*, International Plato Studies 5, Academia, Saint Augustin, 1995

Quine, W.V.O.
: *Set Theory and its Logic*, Harvard University Press, Cambridge, 1963

Rose, Harvey E.
: *Subrecursion: Functions and Hierarchies*, Oxford Logic Guides 9, Clarendon Press, Oxford, 1984

Russell, B. and Whitehead, A.N.,
: *Principia Mathematica*, Cambridge University Press, Cambridge, 1962, second edition 1997

Schoenfield, J.R.
: *Mathematical Logic*, Addison–Wesley, 1967

Spivak, Michael
: *Calculus*, W.A. Benjamin, New York, 1967

Tarski, A.
: "Sur les ensembles finis", *Fundamenta Mathematicae*, **6**, 1924

Takeuti, Gaisi
: *Proof Theory*, in the series *Studies in Logic and the Foundations of Mathematics*, Volume 81, North Holland, Amsterdam, 1975

Vopenka, P.

Mathematics in the Alternative Set Theory, Teubner Verlagsgesellschaft, Leipzig, 1979

William of Ockham

Summa Logicae, translated and with a commentary by Michael J. Loux, *Ockham's Theory of Terms: Part One of the Summa Logicae*, University of Notre Dame Press, South Bend, 1974

Wittgenstein, Ludwig

Tractatus Logico-Philosophicus, translated by D.F. Pears and B.F. McGuiness, Routledge, London, 1995

Philosophical Grammar, Basil Blackwell, Oxford, 1974

Remarks on the Foundations of Mathematics (3rd edition), Basil Blackwell, Oxford, 1978

Wright, Crispin

Frege's Conception of Numbers as Objects, Aberdeen University Press, Aberdeen, 1963

Yessenin-Volpin, A.S.

"The ultra-intuitionistic criticism and the antitraditional program for foundations of mathematics" in *Intuitionism and Proof Theory. Proceedings of the Summer Conference at Buffalo, New York,1968*, edited by A. Kino, John Myhill, and R.E. Vesley, North Holland, Amsterdam, 1970

Zermelo, E.

"Beweiss das jede Menge wohlgeordnet werden kann", *Mathematische Annalen*, **59**, 1904

"Neuer Beweiss für die Möglichkeit einer Wohlordnung", *Mathematische Annalen*, **65**

"Untersuchungen über die Grundlagen der Mengenlehre", *Mathematische Annalen*, **65**, 1908

"Sur les ensembles finis et le principe de l'induction complète", *Acta Mathematica*, **32**, 1909

"Über Grenzzahlen und Mengenbereiche: neue Untersuchungen über die Grundlage der Mengenlehre", *Fundamentica Mathematicae*, **14**, 1930, translated by M. Hallett in *From Kant to Hilbert*, ed. W. Ewald (q.v.)

Index

absolute infinity, 243, 249
Absolute Infinity, Axiom of, 410
abstract objects, 193, 194, 196, 199, 201, 224
abstraction, 202–204
analysing sequence, 165
Archimedes, 52
Aristotle, 1, 14, 30, 33, 36, 40–48
arithmetic
   non–Euclidean, xiv
   Peano axioms for, 159
arithmetica universalis, xv, 59
arithmos, xiv, 21, 70–73, 149, 271, 275
Axiom of Choice, 163
axiomatic definition, 195–206
axiomatic method, 204–205
   fundamental dogma of, 154, 199, 205, 222

Barrow, Isaac, 191
Basic Set Theory, 399, 405
Bell, J.L., 206
Berkeley, G., 195
Bernays, P., 245, 247
Bernays–Gödel set theory, 245
Bishop, E., 270
Borel Determinacy, 188
bounded formula, 407
Bourbaki, N., 99, 107, 108, 205, 220, 387
Brouwer's Principle, 89, 170, 242–251, 384
Brouwer, L.E.J., 270
Burnyeat, Myles, 21

Cantor's Absolute, 176, 241, 250, 261
Cantor's Axiom, 161, 381
Cantor, G., xiv, 47–48, 92, 249, 261, 381
Cantorian finitism, 47, 62, 91, 263, 267, 269, 274
Cantorian Set Theory, 405
cardinality, 134–136
Cartesian product, 129

category, 205
certainty, 104–107
choice function, 163
Cohen, P., 265, 268
Completeness Theorem, 217
concept construction, 210–211
context principle, 27, 226, 227
Continuum Hypothesis, 265
cumulation model, 372
cumulative hierarchy of sets, 170

Decidability for Local Formulas, Axiom Schema of, 407
Dedekind structure, 153
Dedekind, R., 153, 192, 271–273, 382
Dummett, Michael, 19, 40, 49, 225, 230–233, 259, 389, 408

Eilenberg, S., 205
Elements
   Book I, xiv, 160
   Book V, 15, 18
   Book VI, 18
   Book VII, 18, 71
empty set, 74
   as an *arithmos*, 77
equivalence class, 137
equivalence relation, 137
Euclid, 33
Euclidean finite set, 160
Euclidean Finiteness, Axiom of, 277
Euclidean finitism, 47, 61–62, 274, 382–387
Euclidean Set Theory, 406
Eudoxus, 33
Extensional Analysis, Axiom of, 167

Field, Hartry, 31
formal syntax, 207–213
formalism, 218–221, 229–387
formulas

global, 396
local, 396
Foundation, Axiom of
   Cantorian version, 166
   Euclidean version, 298
Frege, G., 22, 26, 28, 45–51, 79–81, 225–232
Friedman, H., 188
function
   global, 78–85
      arithmetical, 301
      defined in a simply infinite system, 331–333
   global propositional, 79
      arithmetical, 301
   local, 130–134

Gödel, K., 107, 185–189, 242, 245, 249, 265, 268
Gentzen, G., 409
global propositional connectives, 406
global propositions, 97
global semantics, 213–221

Hallett, M., 48
Halmos, P., 122
Heath, Sir Thomas, 71
Herbrand's Theorem, 394
Hilbert, D., 107, 151, 199, 228–230, 263
Hilbert–Ackermann Consistency Theorem, 394
Husserl, E., 223, 233

ideal objects, 195, 199, 201, 224
inaccessible point, 186
induction
   along a well-founded partial ordering, 140
   Epsilon, 169
   in a simply infinite system, 328
induction system, 155
Intermediates, 14
intuition (*Anschauung*), 210
isomorphism, 196
isomorphism type, 201
iteration, 158, 184–185, 285

Kant, Immanuel, 210–212
Kleene, S.C., 212, 409
Klein, Jacob, xiii, 18, 21
Kreisel, G., 249, 266
Kronecker, L., 270

Löwenheim–Skolem Theorem, 215
Lawvere, F.W., 206, 392, 393
Lewis, David, 76
lexicographic ordering, 354
Local Definiteness, Axiom Schema of, 404

localisation
   first order
      logical consequence, 217
      logical consistency, 215
      logical validity, 217
   second order, 251–258
      categorical theories, 223
logarithmic exponentials, 309
logical consequence, 214
logical consistency, 214
logical validity, 214

Mac Lane, S., 205, 248
Machover, Moshé, 25
Mahlo's Principle, 186, 405
Mahlo, P., 185
Martin, D., 188
mathematical logic, 207, 208
mathematical objects, 40, 72, 195, 196, 201, 224–236
mathematical structures, 11, 12, 154, 196
Mathematicals, 14, 40
mathesis universalis, xv
measure, 330
minimal parts, 392
Moerdijk, I., 206
morphism, 154
   of Dedekind structures, 154
morphology, 62, 154, 196
Mycielski, Jan, 391, 394

natural numbers
   axiomatic definition of, 153
   origin, 17
nearest neighbour graphs, 392
Nelson, Edward, 387, 391
Newton, Isaac, 17, 191
normal domain, 239
   characteristic of, 240
numeration base, 317
   decomposition with respect to, 318
      over a simpy infinite system, 337
   numeral with respect to, 319
      over a simply infinite system, 337

object, 67–70
One Point Extension Induction
   global version, 278
   local version, 277
operationalism, xvii, 15, 270, 271, 382, 385
operationalist fallacy, xvii, 387
ordered pair, 124–129

Parikh, Rohit, 387
partial function, 140
partial ordering, 137
Peano system, 159

Penrose, R., 259
Plato, 14, 40, 42, 202
Platonism, 14, 42, 202
Poincaré, H., 270
Popham, S., 353, 372
Pritchard, Paul, 21
proof, 99–110
　formalised, 108

Quine, W.V.O., xv, 82

rank
　Cantorian version, 171, 172, 411–414
　Euclidean version, 360
recursion
　on a global function, 170, 175
　　along ω, 180
　　along a simply infinite system
　　　(limited), 333–336
　　along the von Neumann ordinals, 183
　　limited, 175, 311
　on a local function
　　along a simply infinite system, 156
　　along a well-founded partial ordering, 141
　　Epsilon, 169
　　primitive, 289
Regularity, Principle of, 146–149
Reyes, G., 206
Russell, B., 87, 95, 261

satisfiability, 214
scale
　for a species, 331
　universal, 331, 352, 383
Schoenfield, J.R., 208
self-evidence, 10, 123–124, 189
semi-set, 215, 224, 258
set
　cardinal number of, 175
　constituents of, 167
　contrasted with species, 86–87
　definition of, 77–78
　objectivity of, 75
　pure, 168
　support of, 168
　transitive closure of, 168
　ultimate constituents of, 168
simple closure point, 179
Simple Closure Principle, 180
simply infinite system, 192
　Cantorian, 155
　Euclidean, 325–327
　　S-ary expansion of, 338–350
　　Ackermann system over U determined by l, 358
　　cumulative hierarchy over U, 352

hierarchy of S-ary extensions, 350–352
von Neumann, 329, 370
Zermelo, 329, 370
singleton set, 74
smooth infinitesimal analysis, 206
Soundness Theorem, 216
special symbols (definition of)
　Π_1, 97
　Σ_1, 97
　$\cong_c$, 134
　$\leq_c$, 136
　$r_{<a}$, 138
　$r_{\leq a}$, 138
　$\cong_o$, 139
　$\subseteq_S$, 138
　$\leq_o$, 143
　$+_o$, 143
　$\times_o$, 143
　$\epsilon(x)$, 167
　$Constituents(S)$, 167
　$Support(S)$, 168
　$R(S)$, 172
　$rank(S)$, 172
　$Ordertype(r)$, 174
　$V_r(S)$, 174
　$\beth_a$, 176
　$\aleph$, 176
　ZF_1, 237
　ZF_2, 237
　card, 293
　ord, 293
　$\times_c$, 302
　$\exp_c$, 302
　$+_c$, 303
　$-_c$, 303
　$\log_b$, 303
　$\sum_{x \in a} \sigma(x)$, 306
　$\sum_{x < a} \sigma(x)$, 306
　$\prod_{x \in a} \sigma(x)$, 306
　$zg\prod_{x < a} \sigma(x)$, 306
　$lexp_n$, 309
　$\cong_{S\text{-}ary}$, 319, 320
　T is a $\langle d_0, \ldots, d_k \rangle$, 320
　$\leq_{Num(S,r)}$, 321
　$[\ldots n)_S$, 322
　$\mathscr{VN}$, 329
　$\mathscr{Z}$, 329
　$\mathscr{N}[S]$, 338
　$Name_S(a)$, 339
　$Value_S(b)$, 340
　$\mathscr{N}_\mathbf{n}[S]$, 350
　$\mathscr{H}(U)$, 352
　$lex(r)$, 354
　$\mathscr{ACH}(U, l)$, 358
　ZF_0, 371
　$C(M, b)$, 373
　LK_0, 398

LK_{0e}, 399
LK_{0e}^+, 403
LJ_e^*, 406
A^{loc}, 407
A^{bnd}, 407
$\cong_{rank}$, 414
$\leq_{rank}$, 414
species, 85–94
  contrasted with set, 86–87
  definition of, 86
  locally definable, 330
Spengler, Oswald, xvii
Spivak, M., 235
strong axioms of infinity, 250
strong closure point, 181
Strong Closure Principle, 171, 183
Strong Hierarchy Principle
  Cantorian version, 171, 174, 177
  Euclidean version, 365–368
strongly inaccessible number, 186, 240
symbol generated abstraction, 20, 60, 81, 125–128, 196
syncategorematic, 82

Takeuti, G., 398–406
Tarski, A., 274
topos, 205–206
transfinite set, 160
transitive set, 168
truth
  in mathematics, 99–110
  Tarski's definition of, 103

Uniformity, Axiom of, 363
unit, 21, 35, 36, 39, 41
universal validity, 214
universals, problem of, 202

virtual objects, 82, 87
von Neumann natural number, 161
von Neumann ordinal, 144
von Neumann ordinals, 144–146
von Neumann well-orderings, 144
von Neumann–Bernays set theory, 245
Vopenka, P., 215, 387

Weak Hierarchy Principle
  Cantorian version, 171, 172, 177
  Euclidean version, 364–368
weakly inaccessible number, 186
well-foundedness, 139
well-ordering, 139
Weyl, H., 270
Whitehead, A.N., 95
William of Ockham, 82
Wittgenstein, L., 28, 56, 271
Wright, Crispin, 226

Zermelo domain, 240
Zermelo set theory, 238
Zermelo's Principle, 242
Zermelo, E., 76, 163, 185, 237, 242, 274
Zermelo–Fraenkel set theory, 186, 190, 237–242
  formal axioms, 238–239
  quasi-categoricity of, 241

For EU product safety concerns, contact us at Calle de José Abascal, 56–1°,
28003 Madrid, Spain or eugpsr@cambridge.org.

www.ingramcontent.com/pod-product-compliance
Lightning Source LLC
LaVergne TN
LVHW021941060526
838200LV00042B/1890